PHYSICS OF HADRONS AND QCD

PREFACE

The APCTP-RCNP Joint International School on Physics of Hadrons and QCD was held at the Research Center for Nuclear Physics (RCNP), Osaka University on October 12 and 13, 1998. The School is immediately followed by the 1998 YITP Workshop on QCD and Hadron Physics, which was held at the Yukawa Institute for Theoretical Physics, Kyoto University from October 14 to 16, 1998.

The purpose of the School and the Workshop was to study recent topics in QCD and hadron physics from various points of view. The subjects included perturbative and nonperturbative aspects of QCD, chiral effective theory in hadron physics and high temperature and density nuclear matter physics.

Another purpose of the School and Workshop was to enhance communications and collaborations among researchers in the Asia and Oceania region. Nowadays the situation of physics research changes rapidly around us, especially in the Asia and Oceania areas, and many productive collaborations are in progress at personal and institutional levels. Many exciting discussions were exchanged and fruitful collaborations were started among the participants on current interesting topics and future perspectives.

The School consisted of six lectures by the most active physicists invited from Asia, Europe and the US. For the Workshop, 40 invited and contributed talks were given on recent studies in QCD and hadron physics. These proceedings include the lectures and talks presented at the School and Workshop.

The School was hosted by the Asia Pacific Center for Theoretical Physics (APCTP), Seoul, Korea and the Research Center for Nuclear Physics (RCNP) of Osaka University as a joint international school. The Workshop was hosted by the Yukawa Institute for Theoretical Physics (YITP) at Kyoto University as a part of its annual activity.

The organizers are grateful for the generous support given to the School from APCTP and RCNP. We especially thank Prof. Y.M. Cho and Prof. H. Toki for their efforts to realize the Joint International School of the two institutes. We also acknowledge the help of many students and young researchers of RCNP and YITP during the School and Workshop.

Hiroyuki YABU (Tokyo Metropolitan University)
Makoto OKA (Tokyo Institute of Technology)
Tetsuo MATSUI (Yukawa Institute for Theoretical Physics)
Kazunori ITAKURA (Yukawa Institute for Theoretical Physics)

大阪大学
核物理研究センター
RESEARCH CENTER FOR NUCLEAR PHYSICS
OSAKA UNIVERSITY

The APCTP-RCNP Joint International School on Physics of Hadrons and QCD

School Organization

The International Advisory Committee

I.-T. Cheon	(Yonsei University, Seoul)
W. Haxton	(University of Washington, Seattle)
D.-P. Min	(Seoul National University, Taipei)
A. W. Thomas	(University of Adelaide, Adelaide)
K. Yazaki	(Tokyo University, Tokyo)
Z.-Y. Zhang	(Institute of High Energy Physics, Beijin)

The Foreign Liaisons

M.-C. Chu	(Hong Kong University, Hong Kong)
P. C. Hwang	(National Taiwan University, Taipei)
H.-K. Lee	(Hangyang University, Seoul)
S.-H. Lee	(Yonsei University, Seoul)
B.-Y. Park	(Chung-Nam University, Chung-Nam)
A. Williams	(University of Adelaide, Adelaide)

The Organizing Commitee

H. Yabu	(Tokyo Metropolitan University, Tokyo, Chair)
M. Oka	(Tokyo Institute of Technology, Tokyo, Co-chair and representative of RCNP)
T. Matsui	(Yukawa Institute for Theoretical Physics, Kyoto)
T. Hatsuda	(Kyoto University, Kyoto)
K. Itakura	(Yukawa Institute for Theoretical Physics, Kyoto)
J. Kodaira	(University of Hiroshima, Hiroshima)
Y. Koike	(Niigata University, Niigata)
O. Miyamura	(University of Hiroshima, Hiroshima)
T. Sakai	(Research Center for Nuclear Physics, Osaka)
H. Toki	(Research Center for Nuclear Physics, Osaka)

The 1998 YITP-Workshop on QCD and Hadrons Physics

Workshop Organization

The Organizers

H. Yabu	(Tokyo Metropolitan University, Tokyo, Chair)
M. Oka	(Tokyo Institute of Technology, Tokyo)
T. Matsui	(Yukawa Institute for Theoretical Physics, Kyoto)
T. Hatsuda	(Kyoto University, Kyoto)
K. Itakura	(Yukawa Institute for Theoretical Physics, Kyoto)
J. Kodaira	(University of Hiroshima, Hiroshima)
Y. Koike	(Niigata University, Niigata)
O. Miyamura	(University of Hiroshima, Hiroshima)
H. Toki	(Research Center for Nuclear Physics, Osaka)
M.-C. Chu	(Hong Kong University, Hong Kong)
P. C. Hwang	(National Taiwan University, Taipei)
H.-K. Lee	(Hangyang University, Seoul)
S.-H. Lee	(Yonsei University, Seoul)
B.-Y. Park	(Chung-Nam University, Chung-Nam)
A. Williams	(University of Adelaide, Adelaide)

CONTENTS

Short Talks

RANDOM MATRIX THEORY AND QCD AT NONZERO TEMPERATURE AND CHEMICAL POTENTIAL

J.J.M. Verbaarschot
Department of Physics and Astronomy, SUNY Stony Brook, Stony Brook, NY 11790, USA,
e-mail: verbaarschot@nuclear.physics.sunysb.edu

In the first part of these lectures we discuss the infrared limit of the spectrum of the QCD Dirac operator. We discuss the global symmetries of the QCD partition function and show that the Dirac spectrum near zero virtuality is determined by the pattern of spontaneous chiral symmetry breaking of a QCD-like partition function with additional bosonic valence quarks and their super-symmetric partners. We show the existence of an energy scale below which the fluctuations of the QCD Dirac spectrum are given by a chiral Random Matrix Theory (chRMT) with the global symmetries of the QCD partition function. Physically, for valence quark masses below this scale the partition function is dominated by the zero momentum modes. In the theory of disordered systems, this energy scale is known as the Thouless energy. In the second part of these lectures we discuss chRMT as a schematic model for the QCD partition function at nonzero temperature and chemical potential. We discuss novel features resulting from the non-Hermiticity of the Dirac operator. The analysis by Stephanov on the failure of the quenched approximation as well as the phase diagram of the chRMT partition function are discussed. We argue that a localization transition does not occur in the presence of light quarks. Several results will be derived in full detail. We mention the flavor symmetries of the QCD Dirac operator for two colors, the calculation of the valence quark mass dependence of the chiral condensate and the reduction of the chRMT partition function to the finite volume partition function

1 Introduction

The main two phenomena that characterize the low-energy limit of QCD are confinement and chiral symmetry breaking. Because of confinement and the spontaneous breaking of chiral symmetry the low-lying states of QCD are given by the corresponding Goldstone bosons. At low temperatures we expect that the thermodynamics of the QCD partition function is that of a gas of (almost) massless pions. Mainly because of lattice simulations [1,2,3,4] we know that chiral symmetry is restored above a critical temperature of about 140 MeV. The order parameter of the chiral phase transition is the chiral condensate which through the Banks-Casher formula [5] is directly related to the spectral density of the QCD Dirac operator. This is our main motivation for detailed studies of QCD Dirac spectra.

One of the questions we wish to address in these lectures is to what extent the infrared limit of the Dirac spectrum can be obtained analytically. After all, the infrared limit of the QCD partition function is determined by the chiral Lagrangian describing Goldstone modes that become noninteracting at zero momenta. On the other hand we know from many studies of complex systems that the correlations of eigenvalues on the scale of the individual level spacings are universal and are given by Random Matrix Theory. More precisely, this has been formulated as the Bohigas-Giannoni-Schmit conjecture [6] which states that spectral correlations of a classically chaotic system are given by Random Matrix Theory (see Guhr et al. [7] for a comprehensive review).

The natural question one may ask is whether one can reconcile the universal nature of the chiral Lagrangian with the universal spectral correlation given by RMT. The first hint for the existence of a close connection came from the work of Leutwyler and Smilga [8] who found that the existence of a low-energy chiral Lagrangian imposes an infinite family of constraints on the spectrum of the Dirac operator. On the other hand, in the field of Random Matrix Theory it has been known for a long time that the spectral correlations can be described by a Goldstone manifold arising from the spontaneous breaking of symmetry between the advanced and retarded Greens functions [9,10,11,12]. The real breakthrough in this problem came with the introduction of valence quarks [13,14,15,16,17]. The idea is [18] to study a partition function that, in addition to the usual sea-quarks, contains a valence quark and its bosonic super-partner. After differentiation with respect to the source term and equating the fermionic and bosonic valence quark masses, the determinants with the valence quark masses cancel and we have access to the Dirac spectrum weighted by the standard QCD action. As is the case for the usual chiral Lagrangian, at low energies the partition function involving both sea quarks and valence quarks is completely determined by the scheme of chiral symmetry breaking and Lorentz-invariance and has been analyzed in the context of partially quenched chiral perturbation theory [19,20]. However, instead of constraints on QCD-Dirac spectra such as were obtained in the work by Leutwyler and Smilga [8], we are now in a position obtain an exact analytical expression for the Dirac spectrum in the domain of applicability of this so called partially quenched chiral perturbation theory.

As is the case for the usual chiral perturbation theory [21], one can distinguish an important mass-scale. For sufficiently large volumes such that the contributions of the non-Goldstone modes are suppressed and valence quark masses

$$m_v \ll \frac{F^2}{\Sigma L^2} \tag{1}$$

(the pion decay constant is denoted by F, Σ is the chiral condensate and L is the length of the box) the contribution of the nonzero momentum modes factorizes from the partition function and its mass dependence is given by the contribution from the zero momentum modes alone [14,15,16,17]. However, QCD is not the only theory with this zero momentum sector. Any theory with the same pattern of spontaneous symmetry breaking as QCD can be reduced to this zero momentum limit. In particular, chiral Random Matrix Theory [22,23], with randomly distributed matrix elements of the Dirac operator is such a theory. In this case the strong interactions required to achieve a spontaneous breaking of chiral symmetry result from strongly non-local random interactions.

Such a picture is well-know from the theory of disordered mesoscopic systems (a number of recent reviews are available [7,24,25]) In this case, the scale below which Random Matrix Theory is valid is known as the Thouless energy, which is the inverse diffusion time of an electron through the sample [26]

$$E_c = \frac{\hbar D}{L^2}, \tag{2}$$

where D is the diffusion constant and L is the linear dimension of the box. The domain with $\delta E \ll E_c$ is known as the ergodic domain whereas the domain $E_c \ll \delta E \hbar/\tau_e$ (with τ_e is the elastic scattering time) is called the diffusive domain or the Altshuler-Shklovskii domain. In the case of QCD, the diffusion process is that of a quark propagating through the Yang-Mills fields in a 4+1 dimensional space time. The interpretation of spontaneous chiral symmetry breaking as a delocalization transition was made earlier in [27]. By analogy with the Kubo formula, Σ plays the role of the conductivity [27].

By now, this picture of the QCD gauge field configurations as a disordered system has been investigated in many lattice QCD simulations [28,29,30,31,32,33,34,35,36 37,38,39,40,41,42,43,44,45,46] and several instanton liquid simulations [22,47,14,15]. In particular, it has been confirmed that the scale below which chiral Random Matrix Theory is valid is given by the Thouless energy (1) [14,15,34,42,46]. It turns out that eigenvalue correlations are given by chRMT all the way up to this scale. Beyond this energy scale one has to take into account the contributions of the nonzero momentum modes. This can be done simply to one-loop order in chiral perturbation theory. In that way, one obtains analytical results for the valence quark mass dependence [48,16] of the chiral condensate in the Altshuler-Shklovskii domain.

In the first part of these lectures we will show that chRMT is an *exact* theory for the spectrum of the QCD Dirac operator in the ergodic domain. This is achieved by showing that in the range (1) the valence quark mass dependence of the chiral condensate obtained from the low-energy limit of the QCD partition function coincides with the valence quark mass dependence obtained from chiral Random Matrix Theory. In the second part of these lectures we take a different route and use chRMT as a *schematic* model of the chiral phase transition[49,50,51,52,53,54]. In that case we obtain only *qualitative* results for the spectrum of the QCD Dirac operator and the chiral phase transition. For example, such models are the random matrix equivalent of a Landau-Ginzburg functional and give rise to mean field critical exponents. At the tricritical point where the upper critical dimension is three they provide us with a reliable description of the the infrared degrees of freedom [55]. The advantage of using random matrix models as opposed to a Landau-Ginzburg functional is that one has access to the Dirac spectrum. In particular, at nonzero chemical potential, this has provided us with valuable insights in the effects of non-Hermiticity on the properties of the QCD partition function [52,56]. The common ingredient of chRMT and QCD at nonzero chemical potential is that the phase of the fermion determinant due to the non-Hermiticity makes Monte Carlo simulations impossible. Since the chRMT partition function can be evaluated analytically, we are in a position to analyze some of these problems in great detail. In this way, a complete analytical understanding of the failure of the quenched approximation at nonzero chemical potential has been obtained [52].

In the first lecture (section 2) we summarize the main properties of the Euclidean QCD partition function. We discuss in detail the global symmetries and give a brief review of lattice QCD (see lectures by Tom Blum [57] for more details). In the second lecture we analyze the infrared limit of the QCD partition function (section 3). We discuss the domain of validity of the zero momentum limit of the effective chiral partition function and introduce the partially quenched chiral partition function. An important point of this lecture is the discussion of the integration manifold. Explicit calculations of the valence quark mass dependence of the chiral condensate that rely on the structure of this Riemannian super-manifold are presented in section 4. Results are derived in detail both by perturbative and non-perturbative methods. In the third lecture we introduce chiral Random Matrix Theory. We discuss its main properties and emphasize the importance of universality (see section 5). In the second half of this lecture we present lattice QCD results for the statistical properties of Dirac eigenvalues and compare the results with random matrix theory. In the last lecture we introduce chiral Random Matrix Theory as a schematic model for the chiral phase transition. In the framework of this model, we discuss chiral symmetry breaking, quenching at nonzero chemical potential, Yang-Lee zeros in the complex chemical potential plane, the structure of Dirac spectra at nonzero chemical potential and the phase diagram in the mass, chemical potential and temperature plane. Finally, we present general arguments that a localization transition does not occur in QCD with light quarks. Concluding remarks are made in section 8.

2 The QCD Partition Function

The QCD partition function describing strong interactions in a box of volume $V_3 = L^3$ can be expressed in terms of the eigenvalues of the QCD Hamiltonian E_k as

$$Z^{QCD} = \sum_k e^{-\beta E_k}, \tag{3}$$

where β is the inverse temperature. At low temperatures ($\beta \to \infty$) the partition function is dominated by the lightest states of the theory, namely the vacuum state, with an energy density of E_0/V_3, and massless excitations thereof. The partition function Z_{QCD} can be rewritten as Euclidean functional integral over the nonabelian gauge fields A_μ

$$Z^{QCD} = \int dA_\mu \prod_{f=1}^{N_f} \det(D + m_f) e^{-S_{YM}}, \tag{4}$$

where S_{YM} is the Yang-Mills action. We assume that this partition function is properly regularized, for example, by a lattice regularization. The gauge fields A_μ are nonabelian gauge fields in Euclidean space time of volume $V = L^3\beta$. In terms of the generators, T_a, of $SU(N_c)$ they can be written as

$$A_\mu = A_\mu^a \frac{T_a}{2}, \tag{5}$$

The anti-Hermitean Dirac operator is defined by

$$D = \gamma_\mu(\partial_\mu + iA_\mu), \tag{6}$$

where the γ_μ are the Euclidean Dirac matrices with $\{\gamma_\mu, \gamma_\nu\} = 2\delta_{\mu\nu}$. In these lectures we will use the chiral representation of the gamma matrices (with diagonal γ_5).

2.1 Symmetries of the QCD Partition Function

It is well-known that the QCD action is greatly constrained by gauge symmetry, Euclidean Poincaré invariance and renormalizability. These symmetries determine the structure of the Dirac operator (6). In this section we will discuss the global symmetries of the Dirac operator. They play an essential role in its spectral properties in the deepest infrared. In particular, the chiral symmetry, the flavor symmetry and the anti-unitary symmetry of the continuum Dirac operator are discussed.

The chiral symmetry, or the $U_A(1)$ symmetry, can be expressed in terms of the anti-commutation relation

$$\{\gamma_5, D\} = 0. \tag{7}$$

This implies that all nonzero eigenvalues occur in pairs $\pm i\lambda_k$ with eigenfunctions given by ϕ_k and $\gamma_5\phi_k$. If $\lambda_k = 0$ the possibility exists that $\gamma_5\phi_k \sim \phi_k$, so that $\lambda_k = 0$ is an unpaired eigenvalue. According to the Atiyah-Singer theorem, the total number of such zero eigenvalues is a topological invariant, i.e., it does not change under continuous transformations of the gauge field configuration. Indeed, this possibility is realized by the field of an instanton which is a solution of the classical equations of motion. On the other hand, it cannot be excluded that $\lambda_k = 0$ while ϕ_k and $\gamma_5\phi_k$ are linearly independent. However, this imposes additional constraints on the gauge fields which will be violated by infinitesimal deformations. Generically, such situation does not occur.

In a decomposition according to the total number of topological zero modes, the QCD partition function can be written as

$$Z^{QCD} = \sum_\nu e^{i\nu\theta} Z_\nu^{QCD}, \tag{8}$$

where

$$Z_\nu^{QCD} = \langle \prod_f m_f^\nu \prod_k (\lambda_k^2 + m_f^2) \rangle_\nu. \tag{9}$$

Here, $\langle \cdots \rangle_\nu$ denotes the average over gauge-field configurations with topological charge ν weighted by the Yang-Mills action. In we introduce right-handed and left-handed masses as complex conjugated masses we find that the θ dependence of the QCD partition function is only through the combination $me^{i\theta/N_f}$.

A second important global symmetry is the flavor symmetry. This symmetry can be best explained by writing the fermion determinant in the QCD partition function as a functional integral over Grassmann variables,

$$\prod_f \det(D + m_f) = \int d\psi d\bar{\psi} e^{\int d^4x \sum_{f=1}^{N_f} \bar{\psi}^f (D+m_f)\psi^f}. \tag{10}$$

In a chiral basis with $\psi_R = \gamma_5\psi_R$ and $\psi_L = -\gamma_5\psi_L$, this can be rewritten as

$$\prod_f \det(D+m_f) = \int d\psi d\bar{\psi} e^{\int d^4x [\sum_{f=1}^{N_f} \bar{\psi}_R^f D\psi_L^f + \bar{\psi}_L^f D\psi_R^f + \bar{\psi}_R^f m_f \psi_R^f + \bar{\psi}_L^f m_f \psi_L^f]}. \tag{11}$$

For $m_f = 0$ we have the symmetry

$$\begin{aligned} \psi_L &\to U\psi_L, \quad \bar{\psi}_R \to \bar{\psi}_R U^{-1}, \\ \bar{\psi}_L &\to \bar{\psi}_L V^{-1}, \quad \psi_R \to V\psi_R. \end{aligned} \tag{12}$$

The only condition to be imposed on U and V is that their inverse exists. If the number of lefthanded modes is equal to the number of right-handed modes we thus have an invariance under $Gl(N_f)_R \times Gl(N_f)_L$. However, if the number of left-handed modes is not equal to the number of right-handed modes, the axial-symmetry group is broken to an $Sl(N_f)$ subgroup whereas the vector symmetry with $U = V$ remains unbroken. For $m = 0$ the flavor symmetry is thus broken explicitly to $Gl_V(N_f) \times Sl_A(N_f)$ by instantons or the anomaly.

What is much more important, though, is the spontaneous breaking of the axial flavor symmetry. From lattice QCD simulations and phenomenological arguments we know that the expectation value

$$\langle \bar{\psi}\psi \rangle = \langle \bar{\psi}_R \psi_R \rangle + \langle \bar{\psi}_L \psi_L \rangle \approx (240\,MeV)^3 \tag{13}$$

in the vacuum state of QCD instead of the symmetric possibility $\langle \bar{\psi}\psi \rangle = 0$. Phenomenologically, this is known because of the presence of Goldstone modes. The pions are much lighter than the σ mesons. The spontaneous breaking of the axial symmetry also follows from the absence of parity doublets. For example, the pion mass and the δ (or a_0) mass are very different ($m_\pi = 135MeV$ and $m_\delta = 980MeV$).

On easily verifies that $\langle \bar{\psi}\psi \rangle$ is only invariant for $U = V$. The vacuum state thus breaks the chiral symmetry down to $Gl_V(N_f)$. Notice that only the axial symmetries are broken. This is in agreement with the Vafa-Witten theorem [58] which states that vector symmetries cannot be broken in vector-like theories such as QCD. We also observe that the *complete* axial group is broken. The reasons behind this maximum breaking [59] of chiral symmetry are less well understood.

For three or more colors the gauge fields are not related by complex conjugation and there are no anti-unitary symmetries. However, the situation is more interesting for $N_c = 2$ which will be discussed in the next section.

2.2 Special role of $N_c = 2$

The group $SU(2)$ is the only nontrivial special unitary group that is pseudoreal. As will be shown in this section, the consequence is that for $N_c = 2$ the symmetry group of the QCD partition function in the chiral limit is enlarged to $U(2N_f)$.

As before, the fermionic action is obtained by writing the determinant in the partition function as a Gaussian integral over the Grassmann fields ϕ and $\bar{\phi}$. Using the chiral representation of the γ matrices with

$$\gamma_\mu = \begin{pmatrix} 0 & \hat{\sigma}_\mu \\ \hat{\sigma}_\mu^+ & 0 \end{pmatrix}, \tag{14}$$

and $\hat{\sigma}_\mu = (1, i\sigma_k)$ with σ_k the Pauli σ-matrices, the fermionic action can be written as

$$S_F = \int d^4x \sum_{f=1}^{N_f} \begin{pmatrix} \bar{\phi}_R \\ \bar{\phi}_L \end{pmatrix} \begin{pmatrix} m_f & \hat{\sigma}_\mu(\partial_\mu + iA_\mu) \\ \hat{\sigma}_\mu^+(\partial_\mu + iA_\mu) & m_f \end{pmatrix} \begin{pmatrix} \phi_R \\ \phi_L \end{pmatrix}. \tag{15}$$

For $N_c = 2$, we have that $A_\mu^T = -\tau_2 A_\mu \tau_2$ (with $\tau_2 = \sigma_2$ in color space). Combining this with the relation $\sigma_\mu^* = \sigma_2 \sigma_\mu \sigma_2$ we find

$$\bar{\phi}_L^f \hat{\sigma}_\mu^+ (\partial_\mu + iA_\mu)\phi_R^f = \phi_R^f \hat{\sigma}_2 \tau_2 \sigma_\mu (\partial_\mu + iA_\mu) \sigma_2 \tau_2 \bar{\phi}_L^f, \tag{16}$$

where we have used that ∂_μ is anti-Hermitean and that the fermion fields are anti-commuting Grassmann variables. The fermionic action can thus be rewritten as

$$S_F(m_f = 0) = \int d^4x \sum_{f=1}^{N_f} \begin{pmatrix} \bar{\phi}_R^f \\ \sigma_2\tau_2\phi_R^f \end{pmatrix} \begin{pmatrix} \hat{\sigma}_\mu(\partial_\mu + iA_\mu) & 0 \\ 0 & \hat{\sigma}_\mu(\partial_\mu + iA_\mu) \end{pmatrix} \begin{pmatrix} \phi_L^f \\ \sigma_2\tau_2\bar{\phi}_L^f \end{pmatrix} \tag{17}$$

Obviously the symmetry group is $Gl(2N_f)$. If the number of left-handed modes is not equal to the number of right-handed modes because of the anomaly, an axial $U(1)$ is broken explicitly as for three or more colors.

The mass term is given by

$$\begin{aligned} S_m = \int d^4x \sum_{f=1}^{N_f} \Bigg[& \begin{pmatrix} \phi_L^f \\ \sigma_2\tau_2\bar{\phi}_L^f \end{pmatrix} \begin{pmatrix} 0 & -m_f\sigma_2\tau_2 \\ m_f\sigma_2\tau_2 & 0 \end{pmatrix} \begin{pmatrix} \phi_L^f \\ \sigma_2\tau_2\bar{\phi}_L^f \end{pmatrix} \\ + & \begin{pmatrix} \bar{\phi}_R^f \\ \sigma_2\tau_2\phi_R^f \end{pmatrix} \begin{pmatrix} 0 & m_f\sigma_2\tau_2 \\ -m_f\sigma_2\tau_2 & 0 \end{pmatrix} \begin{pmatrix} \bar{\phi}_R^f \\ \sigma_2\tau_2\phi_R^f \end{pmatrix} \Bigg]. \end{aligned} \tag{18}$$

Also in this case we expect maximum spontaneous chiral symmetry breaking consistent with the Vafa-Witten theorem. This means that only the subgroup of $U(2N_f)$ that leaves both $\bar{\phi}_R\phi_R$ and $\bar{\phi}_L\phi_L$ invariant remains unbroken. Therefore, only the subgroup that leaves

$$\begin{pmatrix} \phi_L^f \\ \bar{\sigma}_2\tau_2\bar{\phi}_L^f \end{pmatrix} \begin{pmatrix} 0 & -\sigma_2\tau_2 \\ \sigma_2\tau_2 & 0 \end{pmatrix} \begin{pmatrix} \phi_L^f \\ \sigma_2\tau_2\bar{\phi}_L^f \end{pmatrix} \tag{19}$$

invariant remains unbroken ($\bar{\phi}_R\phi_R$ is invariant under the same transformations). This is the symplectic group $Sp(N_f)$. The Goldstone manifold is thus given by the coset $SU(2N_f)/Sp(N_f)$.

For $N_c = 2$ the Dirac-operator has the anti-unitary symmetry

$$[KC\tau_2, iD] = 0, \tag{20}$$

where K is the complex conjugation operator, $C = \gamma_2\gamma_4$ is the charge conjugation matrix. Because

$$(KC\tau_2)^2 = 1 \tag{21}$$

we can always find a basis such that the Dirac matrix is real for any D. The proof is along the same lines as the proof that time reversal symmetry results in real matrix elements for the Hamiltonian in quantum mechanics.

The anti-unitary symmetry would be violated in the presence of gauge fields B_μ coupling to the axial current (which are not present in QCD). In that case the Dirac operator is given by

$$\gamma_\mu\partial_\mu + i\gamma_\mu A_\mu + \gamma_\mu\gamma_5 B_\mu, \tag{22}$$

which, because of the absence of the factor i in the B_μ-term does not satisfy the commutation relation (20). In other words, the invariance of the Dirac operator under (20) follows from the condition $\gamma_\mu D\gamma_\mu = 2D$ (in four dimensions). For $N_c \geq 3$ this condition does not affect the anti-unitary symmetries, and the the matrix elements of the Dirac operator are arbitrary complex numbers both with and without the B_ν term [60].

2.3 Dirac operator in a Chiral Basis

It is instructive to write the Dirac operator in a chiral basis with n right-handed modes ϕ_k^R and $n+\nu$ left-handed modes ϕ_k^L. Using the anti-commutation relation (7) one can easily derive that

$$\langle\phi_k^R|D|\phi_l^R\rangle = \langle\phi_k^R|\gamma_5 D\gamma_5|\phi_l^R\rangle = -\langle\phi_k^R|D|\phi_l^R\rangle \tag{23}$$

from which it follows that $\langle \phi_k^R | D | \phi_l^R \rangle = 0$. Similarly, $\langle \phi_k^L | D | \phi_l^L \rangle = 0$, resulting in a Dirac Matrix with the structure

$$D = \begin{pmatrix} 0 & T \\ -T^\dagger & 0 \end{pmatrix}, \tag{24}$$

where T is an $n \times (n + \nu)$ matrix. From the inspection of the secular equation one concludes that such matrix has always exactly ν zero eigenvalues. For example, the eigenvalues of the matrix

$$D = \begin{pmatrix} 0 & a & b \\ -a^* & 0 & 0 \\ -b^* & 0 & 0 \end{pmatrix}, \tag{25}$$

are equal to 0 and $\pm i\sqrt{a^*a + b^*b}$. The matrix structure (24) is at the basis of the Random Matrix Theory to be discussed in section 4. A particular realization of a chiral basis is a basis of instanton zero modes which has been used extensively in the instanton liquid model of the QCD vacuum [61,27]

2.4 Lattice QCD

In lattice QCD, the gauge fields represented by unitary matrices on the links of the lattice, i.e.

$$U_{x,x+\mu} = e^{i\frac{T_a}{2} A_\mu^a(x)}, \tag{26}$$

where μ denotes a unit vector in the direction of x_μ. The gluonic lattice action of QCD is the plaquette action given by

$$S = \frac{1}{2} c \sum_P (S_P + S_P^\dagger). \tag{27}$$

For example, the action of an elementary plaquette 1234 is given by

$$S_P = \mathrm{Tr} U_{12} U_{23} U_{43}^{-1} U_{14}^{-1}. \tag{28}$$

This action is invariant under the local gauge transformations

$$U_{kl} \to V_k^{-1} U_{kl} V_l. \tag{29}$$

The fermion fields are represented by Grassmann valued fields on the sites. Under gauge transformations they transform according to

$$\psi_k \to U_{kl} \psi_l. \tag{30}$$

However, a lattice discretization of the Dirac action is more problematic. In a naive lattice discretization we are faced with the fermion doubling problem. This is highlighted by the Nielsen-Ninomiya theorem [62], which states that a lattice discretization of chiral fermions with local interactions is not possible.

There are two widely used discretizations of the fermionic action which deal with the doubling problem at the expense of some of the symmetries of the continuum QCD action: the Kogut-Susskind or staggered action and the Wilson action. The Kogut-Susskind action is defined by

$$\begin{aligned} S^{KS} &= \frac{1}{2}\sum_{n\,\mu} \eta_\mu(n)\bar{\chi}_n(U_{n,n+\mu}\chi_{n+\mu} - U^\dagger_{n-\mu,n}\chi_{n-\mu}) + m\sum_n \bar{\chi}_n\chi_n \\ &= \bar{\chi}_m D^{KS}_{mn}\chi_n, \end{aligned} \tag{31}$$

where $\eta_\mu(n)$ is a phase factor resulting from the diagonalization of the γ-matrices. For $m = 0$ and $n_1 + n_2 + n_3 + n_4$ even, this action is invariant under the transformations

$$\bar{\chi}_n \to e^{i\theta}\bar{\chi}_n, \qquad \chi_{n+\mu} \to e^{-i\theta}\chi_{n+\mu}. \tag{32}$$

For all n_i we have the invariance

$$\bar{\chi}_n \to e^{i\alpha}\bar{\chi}_n, \qquad \chi_n \to e^{-i\alpha}\bar{\chi}_n. \tag{33}$$

For $m \neq 0$ this $U(1) \times U(1)$ symmetry is broken to the second $U(1)$ symmetry only. If we organize the Kogut-Susskind Dirac matrix according to blocks with even and odd $n_1 + n_2 + n_3 + n_4$, one obtains the same block structure as in the chiral representation of the continuum Dirac operator (24). This shows that that the eigenvalues of D^{KS} occur in pairs $\pm\lambda$.

The staggered Dirac operator is anti-Hermitean with eigenvalues $i\lambda_k$ on the imaginary axis. The Dirac operator satisfies the sum-rule

$$\mathrm{Tr} D^{KS} D^{KS} = \mathrm{Tr} D^{KS} D^{KS}\big|_{U=1}. \tag{34}$$

This follows from the observation that U only occurs in the combination UU^{-1} in the left hand side of this equation. An important application of this sum rule is as a test of the accuracy of numerically calculated eigenvalues of the Dirac operator.

The Wilson Dirac action is defined by

$$\begin{aligned} S^W &= -\frac{1}{2}\sum_{n\,\mu} \bar{\psi}^f_n((r - \gamma_\mu)U_{n,n+\mu}\psi_{n+\mu} + (r + \gamma_\mu)U^\dagger_{n-\mu,n}\psi^f_{n-\mu} + \sum_n \bar{\psi}^f_n(m + 4r)\psi^f_n \\ &= \bar{\psi}^f_n D^W_{nm}\psi^f_m. \end{aligned} \tag{35}$$

The term proportional to r was introduced by Wilson to remove the fermion doublers. However, it also destroys the $U_A(1)$ symmetry of D^W, and the eigenvalues do not occur in pairs $\pm\lambda$. On the other hand, this action is invariant under the flavor group $SU_V(N_f)$. The Wilson Dirac operator satisfies the Hermiticity relation

$$D^{W\dagger} = \gamma_5 D^W \gamma_5 \tag{36}$$

from which it follows that $\gamma_5 D^W$ is Hermitean. In the literature Dirac spectra of both D^W with eigenvalues scattered in the complex plane[63,64,65] and $\gamma_5 D^W$ with eigenvalues on the real axis [66,67] have been studied (several reviews appeared recently[68,69]). Below we will restrict ourselves to the Hermitean Wilson Dirac operator. The Wilson Dirac operator satisfies the sum rule

$$\mathrm{Tr} D^{W\dagger} D^W = \mathrm{Tr} D^{W\dagger} D^W \big|_{U=1} . \tag{37}$$

Again this can be seen from the observation that U only occurs in the combination UU^{-1} in the l.h.s. of this equation.

The anti-unitary symmetries of the Wilson Dirac operator are the same as of the continuum Dirac operator. For $N_c = 2$ we thus have

$$[CK\tau_2, iD^W] = 0. \tag{38}$$

For staggered fermions only the eigenvalues of the γ matrices appear in the Dirac operator. Since they are real they do not affect the anti-unitary symmetry. For $N_c = 2$ we thus have

$$[K\tau_2, D^{KS}] = 0. \tag{39}$$

However, $(K\tau_2)^2 = -1$, and it can be shown that it is always possible to find a basis such that the matrix elements of D^{KS} can be organized into real quaternions for any gauge field configuration. The eigenvalues of such matrix are scalar quaternions and are thus doubly degenerate. This can also easily be shown from (39). Because of this relation, if ϕ is an eigenfunction with eigenvalue λ then $K\tau_2\phi$ is also an eigenfunction with eigenvalue λ. But

$$(\phi, K\tau_2\phi) = (K\tau_2\phi, \phi)^* = ((K\tau_2)^2\phi, K\tau_2\phi) = -(\phi, K\tau_2\phi), \tag{40}$$

from which it follows that ϕ and $K\tau_2\phi$ are linearly independent. For a more detailed discussion of the symmetries of the staggered Dirac operator we refer to [10,70,71].

2.5 The chiral phase transition in QCD

It is expected that chiral symmetry will be restored above some critical temperature. This is supported both by naive arguments based on counting the total number of degrees of freedom and by lattice QCD simulations [1]. In the chiral limit, the order parameter for the chiral phase transition is the chiral condensate, $\langle\bar{\psi}\psi\rangle$. For QCD with three color and two massless quarks, it vanishes above a critical temperature of about 140 MeV. The nature of the phase transition is still under dispute. It is not yet clear whether the transition is a second order or a weak first order one [1]. The role of instanton field configurations has to be clarified as well [72,68]. However, with current progress in computational resources, these questions should be answered in the near future. The only remaining fundamental problem is the study of QCD at finite baryon density. Because of the phase of the fermion determinant, this problem cannot be resolved by means of Monte-Carlo simulations. Progress in this area requires new paradigms which make it a particularly challenging area of active research.

2.6 The Banks-Casher relation and microscopic Spectral Density

The order parameter of the chiral phase transition, $\langle\bar{\psi}\psi\rangle$, is nonzero only below the critical temperature or a critical chemical potential. As was shown by Banks and Casher [5], $\langle\bar{\psi}\psi\rangle$ is directly related to the eigenvalue density of the QCD Dirac operator per unit four-volume

$$\Sigma \equiv |\langle\bar{\psi}\psi\rangle| = \lim \frac{\pi\langle\rho(0)\rangle}{V}, \tag{41}$$

where the spectral density of the Dirac operator with eigenvalues $\{\lambda_k\}$ is defined by

$$\rho(\lambda) = \langle\sum_k \delta(\lambda - \lambda_k)\rangle. \tag{42}$$

It is elementary to derive this relation. The chiral condensate is defined as the logarithmic derivative of the the partition function (8). For equal quark masses we obtain,

$$\langle\bar{\psi}\psi\rangle = -\lim \frac{1}{VN_f}\partial_m \log Z^{QCD}(m) = -\lim \frac{1}{V}\langle\sum_k \frac{2m}{\lambda_k^2 + m^2}\rangle. \tag{43}$$

If we express the sum as an integral over the average spectral density, and take the thermodynamic limit before the chiral limit, so that many eigenvalues are less than m, we recover (41). The order of the limits in (41) is important. First we take the thermodynamic limit, next the chiral limit and, finally, the field theory limit. As follows from (43), the sign of $\langle\bar{\psi}\psi\rangle$ changes if m crosses the real axis.

An important consequence of the Bank-Casher formula (41) is that the eigenvalues near zero virtuality are spaced as

$$\Delta\lambda = 1/\rho(0) = \pi/\Sigma V. \tag{44}$$

This should be contrasted with the eigenvalue spectrum of the non-interacting Dirac operator. Then the eigenvalues are those of a free Dirac particle in a box with eigenvalue spacing equal to $\Delta\lambda \sim 1/V^{1/4}$ for the eigenvalues near $\lambda = 0$. Clearly, the presence of gauge fields leads to a strong modification of the spectrum near zero virtuality. Strong interactions result in the coupling of many degrees of freedom leading to extended states and correlated eigenvalues.

Because the eigenvalues near zero are spaced as $\sim 1/\Sigma V$ it is natural to introduce the microscopic variable

$$u = \lambda V\Sigma, \tag{45}$$

and the microscopic spectral density [22]

$$\rho_s(u) = \lim_{V\to\infty} \frac{1}{V\Sigma}\langle\rho(\frac{u}{V\Sigma})\rangle. \tag{46}$$

We expect that this limit exists and converges to a universal function which is determined by the global symmetries of the QCD Dirac operator. The calculation of this universal function from QCD is the main objective of these lectures. We will calculate $\rho_s(u)$ both from the simplest theory in this universality class which is chiral Random Matrix Theory (chRMT) and from the partial quenched chiral Lagrangian which describes the low-energy limit of the QCD partition function. We find that the two results coincide below the Thouless energy.

2.7 Valence Quark mass Dependence of the chiral Condensate

Instead of studying the Dirac spectrum it is often convenient to consider the valence quark mass dependence of the chiral condensate. In terms of the eigenvalues of the Dirac operator it is defined by [73,29,30]

$$\begin{aligned}\Sigma(m_v; m_1, \cdots, m_{N_f}) &= \frac{1}{V}\sum_k \left\langle \frac{1}{i\lambda_k + m_v} \right\rangle \\ &= \frac{1}{V}\int d\lambda \frac{\rho(\lambda; m_1, \cdots, m_{N_f})}{i\lambda + m_v}.\end{aligned} \tag{47}$$

Here, $\langle\cdots\rangle$ denotes an average with respect to the distribution of the eigenvalues.

The relation (47) can then be inverted to give $\rho(\lambda; m_1, \ldots, m_{N_f})$. As mentioned in [16], the spectral density follows from the discontinuity across the imaginary axis,

$$\text{Disc}|_{m_v=i\lambda}\, \Sigma(m_v) = \lim_{\epsilon\to 0} \Sigma(i\lambda+\epsilon) - \Sigma(i\lambda-\epsilon) = 2\pi \sum_k \langle \delta(\lambda+\lambda_k)\rangle = 2\pi\rho(\lambda), \tag{48}$$

where we have suppressed the dependence on the sea-quark masses.

3 Infrared limit of the QCD partition function

3.1 The chiral Lagrangian

For light quarks the low energy limit of QCD is well understood. It is given by the chiral Lagrangian describing the interactions of the pseudoscalar mesons. The reason is that pions are Goldstone bosons which are the only light degrees of freedom in a confining theory such as QCD. To lowest order in the quark masses and the momenta, the chiral Lagrangian is completely dictated by chiral symmetry and Lorentz invariance. In the case of N_f light quarks with chiral symmetry breaking according to $SU_L(N_f)\times SU_R(N_f) \to SU_V(N_f)$ the so called Weinberg Lagrangian is given by [78]

$$\mathcal{L}_{\text{eff}}(U) = \frac{F^2}{4}\,\text{Tr}(\partial_\mu U\, \partial_\mu U^\dagger) - \frac{\Sigma}{2}\,\text{Tr}(\hat{\mathcal{M}}U + \hat{\mathcal{M}}U^\dagger), \tag{49}$$

where F is the pion decay constant, Σ is the chiral condensate and $\mathcal{M}$ is the quark mass matrix. The fields $U(x)$ are $SU(N_f)$ matrices parametrized as

$$U = \exp(i\sqrt{2}\Pi_a t^a/F), \tag{50}$$

with the generators of $SU(N_f)$ normalized according to $\text{Tr}t^a t^b = \delta^{ab}$. The chiral Lagrangian has been used extensively for the description of pion-pion scattering amplitudes.

To lowest order in the pion fields we find for equal quark masses

$$\mathcal{L}_{\text{eff}}(U) = \frac{1}{2}\partial_\mu\Pi^a\partial^\mu\Pi^a + \frac{\Sigma m}{F^2}\Pi^a\Pi^a. \tag{51}$$

This results in the pion propagator $1/(p^2+m_\pi^2)$ with pion mass given by the Gellmann-Oakes-Renner relation

$$m_\pi^2 = \frac{2m\Sigma}{F^2}. \tag{52}$$

In the long-wavelength limit the order of magnitude of the different terms contributing to the action of the chiral Lagrangian is given by [21]

$$S = \int d^4x \mathcal{L}(U) \sim L^{d-2} {\Pi^a_{NZM}}^2 + L^d \frac{\Sigma m}{F^2} ({\Pi^a_{ZM}}^2 + {\Pi^a_{NZM}}^2). \tag{53}$$

Here, Π^a_{ZM} represents the zero momentum modes with no space time dependence, whereas the nonzero momentum modes are denoted by by $\Pi^a_{NZM}(x)$. This decomposition has two immediate consequences. First, for $\frac{\Sigma m}{F^2} \gg \frac{1}{V}$ the fluctuations of the pion fields are small and it is justified to expand U in powers of Π^a. Second, for

$$\frac{\Sigma m}{F^2} \ll \frac{1}{\sqrt{V}} \tag{54}$$

the fluctuations of the zero modes dominate the fluctuations of the nonzero modes, and only the contribution from the zero modes has to be taken into account for the calculation of an observable. In this limit the so called finite volume partition function is given by

$$Z^{\text{eff}}_{N_f}(\mathcal{M}, \theta) \sim \int_{U \in SU(N_f)} dU e^{V \Sigma \mathrm{Re}\, \mathrm{Tr}\, \mathcal{M} U e^{i\theta/N_f}}, \tag{55}$$

where the θ-dependence follows from the dependence of the QCD partition function on the combination [8] $me^{i\theta/N_f}$. We emphasize that any theory with the same pattern of chiral symmetry breaking as QCD can be reduced to the same extreme infrared limit.

3.2 Leutwyler-Smilga Sum Rules

The finite volume partition function in the sector of topological charge ν follows by Fourier inversion according to (8). The partition function for $\nu = 0$ is thus given by (55) with the integration over $SU(N_f)$ replaced by an integral over $U(N_f)$.

The Leutwyler-Smilga sum-rules [8] are obtained by expanding the partition function in powers of m before and after averaging over the gauge field configurations and equating the coefficients. This corresponds to an expansion in powers of m of both the QCD partition function (4) and the finite volume partition function (55) in the sector of topological charge ν. As an example, we consider the coefficients of m^2 in the sector with $\nu = 0$. After performing the group integrals we find the sum-rule

$$\langle {\sum}' \frac{1}{\lambda_k^2} \rangle = \frac{\Sigma^2 V^2}{4N_f}, \tag{56}$$

where the prime indicates that the sum is restricted to nonzero positive eigenvalues.

By equating higher powers of m^2 one can generate an infinite family of sum-rules for the eigenvalues of the Dirac operator. However, they are not sufficient to determine the Dirac spectrum. The reason is that the mass in the propagator also occurs in the fermion-determinant of the QCD partition function. However, as will be shown in the next section, the Dirac spectrum can be obtained from a chiral Lagrangian corresponding to QCD with additional bosonic and fermionic quarks [16]. We conclude that chiral symmetry breaking leads to correlations of the inverse eigenvalues which are determined by the underlying global symmetries.

3.3 The partially Quenched QCD Partition Function

As was mentioned in previous section, the valence quark mass dependence of the chiral condensate cannot be extracted from the QCD partition function. The solution to this problem is to simultaneously introduce yet another quark species of opposite statistics [16]. This corresponds to a Euclidean partition function of the form

$$Z^{\rm pq} = \int dA \, \frac{\det(D + m_{v1})}{\det(D + m_{v2})} \prod_{f=1}^{N_f} \det(D + m_f) \, e^{-S_{YM}} \,, \qquad (57)$$

which we will call the partially quenched or pq-QCD partition function. When $m_{v1} = m_{v2}$ this partition function simply coincides with the original QCD partition function. However, it is now also the generator of a mass-dependent chiral condensate (see (47)) for the additional (say, fermionic) quark species. In the sector of topological charge ν we find

$$\Sigma(m_v; m_1, \cdots, m_{N_f}) = \frac{1}{V} \frac{\partial}{\partial m_{v1}} \bigg|_{m_{v1}=m_{v2}=m_v} \log Z_\nu^{\rm pq}. \qquad (58)$$

Our aim is to find the chiral Lagrangian corresponding to (57). If we are successful, we have succeeded in deriving a partition function for the extreme infrared limit of the spectrum of the QCD Dirac operator. These questions will be addressed in the next sections.

3.4 The infrared limit of QCD

If we don't write the determinants in terms of integrals over complex conjugated variables, the global flavor symmetry of the partially quenched QCD partition function (57) is broken spontaneously according to

$$Gl_R(N_f+1|1) \times Gl_L(N_f+1|1) \to Gl_V(N_f+1|1) \tag{59}$$

with an axial $U(1)$ group broken explicitly by the anomaly. Here, the groups $Gl_R(N_f+1|1)$, $Gl_L(N_f+1|1)$ and $Gl_V(N_f+1|1)$ are super-groups of matrices acting on vectors with N_f+1 fermionic components and one bosonic component. The subscript refers to right-handed (R), left-handed (L) and vector (V), respectively. The latter transformations transform the right-handed and the left-handed fermion fields in the same way. For a confining theory such as QCD the only low-lying modes are the Goldstone modes associated with the spontaneous breaking of chiral symmetry. The quark masses play the role of symmetry breaking fields.

Although the axial supergroup $Gl_A(N_f+1|1)$ is a symmetry group of the pq-QCD action (57) it is not necessarily a symmetry of the QCD partition function. It may be that the symmetry transformations violate the convergence of the integrals in of the partition function. There are no problems for the Grassmann variables. However, the integrations over the bosonic fields are only convergent if the fields are related by complex conjugation. For this reason a $U_A(1)$ transformation in which the bosonic fields ϕ_R and ϕ_L are transformed according to a different phase factor is not a symmetry of the partition function. What is a symmetry of the partition function is the axial transformation

$$\phi_L \to e^s \phi_L, \qquad \phi_L^* \to e^s \phi_L^*, \tag{60}$$

$$\phi_R^* \to e^{-s} \phi_R^* \qquad \phi_R \to e^{-s} \phi_R. \tag{61}$$

Mathematically, this symmetry group is $Gl(1)/U(1)$. Had we restricted ourselves to the unitary subgroup $U(N_f+1|1)$ of $Gl(N_f+1|1)$ from the start, we would have missed this class of symmetry transformations.

Taking into account the chiral anomaly, the chiral symmetry in (57) is broken spontaneously according to (59). The symmetry of the QCD partition function is thus reduced to $Sl_V(N_f+1|1) \oslash Gl_V(1)$ where $\oslash$ denotes the semi-direct product. The Goldstone manifold corresponding to the symmetry breaking pattern (59) is based on the symmetric superspace $Sl_A(N_f+1|1)$. In our effective partition function the terms that break the axial symmetry will be included explicitly resulting in an integration manifold given by $Sl_A(N_f+1|1) \otimes Gl_A(1)$. However, this manifold is not a super-Riemannian manifold and is not suitable as an integration domain for the low energy partition function. As an integration domain we choose the maximum Riemannian submanifold of $Gl_A(N_f+1|1)$. This results in a fermion-fermion block given by the compact domain $U_A(N_f+1)$, whereas the boson-boson block is restricted to the non-compact domain $Gl(1)/U(1)$. Because of the super-trace, this compact/non-compact structure is required for obtaining a positive definite quadratic form for the mass term and the kinetic term of our low energy effective partition function [79]. For a detailed mathematical discussion of this construction we refer to a paper by Zirnbauer [80].

To lowest order in the momenta, the infrared limit of the QCD partition function is uniquely determined by the geometry of the Goldstone manifold and Lorentz invariance. With the singlet field $\Phi_0 \equiv \mathrm{Str}\Phi$ fluctuating about the vacuum angle θ with an amplitude given by the singlet mass we thus obtain the effective partition function [18,19]

$$Z(\theta,\hat{\mathcal{M}}) = \int_{U\in Gl(N_f+1|1)} dU \exp \int d^4x \left[\frac{F^2}{4} \mathrm{Str}(\partial_\mu U\, \partial_\mu U^{-1}) \right. \\ \left. + \frac{\Sigma}{2} \mathrm{Str}(\hat{\mathcal{M}}U + \hat{\mathcal{M}}U^{-1}) + \frac{F^2 m_0^2}{12} (\frac{\sqrt{2}\Phi_0}{F} - \theta)^2 \right]. \tag{62}$$

This partition function contains terms up to order p^2. The kinetic term of the singlet field is subleading and has not been included. The supertrace (Str) and the superdeterminant (Sdet) of a graded matrix with bosonic blocks a and b and fermionic blocks σ and ρ are defined as [81,82]

$$\mathrm{Str}\begin{pmatrix} a & \sigma \\ \rho & b \end{pmatrix} = \mathrm{Tr}a - \mathrm{Tr}b, \tag{63}$$

and

$$\mathrm{Sdet} = \frac{\det(a-\sigma b^{-1}\rho)}{\det b}. \tag{64}$$

Recently, this partition function for $N_f = 0$ was derived from a two sublattice random flux model [83] using the flavor-color transformation introduced by Zirnbauer [84]. In our language, such model is QCD at infinite coupling with $U(N)$ gauge fields and quenched Kogut-Susskind fermions with no phase factor due to the γ matrices.

By a generalization of an argument due to Gasser and Leutwyler [21] to be discussed in the next section it can be shown that for masses in the range (1) this partition function factorizes into a product over zero momentum modes and non-zero momentum modes [14,16] To leading order, the mass dependence of the QCD partition function is thus given by

$$Z_{\rm eff}(\theta, \hat{\mathcal{M}}) = \int_{U \in Gl(N_f+1|1)} dU e^{V\frac{\Sigma}{2}\,{\rm Str}(\hat{\mathcal{M}}U + \hat{\mathcal{M}}U^{-1}) - \frac{F^2 m_0^2 V}{12}\left(\frac{\sqrt{2}\Phi_0}{F} - \theta\right)^2}. \quad (65)$$

The partition function in a sector of topological charge ν follows by Fourier inversion [8]

$$Z_{\rm eff}^{\nu}(\hat{\mathcal{M}}) = \frac{1}{2\pi}\int_0^{2\pi} d\theta e^{-i\nu\theta} Z_{\rm eff}(\theta, \hat{\mathcal{M}}). \quad (66)$$

We perform the integration over θ after a Hubbard-Stratonovitch transformation which linearizes the singlet mass term. Up to a mass independent factor this results in the partition function [16]

$$Z_{\rm eff}^{\nu}(\hat{\mathcal{M}}) = \int_{U \in Gl(N_f+1|1)} dU\, {\rm Sdet}^{\nu} U\, e^{V\frac{\Sigma}{2}\,{\rm Str}(\hat{\mathcal{M}}U + \hat{\mathcal{M}}U^{-1})}. \quad (67)$$

3.5 Domains in (Partially-Quenched) Chiral Perturbation Theory

In chiral perturbation theory the different domains of validity where analyzed in detail by Gasser and Leutwyler [21]. A similar analysis applies to partially quenched chiral perturbation theory. The idea is as follows. The U field can be decomposed as [21]

$$U = U_0 e^{i\psi(x)}. \quad (68)$$

where U_0 is a constant (zero-momentum) field. The kinetic term for the ψ fields is given by

$$\frac{1}{2}\int d^4x \partial_\mu \psi^a \partial_\mu \psi^a \sim L^2 \psi^2. \quad (69)$$

We observe that the magnitude of the fluctuations of the ψ field are of order $1/L$ which justifies a perturbative expansion of $\exp(i\psi(x))$. The fluctuations of the zero modes, on the other hand, are only limited by the mass term

$$\frac{1}{2} V\Sigma {\rm Str}\mathcal{M}(U_0 + U_0^{-1}). \quad (70)$$

For quark masses $m \gg 1/V\Sigma$, the field U_0 fluctuates close to the identity and the U_0 field can be expanded around the identity as well. This is the domain of chiral perturbation theory.

For valence quark masses in the range

$$\frac{1}{V\Sigma} \ll m_v \ll \frac{F^2}{\Sigma L^2} \tag{71}$$

the valence quark mass dependence of the chiral condensate is dominated by the zero momentum modes which can be treated perturbatively. Below we will show that in this domain chiral perturbation theory and random matrix theory coincide. In the theory of disordered mesoscopic systems it is well-known that random matrix theory is valid below an energy scale $\sim 1/L^2$. This will be discussed in the next section.

3.6 Picture from Mesoscopic Physics

In disordered mesoscopic physics it has been found that Random Matrix Theory is valid for eigenvalues separated less than the energy scale E_c (see eq. (2)). In this context E_c is defined as the inverse tunneling time of an electron through the sample. For diffusive motion, the distance Δx a particle diffuses in time interval $\Delta\tau$ is given by

$$(\Delta x)^2 = D\Delta\tau, \tag{72}$$

where D is the diffusion constant. The diffusion time through the sample is thus given by L^2/D resulting in a Thouless energy of

$$E_c = \frac{\hbar D}{L^2}. \tag{73}$$

A second energy scale that enters in mesoscopic physics is $\hbar/\tau_e$, where τ_e is the elastic scattering time. Based one these two scales three different domains for the energy difference, δE, that enters in the two-point correlation function, can be distinguished: the ergodic domain, the diffusive domain or Altshuler-Shklovskii domain and the ballistic domain. For energy differences $\delta E \ll E_c$ eigenvalue correlations are given by RMT. Since for time scales larger than the diffusion time, an initially localized wave packet explores the complete phase space, this regime is known as the ergodic regime. In the diffusive regime defined by $E_c \ll \delta E \ll \hbar/\tau_e$ only part of the phase space is explored by an initially localized wave packet, resulting in the disappearance of eigenvalue correlations. In this paper we don't consider the ballistic regime with $\delta E \gg \hbar/\tau_e$. For an interpretation of the Thouless energy in terms of the spreading width we refer to [85,86]. As was shown in [87,88], the spectral two-point function can be related to the semiclassical return probability which provides a simple intuitive picture of its asymptotic behavior. For other recent studies on this topic we refer to[89,90,91,92,93,94,95]. What has emerged from these studies is that there is a close relation between eigenvalue correlations and localization properties of the wave functions.

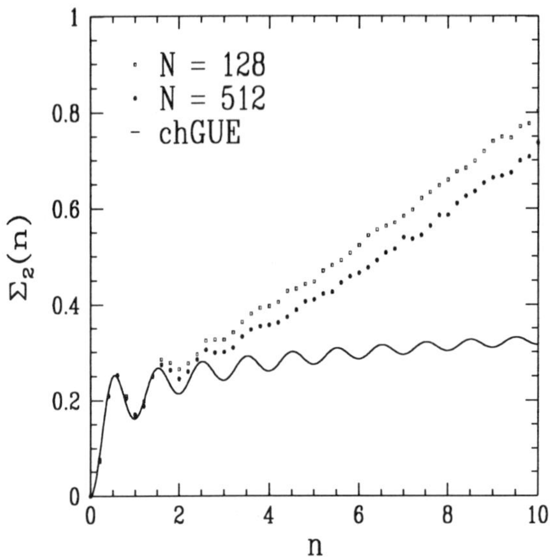

Figure 1: The number variance $\Sigma_2(n)$ versus n approximation for an interval starting at $\lambda = 0$. The total number of instantons is denoted by N.

Based on these ideas we can interpret the Dirac spectrum as the energy levels of a system in 4 Euclidean dimensions and one artificial time dimension. The corresponding classical evolution can be easily derived from the Heisenberg equations of motion

$$\frac{dx}{d\tau} = i[x, D]. \tag{74}$$

According to the Bohigas conjecture the eigenvalue correlations are given by RMT if and only if the corresponding classical motion chaotic. We thus conclude that the classical time evolution of quarks in the Yang-Mills gauge fields is chaotic.

These ideas can be tested by means of lattice QCD [34,35] and instanton liquid [14] simulations. In Fig. 1, we show [14] the number variance $\Sigma_2(n)$, defined as the variance of the number of levels in an interval containing n level on average, versus n for eigenvalues obtained from the Dirac operator in the background of instanton liquid gauge field configurations. The chRMT result, given by the solid curve, is reproduced up to about two level spacings. In units of the average level spacing, $\Delta = 1/\rho(0) = \pi/\Sigma V$, the energy E_c is given by $n_c \equiv E_c/\Delta = F^2L^2/\pi$. For an instanton liquid with instanton density $N/V = 1$ we find that $n_c \approx 0.07\sqrt{N}$. We conclude that chRMT appears to describe the eigenvalue correlations up to the predicted scale.

In mesoscopic physics the Goldstone modes that enter in the theory of impurity scattering are known as the diffusons. The reason is that the Goldstone propagator satisfies a diffusion equation (see for example the book by Efetov [82]). It describes the diffusion of electrons by impurity scattering [10]. Similarly, the pion propagator also satisfies a diffusion equation. This can be seen by Fourier transforming the propagator with regards to the imaginary valence masses $m_v = i\lambda$ and $m_{v'} = i\lambda'$,

$$\Pi(x,\tau) = \int \frac{d^4p}{(2\pi)^4} \int d\lambda d\lambda' \frac{e^{ipx+i\tau(\lambda+\lambda')}}{p^2 + (i\lambda + i\lambda')\Sigma/F^2}. \tag{75}$$

For $(x,\tau) \neq 0$ this propagator satisfies the diffusion equation

$$(D\partial_x^2 - \partial_\tau)\Pi(x,\tau) = 0 \tag{76}$$

with a diffusion constant given by[14,96] $D = F^2/\Sigma$.

4 Calculation of valence quark mass dependence of the chiral condensate from QCD

4.1 Nonperturbative evaluation of $\Sigma(m_v)$

In this section we calculate the valence quark mass dependence of the chiral condensate for the simplest case of $N_f = 0$ and $\nu = 0$ in the domain $m_v \ll F^2/\Sigma L^2$. In this domain the partition function is given by

$$Z(J) = \int_{U\in Gl(1|1)} dU \exp\left[\frac{\Sigma V}{2} \mathrm{Str} \begin{pmatrix} m_v + J & 0 \\ 0 & m_v \end{pmatrix} (U + U^{-1})\right], \tag{77}$$

where the integration is over the maximum super-Riemannian sub-manifold of $Gl(1|1)$. This manifold is parametrized by

$$U = \exp\begin{pmatrix} 0 & \alpha \\ \beta & 0 \end{pmatrix} \begin{pmatrix} e^{i\phi} & 0 \\ 0 & e^s \end{pmatrix} \tag{78}$$

The integration measure is the Haar measure which in terms of this parametrization it is given by

$$\mathrm{Sdet} \frac{\delta U_{kl}}{\delta\phi\, \delta s\, \delta\alpha\, \delta\beta}\ d\alpha d\beta d\phi ds, \tag{79}$$

where $\delta U \equiv U^{-1}dU$.

It is straightforward to calculate the Berezinian going from the variables $\delta U_{11}\, \delta U_{22}\, \delta U_{12}\, \delta U_{21}$ to the variables $\delta\phi\, \delta s\, \delta\alpha\, \delta\beta$. The derivative matrix is given by

$$B = \frac{\delta U_{kl}}{\delta\phi\, \delta s\, \delta\alpha\, \delta\beta} = \begin{pmatrix} i & 0 & \frac{\beta}{2} & \frac{\alpha}{2} \\ 0 & 1 & \frac{\beta}{2} & \frac{\alpha}{2} \\ 0 & 0 & e^{s-i\phi} & 0 \\ 0 & 0 & 0 & e^{-s+i\phi} \end{pmatrix}. \tag{80}$$

Using the definition of the graded determinant one simply finds that $\mathrm{Sdet} B = i$. Up to a constant, the integration measure is thus given by $d\phi ds d\alpha d\beta$. In general, for $N_f \neq 0$, the Berezinian is more complicated [81,17].

We also need

$$\frac{1}{2}(U + U^{-1}) = \begin{pmatrix} (1+\frac{\alpha\beta}{2})\cos\phi & \alpha(e^s - e^{-i\phi}) \\ \beta(e^{i\phi} - e^{-s}) & \cosh s(1 - \frac{\alpha\beta}{2}) \end{pmatrix}. \tag{81}$$

After differentiating with respect to the source term ($\Sigma(m_v) = \partial_J \log Z(J)|_{J=0}/V$) this results in

$$\frac{\Sigma(m_v)}{\Sigma} = \int \frac{d\phi ds d\alpha d\beta}{2\pi} \cos\phi(1 + \frac{\alpha\beta}{2}) e^{x\cos\phi(1+\frac{\alpha\beta}{2}) - x\cosh s(1-\frac{\alpha\beta}{2})}. \tag{82}$$

With the Grassmann integral given by the coefficient of $\alpha\beta$ we obtain

$$\frac{\Sigma(m_v)}{\Sigma} = \int \frac{ds d\phi}{4\pi}[\cos\phi + x(\cos\phi + \cosh s)\cos\phi] e^{x(\cos\phi - \cosh s)}. \tag{83}$$

Now all integrals can be expressed in terms of modified Bessel functions. We find

$$\frac{\Sigma(m_v)}{\Sigma} = I_1(x)K_0(x) + \frac{x}{2}(I_2(x)K_0(x) + I_0(x)K_0(x) + 2I_1(x)K_1(x)). \tag{84}$$

After using the recursion relation for modified Bessel functions,

$$I_2(x) = I_0(x) - \frac{2}{x}I_1(x), \tag{85}$$

we arrive at the final result [29,16,17]

$$\frac{\Sigma(m_v)}{\Sigma} = x(I_0(x)K_0(x) + I_1(x)K_1(x)), \tag{86}$$

where $x = mV\Sigma$.

This calculation can be generalized to arbitrary N_f and arbitrary ν. The calculation for arbitrary N_f is much more complicated, but with a natural generalization of the factorized parametrization, and using some known integrals over the unitary group [97], one arrives at an expression in terms of modified Bessel functions

$$\frac{\Sigma(x)}{\Sigma} = x(I_a(x)K_a(x) + I_{a+1}(x)K_{a-1}(x)), \tag{87}$$

where $a = N_f + |\nu|$. This result is in complete agreement [17] with chRMT to be discussed below.

4.2 Perturbative calculation

The valence quark mass dependence of the chiral condensate follows from (see eq. (62))

$$\Sigma(m_v) = \frac{1}{V}\partial_J \log Z^{\nu}_{\text{eff}}(\hat{\mathcal{M}})\Big|_{J=0}. \tag{88}$$

In the quenched limit this results in (see eq. (67)

$$\Sigma(m_v) = \int_{U(x)\in Gl(1|1)} dU(x)\frac{\Sigma V}{2}\partial_J \text{Str}\mathcal{M}(U(x) + U^{-1}(x))e^{S(U(x))}, \tag{89}$$

Perturbatively to one loop order we can expand U around the identity, and do the usual Gaussian integrations for a flat measure. The trace of the propagator of the Goldstone bosons containing valence quarks is given by

$$\langle \pi_k^2 \rangle = \frac{1}{V}\sum_p \frac{(-1)^L}{p^2 + M_\pi^2}, \tag{90}$$

where L is 1 for a bosonic pion and 1 for a fermionic pion. The valence pion mass, M_π, is given by the Gell-Mann-Oakes-Renner relation

$$M_{vv} = \frac{2m_v\Sigma}{F^2}, \quad \text{or} \quad M_{vs} = \frac{(m_v + m_s)\Sigma}{F^2}, \quad \text{or} \quad M_{ss} = \frac{2m_s\Sigma}{F^2}. \tag{91}$$

The calculation of the valence quark-mass dependence of the chiral condensate using the partially quenched chiral Lagrangian is basically straightforward. The only complication is the "η'"-mass term (the terms proportional to m_0^2 in the Lagrangian). However, the quadratic form in the pion fields can be diagonalized[19] resulting in an analytical expression for the propagator. In the limit that the η' mass is much larger than the mass of the Goldstone bosons, we find[48,16]

$$\frac{\Sigma_v(m_v)}{\Sigma} = 1 - \frac{1}{N_f F^2}\left\{N_f^2 \Delta(M_{sv}^2) - \Delta(M_{vv}^2) + (M_{ss}^2 - M_{vv}^2)\, \partial_{M_{vv}^2} \Delta(M_{vv}^2)\right\}. \tag{92}$$

where the trace of the finite volume pion propagator is given by

$$\Delta(M^2) = \frac{1}{V}\sum_p \frac{1}{p^2 + M^2}. \tag{93}$$

Here, the sum is over momenta in a box of volume L^4. From a careful analysis of the diverging momentum summations it follows that this propagator can be expanded in powers of M^2

$$\Delta(M^2) = \frac{1}{M^2 L^4} - \frac{\beta_1}{L^2} + O(\frac{M^2}{\Lambda^2}), \tag{94}$$

where $\beta_1 = 0.140461$ for a 4-dimensional hyper-cubic box, and Λ is the momentum cutoff[98]. In agreement with a naive inspection of the momentum sum, the propagator is dominated by the zero momentum term if

$$M \ll \frac{1}{L}, \tag{95}$$

or the Compton wavelength of the valence pion is much larger than the size of the box. As discussed in section (3.1) we thus find that for

$$m_v \ll \frac{F^2}{\Sigma L^2} \tag{96}$$

the valence quark mass dependence of the chiral condensate is described by the zero momentum contribution to the trace of the propagator. Let us investigate the zero momentum limit of (92) given by replacing the propagator by the first term in (94). In the chiral limit with $m_s \to 0$ we then find

$$\begin{aligned} \Sigma_v &\sim \Sigma(1 - \frac{N_f}{V M_{sv}^2 F^2}) \\ &\sim \Sigma(1 - \frac{N_f}{m_v V \Sigma}). \end{aligned} \tag{97}$$

The nonperturbative result for $\Sigma(m_v)$ obtained for a fixed value of the topological charge (87) ν instead of at $\theta = 0$ can also be expanded in powers of $1/x \equiv 1/m_v V\Sigma$

$$\frac{\Sigma_{RMT}(x)}{\Sigma} \sim 1 - \frac{N_f + |\nu|}{x} + \frac{|\nu|}{x} + O(\frac{1}{x^3}), \tag{98}$$

where the term $|\nu|/x$ results from the factor $m^{|\nu|}$ in the partition function [99]. We find that to leading order in $1/x$ the valence quark mass dependence of the chiral condensate does not depend on the topological charge and agrees with chiral perturbation theory.

The result for the valence quark mass dependence of the chiral condensate (92) is valid in the diffusive domain as well. In that case it is possible to rederive and to extend results by Smilga and Stern [100] for the slope of the Dirac spectrum.

4.3 Two-point correlation function

Let us do a perturbative calculation of the two-point function in the domain that is dominated by the zero-momentum modes, i.e. for valence quark masses well below $F^2/\Sigma L^2$. The generating function is defined by

$$Z(J, J') = \int_{U\in Gl(2|2)} dU e^{S(U)}, \tag{99}$$

where the action is given by

$$S(U) = \frac{\Sigma V}{2}\mathrm{Str}\begin{pmatrix} m+J & 0 & 0 & 0 \\ 0 & m'+J' & 0 & 0 \\ 0 & 0 & m & 0 \\ 0 & 0 & 0 & m' \end{pmatrix}(U + U^{-1}). \tag{100}$$

The spectral two-point correlation function follows from the partition function by [16]

$$\langle\rho(\lambda)\rho(\lambda')\rangle_c = \frac{1}{4\pi^2}\mathrm{Disc}\partial_J\partial_{J'}\ Z(J, J')|_{J=0,J'=0,m=i\lambda,m'=i\lambda'}\ . \tag{101}$$

We are interested in the discontinuity of the connected part of the correlation function. The only contribution is from mesons containing a valence quark with mass m and a valence quark with mass m'. If we expand $U \equiv \exp i\sqrt{2}\Pi_a T^a/F$ in powers of $\Pi_a T^a/F$ we find that the connected spectral two-point function

$$\langle\rho(\lambda)\rho(\lambda')\rangle_c = \frac{\Sigma^2}{F^4 4\pi^2}\ \mathrm{Disc}|_{m=i\lambda,m'=i\lambda'}\ \frac{1}{M^2_{vv'}}, \tag{102}$$

has been expressed in terms of the discontinuity of the valence pion susceptibility. To obtain this result we have used among others that

$$\mathrm{Str}\begin{pmatrix} 1\,0\,0\,0 \\ 0\,0\,0\,0 \\ 0\,0\,0\,0 \\ 0\,0\,0\,0 \end{pmatrix} T_a T^a = \frac{1}{2}, \tag{103}$$

and the super-symmetry of the partition function. To calculate the discontinuity we wish to remind the reader that

$$M_{vv'} = \frac{(|m| + |m'|)\Sigma}{F^2}, \tag{104}$$

where $|m| = \sqrt{(m^2)}$. We finally find

$$\langle \rho(\lambda)\rho(\lambda') \rangle_c = \frac{1}{2\pi^2}\left(\frac{1}{(\lambda+\lambda')^2} + \frac{1}{(\lambda-\lambda')^2}\right), \tag{105}$$

which agrees with the asymptotic result for the spectral correlation function [101] for the chiral Random Matrix Theory to be discussed next.

5 Chiral Random Matrix Theory

5.1 Definition of the Model

The chiral random matrix partition function with the global symmetries of the QCD partition function as input is defined by [22,23]

$$Z_\beta^\nu(\mathcal{M}) = \int DW \prod_{f=1}^{N_f} \det(\mathcal{D} + m_f) e^{-\frac{N\beta}{4}\Sigma \mathrm{Tr} W^\dagger W}, \tag{106}$$

where

$$\mathcal{D} = \begin{pmatrix} 0 & iW \\ iW^\dagger & 0 \end{pmatrix}, \tag{107}$$

and W is a $n \times m$ matrix with $\nu = |n - m|$ and $N = n + m$. As is the case in QCD, we assume that the equivalent of the topological charge ν does not exceed $\sqrt{N}$, so that, to a good approximation, $n = N/2$. Then the parameter Σ can be identified as the chiral condensate and N as the dimensionless volume of space time (Our units are defined such that the density of the modes $N/V = 1$).

The matrix elements of W are either real ($\beta = 1$, chiral Gaussian Orthogonal Ensemble (chGOE)), complex ($\beta = 2$, chiral Gaussian Unitary Ensemble (chGUE)), or quaternion real ($\beta = 4$, chiral Gaussian Symplectic Ensemble (chGSE)). For QCD with three or more colors and quarks in the fundamental representation the matrix elements of the Dirac operator are complex and we have $\beta = 2$. For $N_c = 2$ and quarks in the fundamental representation the situation is more interesting. As discussed in section (2.4) it is always possible to find a basis for which the matrix elements of the continuum Dirac operator are real for all gauge fields. Then the Dyson index of the corresponding chRMT is $\beta = 1$. For Kogut-Susskind fermions in this case the anti-unitary symmetry is different (see section (2.4)) and the matrix elements of the Dirac operator can be organized into real quaternions corresponding to a chRMT with $\beta = 4$. For gauge fields in the adjoint representation the gauge fields are real resulting in an anti-unitary symmetry also corresponding to the class $\beta = 4$ for any value of N_c.

The reason for choosing a Gaussian distribution of the matrix elements is its mathematically simplicity. This model can be generalized to arbitrary potential

$$\mathrm{Tr} W^\dagger W \rightarrow \mathrm{Tr} V(W^\dagger W) \tag{108}$$

where $V(x)$ is such that the resulting probability distribution is well defined. In section (5.4) we will argue that the interesting properties of chRMT do not depend on the choice of $V(x)$.

Below we will also discuss the invariant or Dyson-Wigner Random Matrix Ensembles. They are defined as ensembles of Hermitean matrices $\{H\}$ with independently Gaussian distributed matrix elements, i.e. with probability distribution given by

$$P(H) \sim e^{-\frac{N\beta}{2}\mathrm{Tr} H^\dagger H}. \tag{109}$$

Depending on the anti-unitary symmetry, the matrix elements are real, complex or quaternion real. They are called the Gaussian Orthogonal Ensemble (GOE), the Gaussian Unitary Ensemble (GUE) and the Gaussian Symplectic Ensemble (GSE), respectively. Each ensemble is characterized by its Dyson index β which is defined as the number of independent variables per matrix element. For the GOE, GUE and the GSE we thus have $\beta = 1$, 2 and 4, respectively.

Together with the Wigner-Dyson ensembles and the superconducting random matrix ensembles [102] the chiral ensembles can be classified according to the Cartan classification of large symmetric spaces [80].

5.2 Symmetries

From the structure of the determinant, it follows immediately that the chiral random matrix partition function has the same global flavor and $U_A(1)$ symmetries as the QCD partition function.

It can be shown that in the domain (1) the random matrix partition function can be mapped onto the effective finite volume partition function [22]. We will discuss such derivation in section (7.3) where we extend this model to finite temperature. The most detailed studies to this model were performed in the quenched limit by means of the supersymmetric method [74,75,76,77]. In that case the chRMT partition function can be mapped onto a super-symmetric nonlinear σ-model.

In this model chiral symmetry is broken spontaneously with chiral condensate given by $\Sigma = \lim_{N\to\infty} \pi\rho(\lambda \to 0)/N$, where N is interpreted as the (dimensionless) volume of space time. For complex matrix elements ($\beta = 2$), which is appropriate for QCD with three or more colors and fundamental fermions, the symmetry breaking pattern is [103] $SU(N_f) \times SU(N_f)/SU(N_f)$. For $\beta = 1$ and 4 the symmetry breaking pattern is $SU(2N_f)/Sp(N_f)$ and $SU(N_f)/O(N_f)$ respectively [103], the same as in QCD [59].

Finally, one of the reasons for the mathematical simplicity of this model is that it has more symmetry than the QCD partition function The chRMT partition function is invariant with respect to the unitary transformation

$$W \to UWV^{-1}. \tag{110}$$

Since an arbitrary complex matrix can be decomposed as

$$W = U\Lambda V^{-1}, \tag{111}$$

with Λ a diagonal matrix and U and V unitary matrices, the chRMT partition function for arbitrary potential $V(\lambda)$ can be expressed in terms of the eigenvalues λ_k as

$$Z_\beta^\nu(\mathcal{M}) = \int d\lambda |\Delta(\lambda_k^2)|^\beta \prod_k \lambda_k^{\beta\nu+\beta-1} \prod_f m_f^\nu \prod_{f,k} (\lambda_k^2 + m_f^2) e^{-\frac{N\beta}{4}\sum_k V(\lambda_k^2)}, \tag{112}$$

where the Vandermonde determinant is defined by

$$\Delta(\lambda_k^2) = \prod_{k<l} (\lambda_k^2 - \lambda_l^2). \tag{113}$$

This result greatly simplifies the mathematical treatment of the chRMT partition function. For example, it allows us to use the orthogonal polynomial method. We wish to point out that for $\beta = 2$ and $m_f = 0$ the ratio of the partition function and $m^{N_f|\nu|}$ depends only on ν through the combination $N_f + |\nu|$. This duality between flavor and topology can be understood more directly via Itzykson-Zuber integrals [104,105].

5.3 Properties of the Random Matrix Model

The average spectral density that can be derived from (106) has the familiar semi-circular shape. As can be easily derived from the orthogonal polynomial method, the microscopic spectral density for the chGUE is given by [106,101,23]

$$\rho_S(u) = \frac{u}{2}\left(J_a^2(u) - J_{a+1}(u)J_{a-1}(u)\right), \tag{114}$$

where $a = N_f + |\nu|$. The valence quark mass dependence of the chiral condensate follows from integration over the microscopic spectral density. If we use the microscopic variable $u = \lambda V\Sigma$ as new integration variable, equation (47) can be rewritten as

$$\frac{\Sigma(m_v)}{\Sigma} = \int_0^\infty du \frac{2u\,du}{u^2 + (m_v V\Sigma)^2} \frac{1}{V\Sigma} \rho(\frac{u}{V\Sigma}). \tag{115}$$

In the thermodynamic limit, the spectral density can be replaced by the microscopic spectral density (114). The integrals over the Bessel functions are known and result in the following expression for the valence quark mass dependence of the chiral condensate [29]

$$\frac{\Sigma(x)}{\Sigma} = x(I_a(x)K_a(x) + I_{a+1}(x)K_{a-1}(x)), \tag{116}$$

where $a = N_f + |\nu|$, $x = m_v V\Sigma$ and I_a and K_a are modified Bessel functions. The spectral correlations near $\lambda = 0$ can also be expressed in terms of Bessel functions [101], whereas in the bulk of the spectrum they are given by the invariant random matrix ensembles [107,108]. The microscopic spectral density and the microscopic spectral correlations can also be derived for $\beta = 1$ and $\beta = 4$ but these two cases are mathematically much more complicated. Many other properties of the chRMT partition function have been calculated. Among others, we mention the distribution of the smallest eigenvalue [109,110] and the microscopic spectral density in the presence of nonzero quark masses [110,111,112,113]. For more discussion of the chRMT partition function we refer to [104].

5.4 *Universality*

The aim of universality studies is to identify observables that are stable against deformations of the random matrix ensemble. Not all observables have the same degree of universality. For example, a semicircular average spectral density is found for random matrix ensembles with independently distributed matrix elements with a finite variance. However, this spectral shape does not occur in nature, and it is thus not surprising that it is only found in a rather narrow class of random matrix ensembles. What is surprising is that the *microscopic* spectral density and the *microscopic* spectral correlators are stable with respect to a much larger class of deformations[114,115,116,117,118,119,120,121,122,123]. Two different types of deformations have been considered, those that maintain the unitary invariance of the partition functions and those that break the unitary invariance.

In the first class, the Gaussian probability distribution is replaced by $P(W) \sim \exp(-N\sum_{k=1}^{\infty} a_k \mathrm{Tr}(W^\dagger W)^k)$. For a potential with only a_1 and a_2 different from zero it was shown [116] that the microscopic spectral density is independent of a_2. A general proof valid for arbitrary potential and all correlation functions was given by Akemann et al. [117]. The essence of the proof is a remarkable generalization of the identity for the Laguerre polynomials, $\lim_{n\to\infty} L_n(x/n) = J_0(2\sqrt{x})$, to orthogonal polynomials determined by an arbitrary potential. It was proved by taking the continuum limit of the recursion relation for orthogonal polynomials.

In the second class, an arbritrary fixed matrix is added to W in the Dirac operator (107). It has been shown that the microscopic spectral density and the microscopic spectral correlations remain unaffected [75,76,77] for parameter values that completely modify the average spectral density.

Microscopic universality for deformations that affect the macroscopic spectral density implies the existence of a scale beyond which universality breaks down. It can be interpreted naturally in terms of the spreading width [85] and is the RMT equivalent of the Thouless energy.

Based on the general form of of the pqChPT partition function one could argue that universality of the microscopic correlators in chRMT is automatic. However, one really has to show the stability of the effective partition function with respect to variations of the distribution of matrix elements. For the Wigner-Dyson ensembles the stability of the saddle-point manifold was demonstrated in [115].

For reasons of mathematical simplicity, most universality proofs have been performed for the $\beta = 2$ ensembles. Recently, we have attempted to prove universality by establishing relations between the kernels for the $\beta = 1$ and

$\beta = 4$ correlation functions and the kernel for the $\beta = 2$ correlation functions. For the Gaussian ensembles such relations are exact identities [124]. However, for an arbitrary potential they are only valid asymptotically [125,126].

There are many other results related to the universality of chRMT. We mention recent results in QCD in three dimensions [117,127,128], results for Dirac operators satisfying the Ginsparg-Wilson relation [129], relations between the microscopic spectral density and partition functions with two additional flavors [130].

6 Chiral RMT and an exact Theory for The Fluctuations of Dirac Eigenvalues

6.1 *Statistical Analysis of Spectra*

Spectra for a wide range of complex quantum systems have been studied both experimentally and numerically (a excellent recent review was given by Guhr, Müller-Groeling and Weidenmüller [7]). One basic observation is that the scale of variations of the average spectral density and the scale of the spectral fluctuations separate. This allows us to unfold the spectrum, i.e. we rescale the spectrum in units of the local average level spacing. Specifically, the unfolded spectrum is given by

$$\lambda_k^{\rm unf} = \int_{-\infty}^{\lambda_k} \langle \rho(\lambda') \rangle d\lambda', \tag{117}$$

with unfolded spectral density

$$\rho_{\rm unf}(\lambda) = \sum_k \delta(\lambda - \lambda_k^{\rm unf}). \tag{118}$$

The fluctuations of the unfolded spectrum can be measured by suitable statistics. We will consider the nearest neighbor spacing distribution, $P(S)$, and moments of the number of levels in an interval containing n levels on average. In particular, we will consider the number variance, $\Sigma_2(n)$, and the first two cumulants, $\gamma_1(n)$ and $\gamma_2(n)$. Another useful statistic is the $\Delta_3(n)$-statistic introduced by Dyson and Mehta [131]. It is related to $\Sigma_2(n)$ via a smoothening kernel. The advantage of this statistic is that its fluctuations as a function of n are greatly reduced. Both $\Sigma_2(n)$ and $\Delta_3(n)$ can be obtained from the pair correlation function.

Analytical results for all spectral correlation functions have been derived for each of the three ensembles [132] via the orthogonal polynomial method. We only quote the most important results. The nearest neighbor spacing distribution, which is known exactly in terms of a power series, is well approximated by

$$P(S) \sim S^{\beta} \exp(-a_{\beta} S^2), \tag{119}$$

where a_β is a constant of order one. The asymptotic behavior of $\Sigma_2(n)$ and $\Delta_3(n)$ is given by

$$\Sigma_2(n) \sim (2/\pi^2\beta) \log n \quad \text{and} \quad \Delta_3(n) \sim \beta\Sigma_2(n)/2. \tag{120}$$

Characteristic features of random matrix correlations are level repulsion at short distances and a strong suppression of fluctuations at large distances.

For uncorrelated eigenvalues the level repulsion is absent and one finds

$$P(S) = \exp(-S), \tag{121}$$

and

$$\Sigma_2(n) = n \quad \text{and} \quad \Delta_3(n) = n/15. \tag{122}$$

6.2 Results for Spectral Correlation Functions

Recently, lattice QCD Dirac spectra were calculated and analyzed by a number of different groups. It was found that spectral correlations below a given scale are given by chRMT, whereas above this scale, deviations from chRMT due to the nonzero momentum excitations were observed. [28,29,30,31,32,33,34,35,36 37,38,39,40,41,42,46,43 22,47,14,15].

The first complete Dirac spectra on relatively large lattices were obtained by Kalkreuter [133] who calculated *all* eigenvalues of the $N_c = 2$ Dirac operator both for Kogut-Susskind (KS) fermions and Wilson fermions for lattices as large as 12^4. In both cases, the Dirac matrix was tri-diagonalized by Cullum's and Willoughby's Lanczos procedure [134] and diagonalized with a standard QL algorithm. This improved algorithm makes it possible to obtain *all* eigenvalues. This allows us to test the accuracy of the eigenvalues by means of sum-rules for the sum of the squares of the eigenvalues of the lattice Dirac operator (see eqs. (34, 37)). Typically, the numerical error in the sum rule is of order 10^{-8}.

Results for the spectral correlations for eigenvalues calculated by Kalkreuter [133] both for KS and Wilson fermions are shown in Fig. 2. The simulations for KS fermions were performed for 4 dynamical flavors with $ma = 0.05$ on a 12^4 lattice (the lattice spacing is denoted by a). The simulations for Wilson fermions were done for two dynamical flavors on an $8^3 \times 12$ lattice.

The results for $\Sigma_2(n)$, $\gamma_1(n)$ and $\gamma_2(n)$ obtained by spectral averaging show an impressive agreement with the RMT predictions. In the context of RMT it has been shown that spectral averages and ensemble averages coincide [135]. This property is known as spectral ergodicity.

Spectra for different values of β have been analyzed as well. It is probably no surprise that random matrix correlations are found at stronger couplings. What is surprising, however, is that even in the weak-coupling domain ($\beta = 2.8$) the eigenvalue correlations are in complete agreement with Random Matrix Theory.

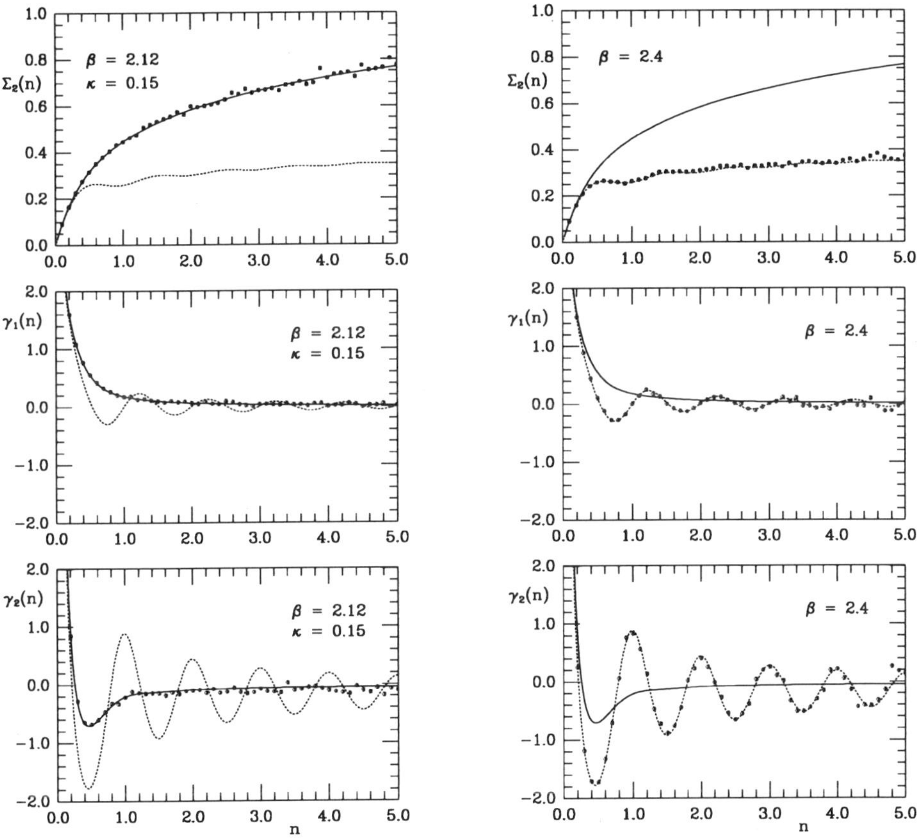

Figure 2: The number variance, $\Sigma_2(n)$ and the first two cumulants, $\gamma_1(n)$ and $\gamma_2(n)$ as a function of n.

In the case of three or more colors with fundamental fermions, both the

Wilson and Kogut-Susskind Dirac operator do not possess any anti-unitary symmetries. Therefore, our conjecture is that in this case the spectral correlations in the bulk of the spectrum of both types of fermions can be described by the GUE. This was recently confirmed for a wide range of β-value both below and above the deconfinement phase transition [39,43].

As far as correlations in the bulk of the spectrum are concerned, in the case of two fundamental colors the continuum theory and Wilson fermions are in the same universality class. It is an interesting question of how spectral correlations of KS fermions evolve in the approach to the continuum limit. Certainly, the Kramers degeneracy of the eigenvalues remains. However, since Kogut-Susskind fermions represent 4 degenerate flavors in the continuum limit, the Dirac eigenvalues should obtain an additional two-fold degeneracy. We are looking forward to more work in this direction.

6.3 Correlations near zero virtuality

Spectral ergodicity cannot be exploited in the study of the microscopic spectral density and, in order to gather sufficient statistics, a large number of independent spectra is required. One way to proceed is to generate instanton-liquid configurations which can be obtained much more cheaply than lattice QCD configurations. Results of such analysis [47] show that for $N_c = 2$ with fundamental fermions the microscopic spectral density is given by the chGOE. For $N_c = 3$ it is given by the chGUE. One could argue that instanton-liquid configurations can be viewed as smoothened lattice QCD configurations. Roughening such configurations will only improve the agreement with Random Matrix Theory. Recently, these expectations were confirmed by lattice QCD simulations [31]. Results for 1416 quenched $SU(2)$ Kogut-Susskind Dirac spectra on a 10^4 lattice are shown in Fig. 3. We show both the distribution of the smallest eigenvalue (left) and the microscopic spectral density (right). The results [136] for the chGSE are represented by the dashed curves.

Agreement of the microscopic spectral density with chRMT for $N_c = 3$ was first demonstrated by means of the valence quark mass dependence of the chiral condensate [29]. Recently, these results were confirmed by the calculation of complete Dirac spectra [41,42,46].

6.4 The valence quark mass dependence of the chiral condensate

The microscopic spectral density near zero virtuality was first studied in terms of the valence quark mass dependence of the chiral condensate. The lattice data for $\Sigma(m_v)$ were obtained by the Columbia group [73] for two dynamical flavors with sea-quark mass $ma = 0.01$ and $N_c = 3$ on a $16^3 \times 4$ lattice. In

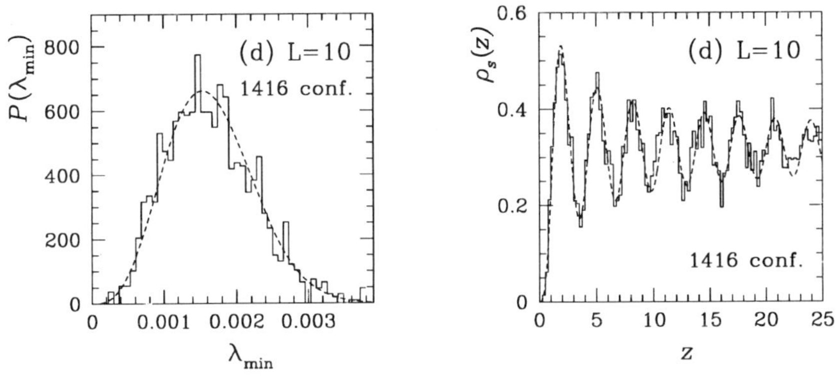

Figure 3: The distribution of the smallest eigenvalue (left) and the microscopic spectral density (right) for two colors and $\beta = 2.0$.

Fig. 4 we plot the ratio $\Sigma(m_v)/\Sigma$ as a function of $x = m_v V\Sigma$ (the 'volume' V is equal to the total number of Dirac eigenvalues), and compare the results with the universal curves obtained from pq-QCD and chRMT (see eq. (87)). We observe that the lattice data for different values of β fall on a single curve. Moreover, in the mesoscopic range this curve coincides with the random matrix prediction for $N_f = \nu = 0$. Of course this is no surprise. In the region where

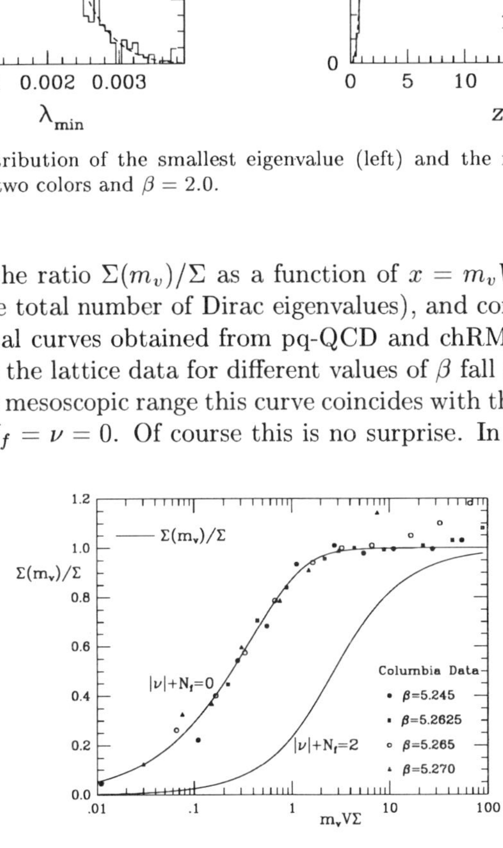

Figure 4: The valence quark mass dependence of the chiral condensate $\Sigma(m)$ plotted as $\Sigma(m_v)/\Sigma$ versus $m_v V\Sigma$. The dots and squares represent lattice results by the Columbia group [73] for values of β as indicated in the label of the figure.

valence quark mass is much less than the current quark mass, the fermion determinant has no bearing on the Dirac spectrum and we have effectively

$N_f = 0$. On a lattice the zero mode states and a much larger number of nonzero modes states are completely mixed, and therefore it is not surprising that the effective topological charge that reproduces the lattice results is equal to zero.

7 Chiral Random Matrix Theory as a Schematic Model for the Dirac Spectrum

7.1 Introduction

Up to now we have mainly focussed on chiral Random Matrix Theory as an exact theory for correlations of QCD Dirac spectra on the microscopic scale. However, in the literature one finds another important application of RMT, namely as a schematic model for a disordered system. Two well-known examples in this category are the Anderson model for localization [137] and RMT applied to the theory of random surfaces in connection with quantum gravity [138]. In this section we will study chiral random matrix theory as a *schematic* model of the chiral phase transition at nonzero temperature and chemical potential.

7.2 chRMT and Nonzero Temperature

In order to obtain a chiral random matrix model for QCD at nonzero temperature we first separate the Dirac operator in two pieces

$$D = \tilde{D} + \gamma_0 \partial_0. \tag{123}$$

In a chiral basis with basis functions given by

$$\psi_{kn}(x) = \phi_k(\vec{x}) e^{-\frac{2\pi i}{\beta}(n+\frac{1}{2})x_0} \tag{124}$$

the matrix corresponding to $\gamma_0 \partial_0$ is block diagonal with nonzero matrix elements given by the Matsubara frequencies. In our model we only keep the lowest Matsubara frequencies $\pm\pi T$. We expect that this is a reasonable approximation in the neighborhood of the critical temperature and beyond. The matrix corresponding to $\tilde{D}$ is replaced by a chiral random matrix.

The chRMT Dirac operator at $T \neq 0$ is thus given by [49,50]

$$D^{RMT} = \begin{pmatrix} 0 & iW \\ iW^\dagger & o \end{pmatrix} + \begin{pmatrix} 0 & i\Omega_T \\ i\Omega_T & 0 \end{pmatrix}, \tag{125}$$

where Ω_T is given by

$$\Omega_T = \begin{pmatrix} \pi T \mathbf{1} & 0 \\ 0 & -\pi T \mathbf{1} \end{pmatrix}, \tag{126}$$

and $\mathbf{1}$ is a unit matrix.

If we note that the probability distribution of the random matrix ensemble is invariant under $W \to UWV^{-1}$ and use the equivalent of $i\sigma_1\sigma_3\sigma_2 = 1$, the random matrix Dirac operator can be simplified to

$$D(T) = \begin{pmatrix} 0 & iW + \pi i T\mathbf{1} \\ iW^\dagger + \pi i T\mathbf{1} & 0 \end{pmatrix}. \tag{127}$$

Notice that the dimension of the identity matrix in eqs. (126) and (127) is different.

Our finite temperature chiral random matrix model is thus given by

$$Z(\nu, N_f) = \int dW \det{}^{N_f}(D(T) + m) e^{-n\Sigma^2 \mathrm{Tr} WW^\dagger}. \tag{128}$$

In the next we will perform a saddle point analysis of this model.

7.3 Analytical Solution

In this section we evaluate the partition function (106) using methods which are standard in the supersymmetric formulation of random matrix theory[11,12]. For simplicity we consider the model for $\beta = 2$ and choose a diagonal mass matrix with equal quark masses. The first step is to perform the average over W by performing a gaussian integral. This leads to a four-fermion interaction. After averaging over the matrix elements of the Dirac operator, the partition function becomes

$$\begin{aligned} Z(\nu, N_f) = \int \mathcal{D}\psi^* \mathcal{D}\psi \exp[&- \frac{1}{n\Sigma^2} \psi^{f\,*}_{L\,k} \psi^f_{R\,i} \psi^{g\,*}_{R\,i} \psi^g_{L\,k} + m\psi^{f\,*}_{R\,i}\psi^f_{R\,i} + m\psi^{f\,*}_{L\,k}\psi^f_{L\,k} \\ &- i\pi T(\psi^{f\,*}_{R\,i}\psi^f_{L\,i} + \psi^{f\,*}_{L\,k}\psi^f_{R\,k})]. \end{aligned} \tag{129}$$

The four-fermion terms can be written as the difference of two squares. Each square can be linearized by

$$\exp(-AQ^2) \sim \int d\sigma \exp(-\frac{\sigma^2}{4A} - iQ\sigma) \ . \tag{130}$$

The different Q variables, can be combined into a single complex $N_f \times N_f$ matrix, A, resulting in

$$\begin{aligned} Z(\nu, N_f) = \int \mathcal{D}A\mathcal{D}\psi\mathcal{D}\psi^* \exp[&-\frac{n\Sigma^2\beta}{2} \mathrm{Tr} AA^\dagger \\ &- i\psi^{f\,*}_{L\,k}\psi^g_{L\,k}(A + m) - i\psi^{f\,*}_{R\,i}\psi^g_{R\,i}(A^\dagger + m) - i\pi T(\psi^{f\,*}_{R\,i}\psi^f_{L\,i} + \psi^{f\,*}_{L\,k}\psi^f_{R\,k})] \ . \end{aligned} \tag{131}$$

Note that the temperature-dependent term can be rewritten as

$$-\frac{i\pi T}{2}\begin{pmatrix}\psi_L\\ \psi_L^*\end{pmatrix}\begin{pmatrix}0 & -\mathbf{1}\\ \mathbf{1} & 0\end{pmatrix}\begin{pmatrix}\psi_R\\ \psi_R^*\end{pmatrix}+(L\longleftrightarrow R). \tag{132}$$

Using this, the fermionic integrals can be performed, and the partition function with $|\nu|$ more left-handed modes than right-handed modes is given by

$$Z(\nu,N_f)=\int \mathcal{D}A\exp[-\frac{n\Sigma^2\beta}{2}\mathrm{Tr}AA^\dagger]\mathrm{det}^{|\nu|}(A+m)\mathrm{det}^n\begin{pmatrix}A+m & -\pi iT\\ -\pi iT & A^\dagger+m\end{pmatrix}. \tag{133}$$

Here, A is an arbitrary complex matrix and m is proportional to the identity matrix.

The chiral condensate is defined by

$$\langle\bar\psi\psi\rangle=-\frac{1}{2nN_f}\partial_m\log Z\ , \tag{134}$$

For $n\to\infty$ it can be evaluated with the aid of a saddle point approximation. The saddle point equations are given by

$$-\frac{n\beta\Sigma^2}{2}A+n(A+m)\left((A^\dagger+m)(A+m)+\pi^2T^2\right)^{-1}=0\ . \tag{135}$$

An arbitrary complex matrix can be diagonalized by performing the decomposition

$$A=U\Lambda V^{-1}, \tag{136}$$

with all eigenvalues positive and U and V unitary matrices. We find that the solution of (135) yields $U=V=1$ with eigenvalues Λ_k given by the positive roots of

$$\Sigma^2\Lambda_k((\Lambda_k+m)^2+\pi^2T^2)-\Lambda_k-m=0\ . \tag{137}$$

In order to calculate the condensate, we express the derivative of the partition function in (3.10) in terms of an average over A,

$$|\langle\bar\psi\psi\rangle|=\frac{1}{2nN_f}\langle\mathrm{Tr}\begin{pmatrix}A & -\pi iT\\ -\pi iT & A^\dagger\end{pmatrix}^{-1}\rangle\ . \tag{138}$$

Below T_c and for $m\to 0$ we find from the saddle point equation,

$$|\langle\bar\psi\psi\rangle|=\Sigma(1-\pi^2T^2\Sigma^2)^{1/2}\ . \tag{139}$$

In the chiral limit we thus find a critical point at

$$T_c = \frac{1}{\pi\Sigma} \quad . \tag{140}$$

Note that the dimensions are matched after including the mode density $N/V = 1$. Here, and elsewhere, our convention is that $N = 2n$. At T_c the solution of the saddle point equation develops a non-analytic dependence on m resulting in the condensate

$$|\langle\bar{\psi}\psi\rangle| = \Sigma^{\frac{4}{3}} m^{\frac{1}{3}} \quad . \tag{141}$$

Therefore, we reproduce the mean field value for the critical exponent $\delta = 3$.

It is also possible to obtain an analytical expression for the spectral density of this model as a function of the critical temperature. It can be obtained by solving the Schwinger-Dyson equations for the resolvent of the model [49] or from the observation that the solution of (135) does not depend on the number of flavors. For $N_f \to 0$ the spectral density is then obtained from the discontinuity of the chiral condensate across the real m-axis [51]. The spectral density thus follows from the solution of a cubic equation. The zero temperature limit is a semicircle whereas the high temperature limit is given by two semi-circles separated by $2\pi T$.

7.4 chRMT at Nonzero Chemical Potential

The chemical potential enters in the continuum QCD partition function as

$$i\gamma_0 A_0 \to i\gamma_0 A_0 + \mu\gamma_0 . \tag{142}$$

resulting in the chRMT Dirac operator [52]

$$D(\mu) = \begin{pmatrix} 0 & iW + \mu \\ iW^\dagger + \mu & 0 \end{pmatrix} . \tag{143}$$

If we notice that $\mu\gamma_0$ has the same reality properties as $\gamma_0\partial_0$, we immediately conclude [151] that these ensembles can be classified according to the same anti-unitary symmetries as the chRMT's for $\mu = 0$. However, the term proportional to μ violates the anti-Hermiticity of the Dirac operator, and typically the eigenvalues of $D(\mu)$ will be scattered in the complex plane.

Many of the problems associated with the presence of a chemical potential are related to the loss of non-Hermiticity. Exactly, these problems are reproduced by the chRMT partition function with Dirac operator given by (143). For example, in this model one can study the problems of the quenched approximation [139,140,141], the structure of the Yang-Lee zeros [142,54] and the problems with the Glasgow method [143]. Below we will discuss the first two applications.

7.5 Yang-Lee Zeros

The QCD partition function

$$Z(m,\mu) = \int \det{}^{N_f}(D + m + \mu\gamma_0)e^{-S_{YM}} \tag{144}$$

is a polynomial in m and μ. For simplicity, let us consider the case $m = 0$. Then [144]

$$Z^{QCD}(m=0,\mu) \sim \prod_k (\mu - \mu_k), \tag{145}$$

and the baryon density is given by

$$n_B = \partial_\mu \log Z(m=0,\mu) = \sum_k \frac{1}{\mu - \mu_k}. \tag{146}$$

Since the zeros are scattered in the complex μ-plane we can interpret the real and imaginary parts of n_B as the electric field at μ due to charges located at μ_k.

Physically, we expect that $n_B = 0$ for $\mu < \mu_c \neq 0$ and jumps to a finite value at $\mu = \mu_c$. We thus expect a first order phase transition at μ_c. According to Gauss law, this is consistent with a homogenous distribution of zeros μ_k along the complex circle $|\mu| = \mu_c$ (Of course in the thermodynamic limit, a subleading number of zeros may be present inside the circle.) We also expect that the baryon number density increases smoothly for $\mu > \mu_c$. This requires a radially symmetric cloud of zeros outside of the radius $|\mu| = \mu_c$.

The simplest possible model for the zeros of the partition function, is one with zeros given by the roots of unity,

$$\mu_k = e^{\frac{2\pi i k}{N}}, \qquad k = 1, \cdots, N. \tag{147}$$

This results in the partition function

$$Z(\mu) = \prod_k (\mu - \mu_k) = \mu^N - 1. \tag{148}$$

For the baryon number density we then obtain

$$\begin{aligned} n_B &= \frac{1}{N}\partial_\mu \log Z = \frac{\mu^{N-1}}{\mu^N - 1} \\ &= \begin{cases} \frac{1}{\mu} & \text{for} \quad |\mu| > 1 \\ 0 & \text{for} \quad |\mu| < 1 \end{cases} \quad \text{for} N \to \infty. \end{aligned} \tag{149}$$

In the next section we will study the baryon number density in the chiral random matrix model at nonzero chemical potential.

7.6 Phases in chRMT at $\mu \neq 0$

The chRMT partition function at $\mu \neq 0$ is obtained from the finite T partition function by the replacement $T \to i\mu$. For $N_f = 1$ this results in the σ model

$$Z(\mu) = \int d\sigma d\sigma^* (|\sigma|^2 - \mu^2)^N e^{-N|\sigma|^2}. \tag{150}$$

For $N \to \infty$, the integrals in this partition partition function can be performed by a saddle point approximation. We find that

$$\bar{\sigma} = 0 \quad \text{for} \quad \mu > \mu_c \quad \Rightarrow Z = \mu^{2N}, \tag{151}$$

$$\bar{\sigma} = \sqrt{1+\mu^2} \quad \text{for} \quad \mu < \mu_c \quad \Rightarrow \quad Z = e^{-N(\mu^2+1)}. \tag{152}$$

The critical point is given by the trancendental equation [52] $\mu_c^2 = \exp(-1 - \mu_c^2)$ which is solved by $\mu_c \approx 0.53$

We find that the baryon number density above the critical point shows the same behavior as in our naive model. However, the baryon density for $\mu < \mu_c$ is nonvanishing which disagrees with our expectation for QCD at zero temperature and finite density. The remedy should be clear. In order to obtain a sharp Fermi-Dirac distribution one has to sum over all Matsubara frequencies [145]. This point was ignored in our model.

A much more disturbing observation is that in the thermodynamic limit the partition function for $\mu < \mu_c$ becomes exponentially small [54,28]. This exponential cancellation is due to the phase of the eigenvalues. It makes accurate numerical simulations forbiddingly difficult.

The random matrix partition function is a polynomial in m and μ and it is straightforward to study the zeros of the partition function in the complex m-plane or the complex μ-plane. To assure our numerical accuracy we have determined the zeros of the polynomials by means of multi-precision algorithms [146] using as much as 1000 significant digits. The results are shown in Fig. 5. The branch points at the end of a cut are the points where two saddle-point solutions coincide. In the present case they are given by zeros of the discriminant of a cubic equation. The line of zeros can also be obtained from a saddle point analysis (see right half of the figure). For more details and results in other fields [147,148] we refer to the original literature.

7.7 Failure of the quenched approximation

For QCD with three or more colors and fermions in the fundamental representation, the fermion determinant is complex. The presence of this phase makes Monte-Carlo simulations impossible. A way out might be the quenched

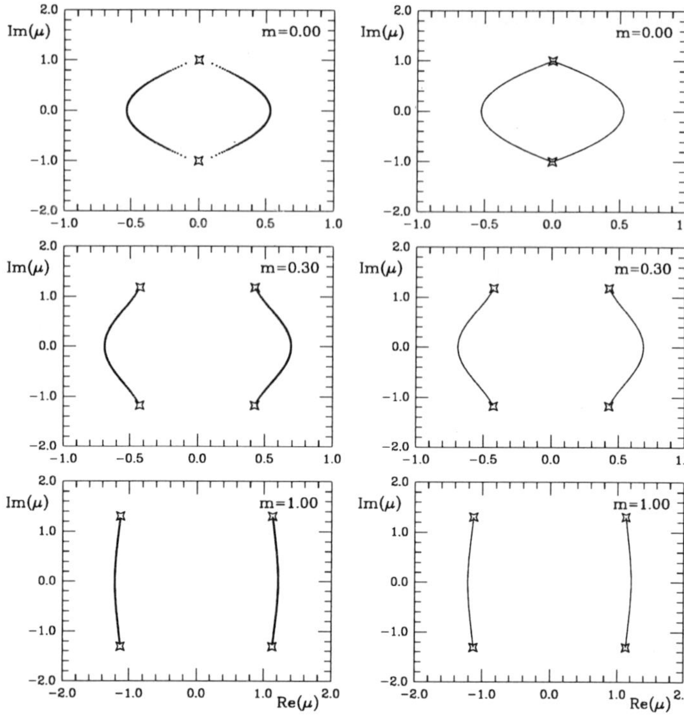

Figure 5: The zeros of the partition function in the complex μ plane (left) and the result obtained from a mean field analysis (right. Results are given for for $m = 0$ (upper), $m = 0.30$ (middle), and $m = 1.0$ (lower) for $n = 192$. The zeros of the discriminant of the cubic equation are denoted by stars. Note that the scale on the x-axis of the lower figure is different.

approximation, i.e. ignore the fermion determinant altogether and hope that it will work. This approach was followed in [139,140] with the conclusion that the critical chemical potential is given by the pion mass instead of the nucleon mass. Obviously, this result is physically wrong!

This problem was first analyzed in chRMT by Stephanov [52]. He showed that the quenched limit is the limit $N_f \to 0$ of a partition functions with fermion determinant

$$|\det(D(\mu) + m)|^{N_f} \tag{153}$$

rather than

$$(\det(D(\mu) + m))^{N_f}. \tag{154}$$

The absolute value can be represented as (let us take $N_f = 2$ for simplicity)

$$\det(D(\mu)+m)\det(D^\dagger(\mu)+m^*) = \int D\psi D\psi^c e^{\bar\psi(D(\mu)+m)\psi+\bar\psi^c(D^\dagger(\mu)+m^*)\psi^c}. \tag{155}$$

Now, Goldstone bosons made out of quarks and conjugate anti-quarks carry a net baryon number resulting in a critical chemical potential given by the pion mass.

The random matrix model corresponding to (155) can be mapped onto a nonlinear σ-model which can be solved by a saddle-point approximation. The exact solution of this model confirms that the critical chemical potential is proportional to $\sqrt{m}$, the same dependence as for the pion mass.

7.8 Triality at $\mu \neq 0$

Let us finally discuss how Dirac spectra at nonzero chemical potential depend on the Dyson index of the matrix elements. We remind the reader that the classification of Dirac operators with chemical potential is the same as for zero chemical potential.

Numerical simulations have been performed for all three classes. A cut along the imaginary axis below a cloud of eigenvalues was found in instanton liquid simulations [149] for $N_c = 2$ at $\mu \neq 0$ which corresponds to $\beta = 1$. In lattice QCD simulations with staggered fermions for $N_c = 2$ [150] a depletion of eigenvalues along the imaginary axis was observed, whereas for $N_c = 3$ the eigenvalue distribution did not show any pronounced features [139].

In the quenched approximation, the spectral properties of the random matrix Dirac operator (143) can easily be studied numerically by diagonalizing a set of matrices with probability distribution (106). In Fig. 6 we show results [151] for the eigenvalues of a few 100×100 matrices for $\mu = 0.15$ (dots). The solid curve represents the analytical result for the boundary of the domain of eigenvalues derived in [52] for $\beta = 2$. However, the method that was used can be extended [151] to $\beta = 1$ and $\beta = 4$ and with the proper scale factors we find exactly the same solution.

For $\beta = 1$ and $\beta = 4$ we observe exactly the same structure as in the previously mentioned (quenched) QCD simulations. We find an accumulation of eigenvalues on the imaginary axis for $\beta = 1$ and a depletion of eigenvalues along this axis for $\beta = 4$. This depletion can be understood as follows. For $\mu = 0$ all eigenvalues are doubly degenerate. This degeneracy is broken at $\mu \neq 0$ which produces the observed repulsion between the eigenvalues.

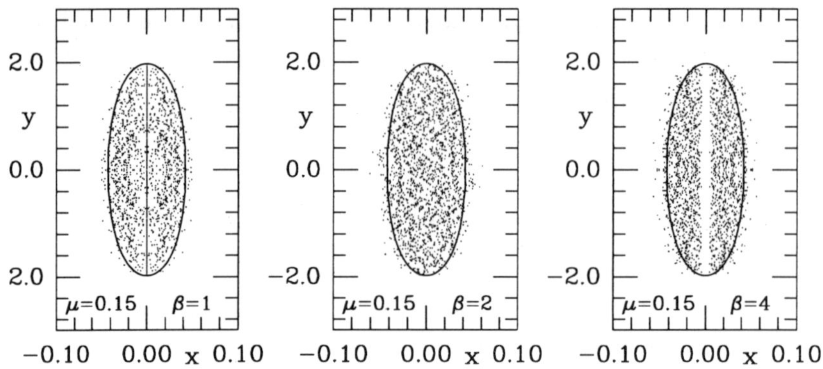

Figure 6: Scatter plot of the real (x), and the imaginary parts (y) of the eigenvalues of the random matrix Dirac operator at nonzero chemical potential. The values of β and μ are given in the labels of the figure. The full curve shows the analytical result for the boundary.

The number of purely imaginary eigenvalues for $\beta = 1$ scales as $\sqrt{N}$ and is thus not visible in a leading order saddle point analysis. Such a $\sqrt{N}$ scaling is typical for the regime of weak non-hermiticity first identified by Fyodorov *et al.* [153]. Using the supersymmetric method of random matrix theory the $\sqrt{N}$ dependence was obtained analytically by Efetov [154]. Also the case $\beta = 4$ was analyzed analytically in [155] with results that are in complete agreement with our numerical simulations. Obviously, more work has to be done in order to arrive at a complete characterization of universal features [156] in the spectrum of nonhermitean matrices.

7.9 The Chiral Phase transition in chRMT

In previous section we have studied a model at nonzero temperature and a model at nonzero chemical potential. These two models can be combined in the following schematic chRMT model [55] for the chiral phase transition,

$$Z = \int DW \det{}^{N_f} \begin{pmatrix} im & W + aT + ib\mu \\ W^\dagger + aT + ib\mu & im \end{pmatrix}. \tag{156}$$

This model can be considered as the matrix equivalent for a Landau-Ginzburg functional. We thus expect to find mean field critical exponents. The advantage over using Landau-Ginzburg theory is that in this case the spectrum of the Dirac operator is accessible. In particular, this might reveal interesting behavior of the Dirac spectrum near the critical point. For example, for $\mu = 0$

we found that the fluctuations of the smallest eigenvalue of the Dirac operator can be used as an order parameter for the chiral phase transition [49].

Also in this case the partition function can be reduced to a σ-model, For the simplest case of $N_f = 1$ it is given by

$$Z(T,\mu) = \int d\sigma e^{-N\Omega(\sigma)}, \tag{157}$$

where

$$\begin{aligned}\Omega(\sigma) = \sigma\sigma^\dagger &- \log((\sigma+m)(\sigma^\dagger+m) - (\mu+iT)^2) \\ &- \log((\sigma+m)(\sigma^\dagger+m) - (\mu-iT)^2).\end{aligned} \tag{158}$$

The chiral condensate is given by the expectation value of σ. The saddle-point equation is a fifth order equation in σ. Therefore this model is very similar to a ϕ^6 Landau-Ginzburg theory. For $m = 0$ we find that σ is real and the saddle point-equation is given by

$$\sigma[\sigma^4 - 2(\mu^2 - T^2 + \frac{1}{2})\sigma^2 + (\mu^2 + T^2)^2 + \mu^2 - T^2] = 0. \tag{159}$$

The critical points occur where one of the solutions of the cubic equation merge with the solution $\sigma = 0$, i.e. along the curve $(\mu^2 + T^2)^2 + \mu^2 - T^2 = 0$. At the tri-critical point three solution merge. This happens if in addition $\mu^2 - T^2 + \frac{1}{2} = 0$.

In Fig. 7 we show the phase diagram in the $\mu T m$ space. In the $m = 0$ plane we observe a line of second order phase transitions and a line of first order phase transitions. They join at the tricritical point. Also joining at the tricritical point is a line of second order phase transitions in the m-directions which is the boundary of the plane of first order transitions in $\mu T m$-space.

7.10 Possibility of a Localization Transition in QCD

From the theory of Anderson localization we know that the eigenvalues corresponding to localized states are statistically independent and thus obey Poisson statistics. Since it has been well established that in QCD at zero temperature the correlations of the eigenvalues of the QCD Dirac operator are given by chiral Random Matrix theory, we conclude that the eigenstates are extended. This also follows from a direct analysis of the Dirac wave functions for gauge field configurations given by a liquid of instantons. However, we wish to point out that studies with Wilson Dirac fermions indicate that eigenfunctions are localized [157]. We do not have an explanation for this discrepancy.

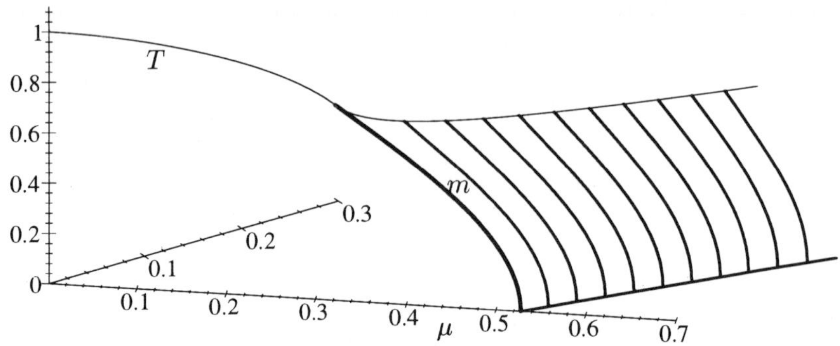

Figure 7: Phase diagram of QCD with two light flavors of mass m as calculated from the random matrix model. The almost parallel curves on the wing surface are cross sections of this surface with m =const planes.

One reason to expect a localization transition with increasing temperature is dimensional reduction. States are more likely to be localized in lower dimensions. For example, in a one-dimensional disordered system all states are localized. One the other hand, because of asymptotic freedom the coupling becomes weaker at higher temperatures. Let us analyze more closely what will happen.

Lattice simulations for the valence quark mass dependence of the chiral condensate were performed for temperatures both below and above the critical temperature [73]. It was found that, in the ergodic domain, all data can be rescaled onto a universal curve given by chRMT. This shows that the states are extended below the critical point.

According to a theoretical argument due to Parisi [158] given in the context of disordered systems a localization transition can only occur in quenched theories. In QCD, this argument can be translated as follows. Let us assume that the eigenfunctions of the Dirac operator are localized. Then the joint eigenvalue distribution factorizes in one-particle distributions $(\lambda^2+m^2)^{N_f}F(\lambda)$. The average spectral density is thus given by

$$\begin{aligned}\rho(\lambda_1) &= \int d\lambda_2 \cdots d\lambda_n \prod_{k=1}^{n}(\lambda_k^2+m^2)^{N_f}F(\lambda_1)\cdots F(\lambda_n)\\ &= (\lambda_1^2+m^2)^{N_f}F(\lambda_1). \qquad (160)\end{aligned}$$

We then find

$$\lim_{\lambda_1 \to 0} \lim_{m \to 0} \lim_{V \to \infty} \rho(\lambda_1) = 0, \tag{161}$$

no spontaneous breaking of chiral symmetry. We thus conclude that in the chiral limit no localization can occur in QCD with light quarks. What happens in quenched theories remains an open question.

8 Concluding Remarks

In these lectures we have presented analytical results for understanding of the infrared limit of the QCD Dirac spectrum. We have shown that the correlations of the Dirac eigenvalues for level spacings below the Thouless energy are given by the zero momentum limit of the partially quenched chiral partition function and agree with results obtained from chiral random matrix theory. A key ingredient in the formulation of the effective theory is the structure of the integration manifold. We have argued that it is given by a super-Riemannian manifold which is characterized by a symbiosis between compact and non-compact degrees of freedom. We have shown that the valence quark mass dependence of the chiral condensate that follows from the zero momentum sector of the effective chiral partition function coincides with the result from chiral Random Matrix Theory. In this formulation, universality is natural. The structure of the chiral Lagrangian only depends on the pattern of chiral symmetry breaking.

An important condition for the validity of the chiral Lagrangian is the requirement that the only low-lying modes are the Goldstone modes associated with the spontaneous breaking of chiral symmetry. Without confinement this would not be the case. On the other hand, according to the Bohigas conjecture, eigenvalue correlations are given by random matrix theory if the corresponding classical system is chaotic. If the classical motion of quarks in 4+1 dimensions is chaotic the eigenvalues of the Dirac operator are correlated according to chRMT. Then necessarily the low-energy dynamics is given is given by the chiral Lagrangian and this can only happen if we have confinement.

Can we reverse this statement? Is the classical motion of a fermion in a confining theory necessarily chaotic? The point I wish to make is that the Bohigas conjecture and confinement are not unrelated. For the understanding of confinement a proof of the Bohigas conjecture might be required. Recent progress in this direction is encouraging[159,160,161,162,163].

9 Acknowledgments

I wish to thank the organizers of the Osaka school and the Kyoto workshop for their hospitality and for organizing such a wonderful meeting. In writing this review I have benefitted greatly from discussions and collaborations with Gernot Akemann, Igor Aleiner, Alexander Altland, Poul Damgaard, Yan Fyodorov, Thomas Guhr, Adam Halasz, Bertram Klein, Andy Jackson, Steffen Meyer, Shinsuke Nishigaki, James Osborn, Andreas Schäfer, Melih Sener, Ben Simons, Robert Shrock, Edward Shuryak, Andrei Smilga, Misha Stephanov, Dominique Toublan, Hans Weidenmüller, Tilo Wettig and Martin Zirnbauer. Dominique Toublan is acknowledged for a critical reading of the manuscript.

References

1. C. DeTar, *Quark-gluon plasma in numerical simulations of QCD*, in *Quark gluon plasma 2*, R. Hwa ed., World Scientific 1995.
2. A. Ukawa, Nucl. Phys. Proc. Suppl. **53** (1997) 106.
3. A. Smilga, Phys. Rep. **291** (1997) 1.
4. F.Karsch, Nucl. Phys. Proc. Suppl. **60A** (1998) 169.
5. T. Banks and A. Casher, Nucl. Phys. **B169** (1980) 103.
6. O. Bohigas, M. Giannoni, Lecture notes in Physics **209** (1984) 1; O. Bohigas, M. Giannoni and C. Schmit, Phys. Rev. Lett. **52** (1984) 1.
7. T. Guhr, A. Müller-Groeling and H.A. Weidenmüller, Phys. Rep. **299** (1998) 189.
8. H. Leutwyler and A. Smilga, Phys. Rev. **D46** (1992) 5607.
9. L. Schäfer and F. Wegner, Z. Phys. **B38** (1980) 113.
10. A.J. McKane and M. Stone, Ann. Phys. (NY) **131** (1981) 36.
11. K. Efetov, Adv. Phys. **32**, 53 (1983).
12. J.J.M. Verbaarschot, H.A. Weidenmüller, and M.R. Zirnbauer, Phys. Rep. **129**, 367 (1985).
13. J.J.M. Verbaarschot in *Continuous Advances in QCD*, A.V. Smilga ed., p. 195, World Scientific, 1994 (hep-th/9405006).
14. J.C. Osborn and J.J.M. Verbaarschot, Phys. Rev. Lett. **81** (1998) 268.
15. J.C. Osborn and J.J.M. Verbaarschot, Nucl. Phys. **B525** (1998) 738.
16. J. Osborn, D. Toublan and J. Verbaarschot, Nucl. Phys. **B** (in press) hep-th/9806110.
17. P. Damgaard, J. Osborn, D. Toublan and J. Verbaarschot, hep-th/9811212.
18. A. Morel, J. Physique **48** (1987) 1111.
19. C. Bernard and M. Golterman, Phys. Rev. D49 (1994) 486; C. Bernard and M. Golterman, hep-lat/9311070.

20. S. Sharpe, Phys. Rev. D 56 (1997) 7052.
21. J. Gasser and H. Leutwyler, Phys. Lett. **188B** (1987) 477.
22. E.V. Shuryak and J.J.M. Verbaarschot, Nucl. Phys. **A560** (1993) 306.
23. J.J.M. Verbaarschot, Phys. Rev. Lett. **72** (1994) 2531; Phys. Lett. **B329** (1994) 351.
24. C.W.J. Beenakker, Rev. Mod. Phys. **69** (1997) 731.
25. G. Montambaux, in *Quantum Fluctuations, Les Houches, Session LXIII*, E. Giacobino, S. Reynaud and J. Zinn-Justin, eds., Elsevier Science, 1995, cond-mat/9602071.
26. B.L. Altshuler, I.Kh. Zharekeshev, S.A. Kotochigova and B.I. Shklovskii, Zh. Eksp. Teor. Fiz. **94** (1988) 343.
27. E. Shuryak, Nucl. Phys. **B302** (1988) 599; D. Diakonov, *Talk given at International School of Physics, 'Enrico Fermi', Course 80: Selected Topics in Nonperturbative QCD*, Varenna, 1995, hep-ph/9602375.
28. M.A. Halasz and J.J.M. Verbaarschot, Phys. Rev. Lett. **74** (1995) 3920; M.A. Halasz, T. Kalkreuter and J.J.M. Verbaarschot, Nucl. Phys. Proc. Suppl. **53** (1997) 266.
29. J.J.M. Verbaarschot, Phys. Lett. **B368** (1996) 137.
30. J.J.M. Verbaarschot, in *Nonperturbative Approaches to Quantum Chromodynamics*, D. Diakonov, ed., Gatchina 1995.
31. M.E. Berbenni-Bitsch, S. Meyer, A. Schäfer, J.J.M. Verbaarschot and T. Wettig, Phys. Rev. Lett. **80** (1998) 1146.
32. J.Z. Ma, T. Guhr and T. Wettig, Eur. Phys. J. **A2** (1998) 87.
33. M.E. Berbenni-Bitsch, A.D. Jackson, S. Meyer, A. Schafer, J.J.M. Verbaarschot and T. Wettig, Nucl. Phys. Proc. Suppl. **63** (1998) 820.
34. M.E. Berbenni-Bitsch, M. Gockeler, T. Guhr, A.D. Jackson, J.Z. Ma, S. Meyer, A. Schäfer, H.A. Weidenmüller, T. Wettig and T. Wilke, Phys. Lett. **B438** (1998) 14.
35. T. Guhr, J.Z. Ma, S. Meyer and T. Wilke, *Statistical analysis and the equivalent of a Thouless energy in lattice QCD Dirac spectra*, hep-lat/9806003.
36. M.E. Berbenni-Bitsch, S. Meyer and T. Wettig, Phys. Rev. **D58** (1998) 071502.
37. H. Markum, R. Pullirsch, K. Rabitsch and T. Wettig, Nucl. Phys. Proc. Suppl. **63** (1998) 832.
38. R. Pullirsch, H. Markum, K. Rabitsch and T. Wettig, hep-lat/9809057.
39. R. Pullirsch, K. Rabitsch, T. Wettig and H. Markum, Phys. Lett. **B427** (1998) 119.
40. P.H. Damgaard, U.M. Heller, A. Krasnitz and T. Madsen, Phys. Lett. **B440** (1998) 129.

41. P.H. Damgaard, U.M. Heller and A. Krasnitz, hep-lat/981006.
42. M. Gockeler, H. Hehl, P.E.L. Rakow, A. Schafer and T. Wettig, Phys. Rev. **D** (in press) hep-lat/9811018.
43. B.A. Berg, H. Markum, R. Pullirsch, hep-lat/9812010.
44. R. Janik, M. Nowak, G. Papp and I. Zahed, Phys. Lett. B389 (1996) 341.
45. H. Hehl and A. Schafer, hep-ph/9806372.
46. M.E. Berbenni-Bitsch, M. Gockeler, H. Hehl, S. Meyer, P.E.L. Rakow, A. Schäfer and T. Wettig, hep-lat/9901013.
47. J.J.M. Verbaarschot, Nucl. Phys. **B427** (1994) 534.
48. M.F.L. Golterman and K.C. Leung, hep-lat/9711033; M.F.L. Golterman, Acta Phys. Polon. **B25** (1994).
49. A.D. Jackson and J.J.M. Verbaarschot, Phys. Rev. **D53** (1996) 7223.
50. T. Wettig, A. Schäfer and H. Weidenmüller, Phys. Lett. **B367** (1996) 28.
51. M. Stephanov, Phys. Lett. **B275** (1996) 249; Nucl. Phys. Proc. Suppl. **53** (1997) 469.
52. M. Stephanov, Phys. Rev. Lett. **76** (1996) 4472.
53. R. Janik, M. Nowak, G. Papp and I. Zahed, Acta Phys. Polon. **B28** (1997) 2949. hep-ph/9708418; R. Janik, M.A. Nowak, G. Papp, J. Wambach, and I. Zahed, Phys. Rev. **E55** (1997) 4100; R. Janik, M.A. Nowak, G. Papp and I. Zahed, Nucl. Phys. B501 (1997) 603.
54. M.A. Halasz, A.D. Jackson and J.J.M. Verbaarschot, Phys. Lett. **B395** (1997) 293; Phys. Rev. **D56** (1997) 5140.
55. M. Halasz, A. Jackson, R. Shrock, M. Stephanov and J. Verbaarschot, Phys. Rev. **D58** (1998) 096007.
56. J. Feinberg and A. Zee, Nucl. Phys. **B504** (1997) 579; Nucl. Phys. **B501** (1997) 643.
57. T. Blum, this volume.
58. C. Vafa and E. Witten, Nucl. Phys. **B234** (1984) 173.
59. M. Peskin, Nucl. Phys. **B175** (1980) 197; S. Dimopoulos, Nucl. Phys. **B168** (1980) 69; M. Vysotskii, Y. Kogan and M. Shifman, Sov. J. Nucl. Phys. **42** (1985) 318; D.I. Diakonov and V.Yu. Petrov, Lecture notes in physics, **417**, Springer 1993.
60. J. Stern, we thank him for raising this point.
61. D.I. Diakonov and V. Yu. Petrov, Nucl. Phys. **B272** (1986) 457.
62. H.B. Nielsen and M. Ninomiya, Nucl. Phys. **185** (1981) 20.
63. W. Bardeen, A. Duncan, E. Eichten, G. Hockney and H. Thacker, Phys. Rev. **D57** (1998) 1633. and H. Thacker, *Quenched approximation artifacts: a detailed study in two-dimensional QED*, hep-lat/9705002.

64. C.R. Gattringer, I. Hip and C.B. Lang, Nucl. Phys. **B508** (1997) 329.
65. C. Gattringer, I. Hip and C. Lang, Nucl. Phys. **B508** (1997) 329.
66. C. Gattringer and I. Hip, Nucl. Phys. **B536** (1998) 363.
67. F. Farchioni, I. Hip, C.B. Lang, M. Wohlgenannt, hep-lat/9812018.
68. J. Negele, hep-lat/9810053.
69. R. Narayanan, hep-lat/9810045,
70. S. Hands and M. Teper, Nucl. Phys. **B347** (1990) 819.
71. J.B. Kogut, J.-F. Lagae and D.K. Sinclair, Nucl. Phys. Proc. Suppl. **63** (1998) 433.
72. T. Schäfer and E. Shuryak, Rev. Mod. Phys. **70** (1998) 323.
73. S. Chandrasekharan, Nucl. Phys. Proc. Suppl. **42** (1995) 475; S. Chandrasekharan and N. Christ, Nucl. Phys. Proc. Suppl. **42** (1996) 527; N. Christ, Lattice 1996.
74. A.V. Andreev, B.D. Simons, and N. Taniguchi, Nucl. Phys **B432 [FS]** (1994) 487.
75. A.D. Jackson, M.K. Sener and J.J.M. Verbaarschot, Nucl. Phys. **B479** (1996) 707.
76. T. Guhr and T. Wettig, Nucl. Phys. **B506** (1997) 589.
77. A.D. Jackson, M.K. Sener and J.J.M. Verbaarschot, Nucl. Phys. **B506** (1997) 612.
78. J. Gasser and H. Leutwyler, Ann. Phys. **158**, 142 (1984); J. Gasser and H. Leutwyler, Nucl. Phys. B**250**, 465 (1985).
79. H. Leutwyler, Ann. Phys. **235** (1994) 165.
80. M. Zirnbauer, J. Math. Phys. **37** (1996) 4986.
81. F.A. Berezin, *Introduction to superanalysis*, Reidel, Dordrecht (1987).
82. K.B. Efetov, *Supersymmetry in disorder and chaos*, Cambridge University Press, (1997); K.B. Efetov, Adv. Phys. **32** (1983) 53.
83. B. Simons and A. Altland, cond-mat/9811134.
84. M.R. Zirnbauer, J. Math. Phys. **38** (1997) 2007; J. Phys. **A 29** (1996) 7113.
85. H.A. Weidenmüller, Nucl. Phys. A 518 (1990) 1.
86. T. Guhr and A. Mueller-Groeling, cond-mat/9702113, J. Math. Phys. (in press).
87. N. Argaman, Y. Imry and U. Smilansky, Phys. Rev. **B47** (1993) 4440.
88. V.E. Kravtsov, in *Correlated Fermions and Transport in Mesoscopic Systems*, Les Arcs, 1996, cond/mat9603166.
89. A. Altland and Y. Gefen, Phys. Rev. Lett. **71** (1993) 3339.
90. D. Braun and G. Montambaux, Phys. Rev. **52** (1995) 13903.
91. A.G. Aronov and A.D. Mirlin, Phys. Rev. **B51** (1995) 6131.
92. Y.V. Fyodorov and A.D. Mirlin, Phys. Rev. **B51** (1995) 13403.

93. V.E. Kravtsov, I.V. Lerner, B.L. Altshuler and A.G. Aronov, Phys. Rev. Lett. **72** (1994) 888.
94. A. Altland, Y. Gefen and G. Montambaux, Phys. Rev. lett. **76** (1996) 1130.
95. J.T. Chalker, V.E. Kravtsov, I.V. Lerner, JETP Lett. 64 (1996) 386.
96. R. Janik, M.A. Nowak, G. Papp and I. Zahed, Phys. Rev. Lett. **81** (1998) 264.
97. R. Brower, P. Rossi and C-I. Tan, Nucl.Phys. **B190** [FS3] (1981) 699; R. Brower and M. Nauenberg, **B180** [FS2] (1981) 221.
98. P. Hasenfratz and H. Leutwyler, Nucl. Phys. B343 (1990) 241.
99. P.H. Damgaard, Phys. Lett. **B425** (1998) 151.
100. A. Smilga and J. Stern, Phys. Lett. **B318** (1993) 531.
101. J.J.M. Verbaarschot and I. Zahed, Phys. Rev. Lett. **70** (1993) 3852.
102. A. Altland, M.R. Zirnbauer, Phys. Rev. Lett. **76** (1996) 3420.
103. A. Smilga and J.J.M. Verbaarschot, Phys. Rev. **D51** (1995) 829.
104. J.J.M. Verbaarschot, *Lectures given at NATO Advanced Study Institute on Confinement, Duality and Nonperturbative Aspects of QCD*, Cambridge, 1997, hep-th/9710114.
105. T. Akuzawa and M. Wadati, J. Phys. Soc. Jap. **67** (1998) 2151.
106. K. Slevin and T. Nagao, Phys. Rev. Lett. **70** (1993) 635.
107. D. Fox and P. Kahn, Phys. Rev. **134** (1964) B1152; (1965) 228.
108. T. Nagao and M. Wadati, J. Phys. Soc. Japan **60** (1991) 2998; J. Phys. Soc. Jap. **60** (1991) 3298; J. Phys. Soc. Jap. bf 61 (1992) 78; J. Phys. Soc. Jap. bf 61 (1992) 1910. **61** (1992) 78, 1910.
109. P. Forrester, Nucl. Phys. **B[FS]402** (1993) 709.
110. S.M. Nishigaki, P.H. Damgaard and T. Wettig, hep-th/9803007.
111. P.H. Damgaard and S.M. Nishigaki, Nucl. Phys. **B518** (1998) 495.
112. T. Wilke, T. Guhr and T. Wettig, Phys. Rev. **D57** (1998) 6486; B. Seif, T. Wettig and T. Guhr, hep-th/9811044.
113. J. Jurkiewicz, M. Nowak and I. Zahed, Nucl. Phys. **B478** (1996) 605; Erratum-ibid. **B513** (1998) 759.
114. M.J. Bowick and E. Brézin, Phys. Lett. **B 268**, 21 (1991).
115. G. Hackenbroich and H.A. Weidenmüller, Phys. Rev. Lett. **74** (1995) 4118.
116. E. Brézin, S. Hikami and A. Zee, Nucl. Phys. **B464** (1996) 411.
117. G. Akemann, P. Damgaard, U. Magnea and S. Nishigaki, Nucl. Phys. **B 487[FS]** (1997) 721.
118. E. Kanzieper and V. Freilikher, Phys. Rev. Lett. **78**, 3806 (1997); Phys. Rev. **E 55**, 3712 (1997).
119. V. Freilikher, E. Kanzieper and I. Yurkevich, Phys. Rev. **E53**, 2200

(1996).
120. P. Zinn-Justin, Nucl. Phys. **B497** (1997) 725.
121. B. Eynard, Nucl. Phys. **B506**, 633 (1997).
122. E. Brézin and S. Hikami, Phys. Rev. **E56**, 264 (1997).
123. C.W.J. Beenakker, Phys. Rev. Lett. **70**, 1155 (1993).
124. J.J.M. Verbaarschot, Nucl. Phys. B426 (1994) 559.
125. M.K. Sener and J.J.M. Verbaarschot, hep-th/9801042.
126. H. Widom, solv-int/9804005.
127. P.H. Damgaard, S.M. Nishigaki ,Phys. Rev. **D57** (1998) 5299.
128. J. Christiansen, hep-th/9809194.
129. K. Splittorff and A.D. Jackson, hep-lat/9805018. K. Splittorf, hep-th/9810248.
130. P.H. Damgaard, Phys. Lett. **B424** (1998) 322; G. Akemann and P.H. Damgaard, Phys. Lett. **B432** (1998) 390/ G. Akemann and P.H. Damgaard, Nucl. Phys. **B528** (1998) 411.
131. F. Dyson and M. Mehta, J. Math. Phys. **4** (1963) 701.
132. M. Mehta, *Random Matrices*, Academic Press, San Diego 1991.
133. T. Kalkreuter, Phys. Lett. **B276** (1992) 485; Phys. Rev. **D48** (1993) 1; Comp. Phys. Comm. **95** (1996) 1.
134. J. Cullum and R.A. Willoughby, J. Comp. Phys. **44** (1981) 329.
135. A. Pandey, Ann. Phys. **134** (1981) 119.
136. T. Nagao and P.J. Forrester, Nucl. Phys. **B435** (1995) 401.
137. P.W. Anderson, Phys. Rev. **109** (1958) 1492.
138. P. Di Francesco, P. Ginsparg, and J. Zinn-Justin, Phys. Rep. **254** (1995) 1.
139. I. Barbour, N. Behihil, E. Dagotto, F. Karsch, A. Moreo, M. Stone and H. Wyld, Nucl. Phys. **B275** (1986) 296; M.-P. Lombardo, J.B. Kogut and D.K. Sinclair, Phys. Rev. **D54** (1996) 2303.
140. I. M. Barbour, S. E. Morrison, E. G. Klepfish, J. B. Kogut and M.-P. Lombardo, *Results on Finite Density QCD*, hep-lat/9705042.
141. I.M. Barbour, S. Morrison and J. Kogut, *Lattice gauge theory simulation at nonzero chemical potential in the chiral limit*, hep-lat/9612012.
142. I. Barbour, A. Bell, M. Bernaschi, G. Salina and A. Vladikas, Nucl. Phys. B386 (1992) 683.
143. M.A. Halasz, Nucl. Phys. **A642** (1998) 324.
144. J. Vink, Nucl. Phys. **B323** (1989) 399.
145. J. Kapusta, *Thermal Field Theory*, Cambridge University Press.
146. D. Bailey, *A Fortran-90 Based Multiprecision System*, NASA Ames RNR Technical Report RNR-94-013.
147. C.N. Yang and T.D. Lee, Phys. Rev. **87** (1952) 104, 410.

148. V. Matteev and R. Shrock, J. Phys. A: Math. Gen. **28** (1995) 5235.
149. Th. Schäfer, Phys. Rev. **D57** (1998) 3950.
150. C. Baillie, K. Bowler, P. Gibbs, I. Barbour and M. Rafique, Phys. Lett. **197B** (1987) 195.
151. M. Halasz, J. Osborn and J. Verbaarschot, Phys. Rev. **D56** (1997) 7059.
152. Y. Fyodorov and H. Sommers, JETP Lett. **63** (1996), 1026.
153. Y. Fyodorov, B. Khoruzhenko and H. Sommers, Phys. Lett. **A 226** (1997) 46.
154. K. Efetov, Phys. Rev. Lett. **79** (1997) 491; Phys. Rev. **B 56** (1997) 9630.
155. A.V. Kolesnikov and K.B. Efetov, cond-mat/9809173.
156. Y. Fyodorov, B. Khoruzhenko and H. Sommers, Phys. Rev. Lett. **79** (1997) 557.
157. K. Jansen, C. Liu, H. Simma and D. Smith, Nucl. Phys. Proc. Supp. **53**, 262 (1997).
158. G. Parisi, J. Phys. A: Math. Gen. **14** (1981) 735.
159. A.V. Andreev, O. Agam, B.D. Simons and B.L. Altshuler, Nucl. Phys. **B482** (1996) 536.
160. A. Altland and M. Zirnbauer, Phys. Rev. Lett. **77** (1996) 4536.
161. A.V. Andreev, B.D. Simons, and N. Taniguchi, Nucl. Phys **B432** **[FS]** (1994) 487.
162. M.R. Zirnbauer, chao-dyn/9810016.
163. M.R. Zirnbauer, chao-dyn/9812023.

LIGHT-FRONT QUANTUM CHROMODYNAMICS

ROBERT J. PERRY
Physics Department, The Ohio State University, Columbus, OH 43210, USA
E-mail: perry.6@osu.edu

Light-front quantum chromodynamics may lead to an accurate constituent approximation for the low-energy properties of hadrons. This requires a cutoff that violates explicit gauge invariance and Lorentz covariance, leading to the calculation of a renormalized QCD hamiltonian using a similarity renormalization group. Renormalization repairs broken symmetries, and in light-front field theory it moves dynamical effects that usually require large numbers of partons to few-body effective interactions. This has been shown to work in QED through dominant contributions to the Lamb shift. In QCD logarithmic confinement arises at second order, and initial bound state calculations produce reasonable results.

1 Motivation and Outline

Quantum chromodynamics (QCD) is the fundamental theory of the strong interaction, but our understanding of QCD and our ability to use it to solve low-energy problems where the interactions are truly strong falls far short of our accomplishments in the study of quantum electrodynamics (QED). Both are gauge theories, but the fact that gluons carry color charge drastically complicates QCD. Couplings in QED are weak at low energies and photons are nearly free, so that the interactions their exchange produces are readily approximated. As a result low energy bound states can be accurately described using a small number of constituents whose interactions appear in the hamiltonian at second order, and the vacuum has no effect on bound states in light-front QED.

The study of QCD and strong interaction phenomenology leads to little hope of deriving similar approximations. Interactions in QCD are strong at low energies and the color-charged gluons interact strongly while mediating interactions. The vacuum is supposed to have a complicated structure to which conferences are devoted. Its structure is assumed to be responsible for essential aspects of the theory such as confinement and chiral symmetry breaking in all widely respected treatments of the theory, and it is hard to imagine an accurate description of individual hadrons that does not also include a complicated description of the vacuum in which they reside. Nonetheless, we advocate an approach in which an accurate description of hadrons resembles the accurate description of atoms in QED.[1,2]

We start with the heretical conjecture that *a constituent picture of hadrons can be derived from QCD.* This conjecture guides our calculations, but the approach I describe is completely fixed by QCD and the constituent picture will fail if it is inadequate.[a]

If a constituent approximation is accurate, we can study the low-energy properties of hadrons (*e.g.*, mesons) by solving a relativistic Schrödinger equation:

$$H_\Lambda \mid \Psi_\Lambda\rangle = E \mid \Psi_\Lambda\rangle, \tag{1}$$

with,

$$\mid \Psi_\Lambda\rangle = \phi^\Lambda_{q\bar{q}} \mid q\bar{q}\rangle + \phi^\Lambda_{q\bar{q}g} \mid q\bar{q}g\rangle + \cdots, \tag{2}$$

where I use shorthand notation for the Fock space components of the state. The exact state vector includes an infinite number of terms. In a constituent approximation we truncate this series, adding terms to improve the approximation. We derive the hamiltonian from QCD, so we must allow for the possibility of constituent gluons. I have indicated that the hamiltonian and the state both depend on a cutoff, Λ, which is critical for the approximation.

This approach has no chance of working without a *renormalization scheme tailored to light-front hamiltonian field theory.* Much of our work has focused on the development of such a renormalization scheme.[3–8]

Consider the conditions under which it might be possible to truncate the above Fock space series without making an arbitrarily large error in the eigenvalue. I focus on the eigenvalue, because it is certainly not possible to approximate all observable properties of hadrons (*e.g.*, wee parton structure functions) this way. For this approximation to be valid, *all* many-body states must approximately decouple from the dominant few-body components.

We know that even in perturbation theory, high energy many-body states do not decouple from few-body states. In fact, the errors from simply discarding high energy states are infinite. In second-order perturbation theory, for example, high energy photons contribute an arbitrarily large shift to the mass of an electron. This second-order effect is illustrated in Fig. 1. The solution to this problem is well-known, renormalization. Renormalization moves the effects of high energy components in the state to effective interactions in the hamiltonian.

It is difficult to see how a constituent approximation can emerge in a hamiltonian calculation using any regularization scheme without a cutoff that

---

[a]The description of light hadrons requires refinements to the simple approximations I describe, because of chiral symmetry breaking.

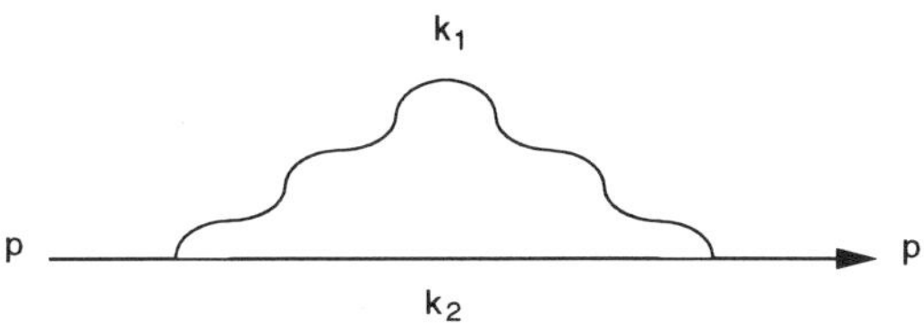

Figure 1. The second-order shift in the self-energy of a bare electron due to mixing with electron-photon states.

either removes high energy degrees of freedom or removes direct couplings between low and high energy degrees of freedom.

In the best case scenario we expect the cutoff to act like a resolution. If the cutoff is increased to an arbitrarily large value, the resolution increases and instead of seeing a few constituents we resolve the substructure of the constituents and the few-body approximation breaks down. As the cutoff is lowered, this substructure is removed from the state vectors, and the renormalization procedure replaces it with effective interactions in the hamiltonian. Any "cutoff" that does not remove this substructure from the states is of no use to us.

This point is well-illustrated by the QED calculations discussed below.[2,9,10] There is a window into which the cutoff must be lowered for the constituent approximation to work. If the cutoff is raised atomic states start to include photons, and as the cutoff is raised further they start to include additional photons and electron-positron pairs. After the cutoff is lowered to a value that can be self-consistently determined *a-posteriori*, photons and pairs are removed from the states and replaced by the Coulomb interaction and relativistic corrections in the hamiltonian. The cutoff cannot be lowered too far using a perturbative renormalization group, hence the window.

Thus, if we remove high energy degrees of freedom, or coupling to high energy degrees of freedom, we should encounter self-energy corrections leading to effective one-body operators, vertex corrections leading to effective vertices, and exchange effects leading to explicit many-body interactions not found in the canonical hamiltonian. We naively expect these operators to be local when acting on low energy states, because simple uncertainty principle arguments indicate that high energy virtual particles cannot propagate very far.

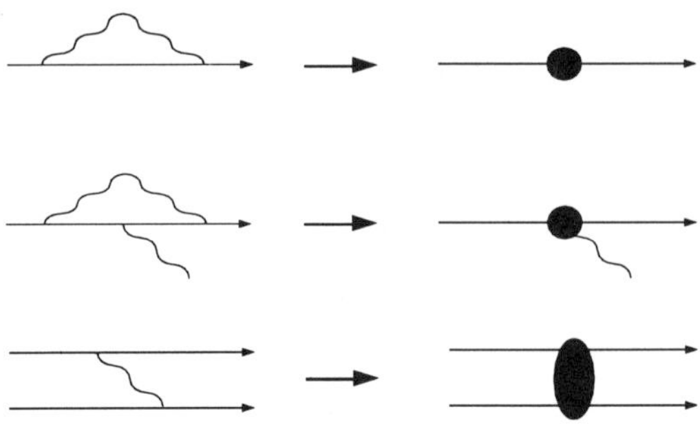

Figure 2. Mixing of low-energy few-body states with high-energy many-body states alters the dispersion relation for single particles, alters emission and absorption vertices, and produces approximately local few-body interactions.

Unfortunately these arguments break down in light-front coordinates, and at best we can maintain transverse locality.[7]

In this article I give a brief overview of the new renormalization techniques we employ, and in a separate article in these Proceedings [11] Sérgio Szpigel and I illustrate them using the two-dimensional δ-function potential.

Low energy many-body states do not typically decouple from low energy few-body states. The worst of these low energy many-body states is the vacuum. This is what drives us to use light-front coordinates.[12] Fig. 3 shows a pair of particles being produced out of the vacuum in equal-time coordinates t and z. The transverse components x and y are not shown, because they are the same in equal-time and light-front coordinates. The figure also shows the light-front time axis,

$$x^+ = t + z \; , \tag{3}$$

and the light-front longitudinal spatial axis,

$$x^- = t - z \; . \tag{4}$$

In equal-time coordinates it is kinematically possible for virtual pairs to be produced from the vacuum, as long as their momenta sum to zero so that three-momentum is conserved. Because of this, the state vector for a proton includes an arbitrarily large number of particles that are disconnected from the proton. The only constraint imposed by relativity is that particle velocities be less than or equal to that of light.

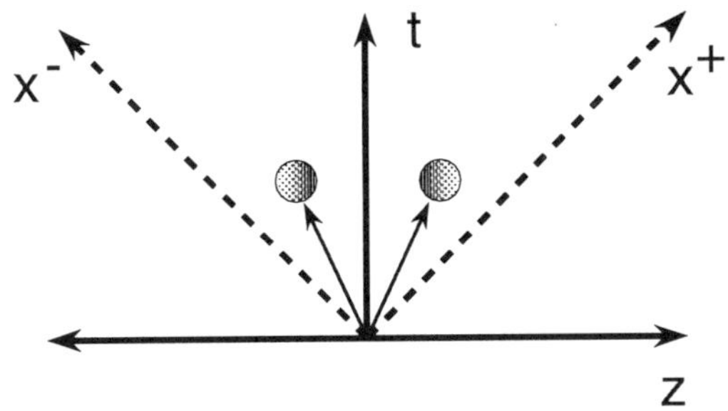

Figure 3. Light-front coordinates. Light-front 'time' is $x^+ = t + z$, and light-front longitudinal position is $x^- = t - z$.

In light-front coordinates, however, we see that all allowed trajectories lie in the first quadrant. In other words, light-front longitudinal momentum, p^+ (conjugate to x^- since $a \cdot b = \frac{1}{2}(a^+b^- + a^-b^+) - \mathbf{a}_\perp \cdot \mathbf{b}_\perp$), is always positive,

$$p^+ \geq 0 \,. \tag{5}$$

We exclude particle modes with $p^+ = 0$, forcing the vacuum to be trivial because it is the only state with $p^+ = 0$. Moreover, the light-front energy of a free particle of mass m is

$$p^- = \frac{\mathbf{p}_\perp^2 + m^2}{p^+} \,. \tag{6}$$

This implies that all free particles with zero longitudinal momentum have infinite energy, unless their mass and transverse momentum are zero.

Is the vacuum really trivial? What about confinement? What about chiral symmetry breaking? What about instantons? What about the job security of theorists who study the vacuum?

I simply discard all $p^+ = 0$ degrees of freedom and replace their effects using a renormalization procedure that does not require any explicit reference to zero modes. Thus the vacuum in our formalism is trivial. We are forced to work in the "hidden symmetry phase" of the theory, and to introduce effective interactions that reproduce all effects associated with the vacuum in other formalisms.[1,13,14] The simplest example of this approach is provided by a scalar field theory with spontaneous symmetry breaking. It is possible to shift the scalar field and deal explicitly with a theory containing symmetry breaking interactions. In the simplest case ϕ^3 is the only relevant or marginal symmetry breaking interaction, and one can simply tune this coupling to the value corresponding to spontaneous rather than explicit symmetry breaking.

The use of a symmetry-breaking cutoff and the removal of zero-modes leads to a large number of operators not found in the canonical QCD hamiltonian. This complicates the renormalization procedure, but it may lead to tremendous simplifications in the final nonperturbative problem. For example, few-body operators must produce confinement manifestly!

Confinement cannot require particle creation and annihilation, flux tubes, etc. This is easily seen using a variational argument. Consider a color neutral quark-antiquark pair that are separated by a distance R, which is slowly increased to infinity. Moreover, to see the simplest form of confinement assume that there are no light quarks, so that the energy should increase indefinitely as they are separated if the theory possesses confinement. At each separation the gluon components of the state adjust themselves to minimize the energy. But this means that the expectation value of the hamiltonian for a state with no gluons must exceed the energy of the state with gluons, and therefore must diverge even more rapidly than the energy of the true ground state. This means that there must be a two-body confining interaction in the hamiltonian. If the renormalization procedure is unable to produce such confining two-body interactions, it is invalid.

1.1 Simple Strategy

I want to outline a conceptually simple strategy for bound state calculations.[2] The first step is to use a perturbative similarity renormalization group [3,4] and coupling coherence [6,7] to find the renormalized hamiltonian as an expansion in powers of the canonical coupling:

$$H^\Lambda = h_0 + g_\Lambda h_1^\Lambda + g_\Lambda^2 h_2^\Lambda + \cdots . \tag{7}$$

We compute this series to a finite order and to date have not required any *ad hoc* assumptions to uniquely fix the hamiltonian. No operators are added to the hamiltonian by hand, so it is completely determined by the underlying theory to this order. This step is illustrated in a separate article in these Proceedings.[11]

The second step is to employ bound state perturbation theory to solve the eigenvalue problem. The complete hamiltonian contains every interaction (although each is cut off) contained in the canonical hamiltonian, and many more. We separate the hamiltonian,

$$H^\Lambda = \mathcal{H}_0^\Lambda + \mathcal{V}^\Lambda \;, \tag{8}$$

treating $\mathcal{H}_0^\Lambda$ nonperturbatively and computing the effects of $\mathcal{V}^\Lambda$ in bound state perturbation theory. We must choose $\mathcal{H}_0^\Lambda$ and Λ so that $\mathcal{H}_0^\Lambda$ is tractable and to minimize corrections from higher orders of $\mathcal{V}^\Lambda$ within a constituent approximation.

If a constituent approximation is valid *after* Λ is lowered to a critical value that must be determined, we may be able to move all particle creation and annihilation to $\mathcal{V}^\Lambda$. $\mathcal{H}_0^\Lambda$ includes many-body interactions that do not change particle number, and these interactions should be primarily responsible for the constituent bound state structure.

There are several obvious flaws in this strategy. Chiral symmetry-breaking operators, which must be included in the hamiltonian since we work entirely in the hidden symmetry phase of the theory, do not appear at any finite order in the coupling. There is only one relevant chiral symmetry breaking operator, and it appears in the canonical hamiltonian when quarks are massive (spin-flip gluon emission by quarks) although it can acquire non-canonical dependence on longitudinal momenta since there is no longitudinal locality. This operator must simply be added if quarks are massless and tuned to fit spectra or fixed by a non-perturbative renormalization procedure.[1,16,17] In addition, there are perturbative errors in the strengths of all operators. We know from simple scaling arguments [18] that when Λ is in the perturbative scaling regime:

- small errors in relevant operators exponentiate in the output,
- small errors in marginal operators produce comparable errors in output,
- small errors in irrelevant operators tend to decrease exponentially in the output.

This means that even if a relevant operator appears (*e.g.*, a constituent quark or gluon mass operator), we may need to tune its strength rather than use its perturbative value to obtain reasonable results. We have not had to do this, but we have recently studied some of the effects of tuning a gluon mass operator.[19]

To date this strategy has produced well-known results in QED [2,9,10] through the Lamb shift, and reasonable results for heavy quark bound states in QCD.[20,21,19] All of these calculations rely on a nonrelativistic reduction of the effective hamiltonian, which leads to drastic simplifications. Glueball calculations are being completed [22] and will be the first fully relativistic QCD calculations in this approach.

The best place to begin the study of light-front field theory and to see how a constituent approximation can arise in light-front gauge theories is the Schwinger model, massless QED in $1+1$ dimensions.[23] This model can be solved analytically,[24] and its physical content is remarkably simple. There is only one physical particle, a massive neutral scalar particle with no self-interactions. The Fock space content of the physical states depends crucially on the coordinate system and gauge, and it is only in light-front coordinates that a simple constituent picture emerges.[25,2] In light-front field theory the physical particle is a bound state of a single electron-positron pair, with a wave function that is constant in the longitudinal fraction carried by either particle.

The Schwinger model does not display the renormalization problems that must be solved in QED_{3+1} and QCD_{3+1}. Before turning to these theories, I discuss the renormalization machinery that has been developed for light-front hamiltonians.

2 Light-Front Renormalization Group

In $3+1$ dimensions we must introduce a cutoff, Λ, and we never perform explicit bound state calculations with Λ anywhere near its continuum limit. In fact, we want to let Λ become as small as possible. In my opinion, any strategy for solving light-front QCD that requires the cutoff to explicitly approach infinity in the nonperturbative part of the calculation is useless. Therefore, we must set up and solve

$$P^-_\Lambda \mid \Psi_\Lambda(P)\rangle = \frac{\mathbf{P}_\perp^2 + M^2}{P^+} \mid \Psi_\Lambda(P)\rangle \; . \tag{9}$$

Physical results, such as the mass, M, can not depend on the arbitrary cutoff, Λ. This means that P^-_Λ and $\mid \Psi_\Lambda\rangle$ must depend on the cutoff in such a way that $\langle \Psi_\Lambda \mid P^-_\Lambda \mid \Psi_\Lambda\rangle$ does not. Wilson based the derivation of his renormalization group on this observation,[26,27,28,18] and we modify Wilson's renormalization group to compute P^-_Λ.

It is difficult to even talk about how the hamiltonian depends on the cutoff without having a means of changing the cutoff. If we can change the cutoff, we can explicitly watch the hamiltonian's cutoff dependence change and fix its cutoff dependence by insisting that this change satisfy certain requirements (*e.g.*, that the limit in which the cutoff is taken to infinity exists).

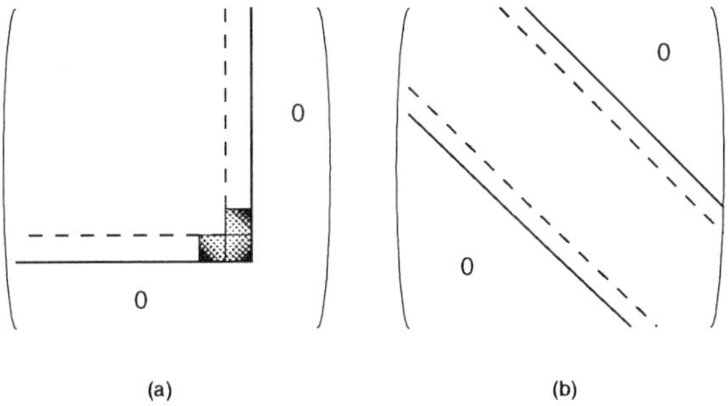

Figure 4. Two ways to run a cutoff on free energy. In (a) a cutoff on the magnitude of the energy is lowered from the solid to the dashed lines, with problems resulting from the removed shaded region. In (b) a cutoff on how far off diagonal matrix elements appear is lowered from the dashed to the solid lines.

We introduce an operator that changes the cutoff,

$$H(\Lambda_1) = T[H(\Lambda_0)] , \tag{10}$$

where I assume that $\Lambda_1 < \Lambda_0$. To simplify the notation, I let $H(\Lambda_l) = H_l$. To renormalize the hamiltonian we study the properties of the transformation.

Fig. 4 displays two generic cutoffs that might be used. Traditionally theorists have used cutoffs that remove high energy states, as shown in Figure 4a. This is the type of cutoff Wilson employed in his initial work [26] and I have studied its use in light-front field theory.[7] When a cutoff on energies is reduced, all effects of couplings eliminated must be moved to effective operators. When these effective operators are computed perturbatively they involve products of matrix elements divided by energy denominators. Expressions closely resemble those encountered in standard perturbation theory, with the second-order operator involving terms of the form

$$\delta V_{ij} \sim \frac{\langle \phi_i \mid V \mid \phi_k \rangle \langle \phi_k \mid V \mid \phi_j \rangle}{\epsilon_i - \epsilon_k} . \tag{11}$$

This new effective interaction replaces missing couplings, so the states ϕ_i and ϕ_j are retained and the state ϕ_k is one of the states removed. The problem comes from the shaded, lower right-hand corner of the matrix, where the energy denominator vanishes for states at the corner of the remaining matrix. In this corner we should use nearly degenerate perturbation theory rather than perturbation theory, but to do this requires solving high energy many-body problems nonperturbatively before solving the low energy few-body problems.

An alternative cutoff, which does not actually remove any states and that can be run by a similarity transformation[b] is shown in Fig. 4b. This cutoff removes couplings between states whose free energy differs by more than the cutoff. The advantage of this cutoff is that the effective operators resulting from it contain energy denominators that are never smaller than the cutoff, so that a perturbative approximation for the effective hamiltonian may work well.

Before discussing the similarity transformation, which runs a cutoff on off-diagonal matrix elements, I need to introduce the classification of operators as relevant, marginal, and irrelevant. This classification is not necessary to use the similarity renormalization group, but it is necessary for the definition of coupling coherence, and it is essential for a full understanding of the renormalization group. It starts with the definition of a fixed point, which is a hamiltonian that is left invariant by the transformation,

$$H^* = T[H^*] \,. \tag{12}$$

Obviously a fixed point is a renormalized hamiltonian, because the cutoff fixed point hamiltonian is identical to the fixed point hamiltonian with infinite cutoff.

Consider the immediate 'neighborhood' of the fixed point, and assume that the trajectory remains in this neighborhood. This assumption must be justified *a posteriori*, but if it is true we should write

$$H_l = H^* + \delta H_l \,, \tag{13}$$

and consider the trajectory of small deviations δH_l.

As long as δH_l is 'sufficiently small,' we can use a perturbative expansion in powers of δH_l, which leads us to consider

$$\delta H_{l+1} = L \cdot \delta H_l + N[\delta H_l] \,. \tag{14}$$

Here L is the linear approximation of the full transformation in the neighborhood of the fixed point, and $N[\delta H_l]$ contains all contributions to δH_{l+1} of $\mathcal{O}(\delta H_l^2)$ and higher.

The object of the renormalization group calculation is to compute trajectories and this requires a representation for δH_l. The problem of computing trajectories is one of the most common in physics, and a convenient basis

---

[b]In deference to the original work I call this a similarity transformation even though in all cases of interest to us it is a unitary transformation.

for the representation of δH_l is provided by the eigenoperators of L, since L dominates the transformation near the fixed point. These eigenoperators and their eigenvalues are found by solving

$$L \cdot O_m = \lambda_m O_m \; . \tag{15}$$

For light-front field theory this linear transformation is a scaling of the transverse coordinate, the eigenoperators are products of field operators and transverse derivatives, and the eigenvalues are determined by the transverse dimension of the operator. All operators can include both powers and inverse powers of longitudinal derivatives because there is no longitudinal locality.[7]

Using the eigenoperators of L as a basis we can represent δH_l,

$$\delta H_l = \sum_{m \in R} \mu_{m_l} O_m + \sum_{m \in M} g_{m_l} O_m + \sum_{m \in I} w_{m_l} O_m \; . \tag{16}$$

Here the operators O_m with $m \in R$ are *relevant* (*i.e.*, $\lambda_m > 1$), the operators O_m with $m \in M$ are *marginal* (*i.e.*, $\lambda_m = 1$), and the operators with $m \in I$ are *irrelevant* (*i.e.*, $\lambda_m < 1$). The motivation behind this nomenclature is made clear by considering repeated application of L, which causes the relevant operators to grow exponentially, the marginal operators to remain unchanged in strength, and the irrelevant operators to decrease in magnitude exponentially.

2.1 Similarity transformation

Stan Głazek and Ken Wilson studied the problem of small energy denominators that typically appear in effective interactions,[27] and realized that a similarity transformation that runs a different form of cutoff (as discussed above) avoids this problem.[3] Independently, Wegner[4] developed a similarity transformation that is easier to use than those first studied by Głazek and Wilson. Recently Walhout has developed a transformation that may be well-suited to analytical and higher-order calculations.[29]

Many details and a simple example of the use of Wegner's similarity transformation can be found elsewhere in these Proceedings.[11] Here I summarize the details needed to complete QED and QCD calculations through second order. Consider a hamiltonian, $H_s = h + V_s$, where h is a free hamiltonian that may contain mass terms. The dependence of H on the cutoff, where s is the inverse of the cutoff squared, is given by

$$\frac{dH_s}{ds} = [H_s, [H_s, h] \; . \tag{17}$$

This equation is most easily analyzed in terms of matrix elements between eigenstates of h,

$$h|\phi_i\rangle = \epsilon_i|\phi_i\rangle \, . \tag{18}$$

A reduced interaction, $\overline{V}_s$, can be defined,

$$V_{sij} = e^{-s\Delta_{ij}^2}\, \overline{V}_{sij} \, , \tag{19}$$

where $\Delta_{ij} = \epsilon_i - \epsilon_j$. A gaussian cutoff factor that clearly forces the interactions towards the diagonal is isolated, and the equation for $\overline{V}_s$ starts at second order,

$$\frac{d\overline{V}_{sij}}{ds} = \sum_k (\Delta_{ik} + \Delta_{jk})\, \overline{V}_{sik}\, \overline{V}_{skj}\, e^{-2s\Delta_{ik}\Delta_{jk}} \, , \tag{20}$$

where I use $\Delta_{ij}^2 - \Delta_{ik}^2 - \Delta_{jk}^2 = -2\Delta_{ik}\Delta_{jk}$.

This is a first-order differential equation and it is exact. Its solution is completely determined when a complete set of boundary conditions are specified. In its simplest form the boundary conditions would be given by specifying H_{s_0} using a single value of s_0 for all matrix elements, with $s_0 = 0$ corresponding to an infinite cutoff. As discussed in the simple δ-function example,[11] divergences prevent us from using $s_0 = 0$ to specify boundary conditions for relevant and marginal operators, but we are free to choose different values of s_0 for different operators and this freedom is crucial. The boundary conditions are given by coupling coherence, which I discuss in the next section.

For QED and QCD we can approximate $\overline{V}_s$ using a perturbative expansion in powers of a single running coupling constant,

$$\overline{V}_s = g_s \overline{V}^{(1)} + g_s^2 \overline{V}_s^{(2)} + \cdots \, . \tag{21}$$

In both QED and QCD the coupling does not begin to run until third order, and the first order term, $\overline{V}^{(1)}$, is given by the canonical hamiltonian and has no dependence on s other than that of the running coupling. To second order we have

$$\frac{d\overline{V}^{(2)}_{sij}}{ds} = \sum_k (\Delta_{ik} + \Delta_{jk})\, \overline{V}^{(1)}_{ik}\, \overline{V}^{(1)}_{kj}\, e^{-2s\Delta_{ik}\Delta_{jk}} \, . \tag{22}$$

Integrating from s_0 to s, we obtain

$$\overline{V}^{(2)}_{sij} = \frac{1}{2} \sum_k \overline{V}^{(1)}_{ik}\, \overline{V}^{(1)}_{kj} \left(\frac{1}{\Delta_{ik}} + \frac{1}{\Delta_{jk}} \right) \times$$

$$\times \left[e^{-2s_0\Delta_{ik}\Delta_{jk}} - e^{-2s\Delta_{ik}\Delta_{jk}} \right] \, . \tag{23}$$

This simple equation is sufficient to compute the renormalized QED and QCD hamiltonians through second order the canonical couplings.

In order to use the concept of fixed points in the similarity renormalization group, we need to modify the transformation slightly so that it is possible to get a fixed point with interactions. The free hamiltonian h is obviously a fixed point, but Eq. (20) clearly shows that the linearized transformation is given by Eq. (19). If we want to have fixed points for which V_s is not zero, we need to modify the transformation so that the gaussian cutoff factor in Eq. (19) does not change as s changes. This is most easily accomplished in the traditional manner by rescaling variables so that the eigenvalues in Eq. (19) absorb the change in s. For example, in massless light-front field theory,

$$\epsilon_i = p_i^- = \frac{p_{\perp i}^2}{p_i^+} \,. \tag{24}$$

We can absorb a change in s by rescaling all transverse momenta, and this leads to an operator classification based on the transverse engineering dimension of operators. There are other interesting possibilities,[7] but the result is that transverse local operators (i.e., no inverse powers of transverse momenta) that vanish when all transverse momenta are taken to zero are irrelevant. The electron-photon and quark-gluon couplings are marginal, except for a spin-flip piece that is relevant and breaks light-front chiral symmetry.[1,16]

The interesting new feature of light-front field theory that is not encountered in equal-time or euclidean field theory is that there is no longitudinal locality, so that the longitudinal dimension of operators does not affect their classification. A relevant or marginal operator can contain a function of longitudinal momentum fractions, and these functions inevitably appear in renormalized light-front hamiltonians.[7] This means that there are effectively an infinite number of relevant and marginal operators in renormalized light-front hamiltonians, and such a situation is usually regarded as a disaster for renormalization group treatments because it indicates that there may be an infinite number of free physical parameters. Ken Wilson and I developed coupling coherence to deal with this problem.[6,7]

2.2 Coupling coherence

The basic mathematical idea behind coupling coherence was first formulated by Oehme, Sibold, and Zimmerman.[30] They were interested in field theories where many couplings appear, such as the standard model, and wanted to find some means of reducing the number of couplings.

The puzzle that led us to the same results is how to reconcile our knowledge from covariant formulations of QCD that only one running coupling constant characterizes the renormalized theory with the appearance of new counterterms and functions required by the light-front formulation. What happens in perturbation theory when there are effectively an infinite number of relevant and marginal operators? In particular, does the solution of the perturbative renormalization group equations require an infinite number of independent counterterms (*i.e.*, independent functions of the cutoff)? Coupling coherence provides the conditions under which a finite number of running variables determines the renormalization group trajectory of the renormalized hamiltonian. To leading nontrivial orders these conditions are satisfied by the counterterms introduced to restore Lorentz covariance in scalar field theory and gauge invariance in light-front gauge theories. In fact, the conditions can apparently be used to determine all counterterms in the hamiltonian, including relevant and marginal operators that contain functions of longitudinal momentum fractions; and with no direct reference to Lorentz covariance, this symmetry seems to be restored to observables.[7]

A coupling-coherent hamiltonian is analogous to a fixed point hamiltonian, but instead of reproducing itself exactly it reproduces itself in form with a limited number of independent running couplings. If g_Λ is the only independent coupling in a theory, in a coupling-coherent hamiltonian *all other couplings are invariant functions of* g_Λ, $f_i(g_\Lambda)$. The extra couplings $f_i(g_\Lambda)$ depend on the cutoff only through their dependence on the running coupling g_Λ, and in general we demand $f_i(0) = 0$. This boundary condition on the dependent couplings is motivated in our calculations by the fact that it is the combination of the cutoff and the interactions that force us to add the counterterms we seek, so the counterterms should vanish when the interactions are turned off.

Let me illustrate the idea with a simple example with a finite number of relevant and marginal operators *ab initio*, and use coupling coherence to discover when only one or two of these may independently run with the cutoff. Such conditions are met when an underlying symmetry exists, although this is not necessary.

Consider a theory in which two scalar fields interact,

$$V(\phi) = \frac{\lambda_1}{4!}\phi_1^4 + \frac{\lambda_2}{4!}\phi_2^4 + \frac{\lambda_3}{4!}\phi_1^2\phi_2^2 \,. \tag{25}$$

Under what conditions are there fewer than three independent running coupling constants? We can use a simple cutoff on Euclidean momenta, $q^2 < \Lambda^2$.

Letting $t = \ln(\Lambda/\Lambda_0)$, the Gell-Mann–Low equations are

$$\frac{\partial \lambda_1}{\partial t} = 3\zeta\lambda_1^2 + \frac{1}{12}\zeta\lambda_3^2 + \mathcal{O}(2 \text{ loop}) \,, \tag{26}$$

$$\frac{\partial \lambda_2}{\partial t} = 3\zeta\lambda_2^2 + \frac{1}{12}\zeta\lambda_3^2 + \mathcal{O}(2 \text{ loop}) \,, \tag{27}$$

$$\frac{\partial \lambda_3}{\partial t} = \frac{2}{3}\zeta\lambda_3^2 + \zeta\lambda_1\lambda_3 + \zeta\lambda_2\lambda_3 + \mathcal{O}(2 \text{ loop}) \,; \tag{28}$$

where $\zeta = \hbar/(16\pi^2)$. It is not important at this point to understand how these equations are derived.

Assume that there is only one independent variable, $\tilde{\lambda} = \lambda_1$, so that λ_2 and λ_3 are functions of $\tilde{\lambda}$. In this case Eqs. (27) and (28) become,

$$\left(3\tilde{\lambda}^2 + \frac{1}{12}\lambda_3^2\right) \frac{\partial \lambda_2}{\partial \tilde{\lambda}} = 3\lambda_2^2 + \frac{1}{12}\lambda_3^2 \,, \tag{29}$$

$$\left(3\tilde{\lambda}^2 + \frac{1}{12}\lambda_3^2\right) \frac{\partial \lambda_3}{\partial \tilde{\lambda}} = \frac{2}{3}\lambda_3^2 + \tilde{\lambda}\lambda_3 + \lambda_2\lambda_3 \,. \tag{30}$$

The only non-trivial solutions are $\lambda_2 = \tilde{\lambda}$, and either $\lambda_3 = 2\tilde{\lambda}$ or $\lambda_3 = 6\tilde{\lambda}$. If $\lambda_3 = 2\tilde{\lambda}$,

$$V(\phi) = \frac{\tilde{\lambda}}{4!}\left(\phi_1^2 + \phi_2^2\right)^2 \,, \tag{31}$$

and we find the $O(2)$ symmetric theory. If $\lambda_3 = 6\tilde{\lambda}$,

$$V(\phi) = \frac{\tilde{\lambda}}{2 \cdot 4!}\left[\left(\phi_1 + \phi_2\right)^4 + \left(\phi_1 - \phi_2\right)^4\right] \,, \tag{32}$$

and we find two decoupled scalar fields. Therefore, λ_2 and λ_3 do not run independently with the cutoff if there is a symmetry that relates their strength to λ_1.

The condition that a limited number of variables run with the cutoff does not only reveal symmetries broken by the regulator, it may also be used to uncover symmetries that are broken by the vacuum. For example, it is straightforward to show that in a scalar theory with a ϕ^3 coupling, this coupling can be fixed as a function of the ϕ^2 and ϕ^4 couplings only if the symmetry is spontaneously broken rather than explicitly broken.[6]

For the QED and QCD calculations, I need to compute the hamiltonian to second order, while the canonical coupling runs at third order. In this case we can use Eq.(23), with $s_0 \to \infty$ for relevant operators and $s_0 \to 0$ for irrelevant operators, and with the bare coupling e replaced by the running coupling e_s or e_Λ. The more interesting case of marginal operators can be avoided at first.[7,8]

3 Light-Front QED and QCD

Various forms of the canonical light-front QED and QCD hamiltonians can be found in several articles.[31,32] Following Brodsky and Lepage, I have displayed these hamiltonians elsewhere.[2,33]

In light-cone gauge and using light-front coordinates, it is possible to explicitly eliminate all unphysical degrees of freedom and write the hamiltonian in terms of two-component fermions and transverse gluons. Any ambiguities in the procedure that come from the zero-mode problem or normal-ordering are resolved by coupling coherence, so the renormalized hamiltonian is apparently uniquely determined order-by-order in the running coupling.

3.1 Light-front QED

In this section I follow the strategy outlined in the first section to compute the positronium spectrum. I outline the calculation through the leading order Bohr results [2] and indicate how higher order calculations proceed.[9,10]

The first step is to compute a renormalized cutoff hamiltonian as a power series in the running coupling e_Λ,

$$H^\Lambda_N = h_0 + e_\Lambda h_1 + e_\Lambda^2 h_2 + \cdots + e_\Lambda^N h_N \; . \tag{33}$$

Having obtained the hamiltonian to some order in e_Λ, the next step is to split it into two parts,

$$H^\Lambda = \mathcal{H}_0^\Lambda + \mathcal{V}^\Lambda \; . \tag{34}$$

$\mathcal{H}_0^\Lambda$ must be accurately solved non-perturbatively, producing a zeroth order approximation for the eigenvalues and eigenstates. The greatest ambiguities in the calculation appear in the choice of $\mathcal{H}_0^\Lambda$, which requires one of science's most powerful computational tools, trial and error.

In QED and QCD I *conjecture* that for sufficiently small Λ all interactions in $\mathcal{H}_0^\Lambda$ preserve particle number, with all interactions that involve particle creation and annihilation in $\mathcal{V}^\Lambda$. Corrections from $\mathcal{V}^\Lambda$ are then computed in bound state perturbation theory.

Since $\mathcal{H}_0^\Lambda$ is assumed to include interactions that preserve particle number, the zeroth order positronium ground state is a pure electron-positron state. We only need one- and two-body interactions; *i.e.*, the electron self-energy and the electron-positron interaction. The hamiltonian is computed to second order using Eq. (23). We must specify s_0 in Eq. (23), which corresponds to the inverse cutoff squared at which boundary conditions are placed on the hamiltonian. Coupling coherence leads to the prescription that $s_0 \to 0$ for irrelevant operators and $s_0 \to \infty$ for relevant operators.

Bare electron mixing with electron-photon states leads to a self-energy (see Fig. 1):

$$\Sigma^{\Lambda}_{coh}(p) = \frac{e_\Lambda^2}{8\pi^2 p^+}\Bigg\{2y\Lambda^2 \ln\left(\frac{y^2\Lambda^2}{(y\Lambda^2+m^2)\epsilon}\right) - \frac{3}{2}y\Lambda^2 + \frac{1}{2}\frac{ym^2\Lambda^2}{y\Lambda^2+m^2} + 3m^2\ln\left(\frac{m^2}{y\Lambda^2+m^2}\right)\Bigg\} + \mathcal{O}(\epsilon/y)\;; \qquad (35)$$

where y is the fraction of longitudinal momentum carried by the electron, $y = p^+/P^+$. To simplify the discussion I have replaced the gaussian cutoff factors that appear in all integrals with step functions, and completed the integrals analytically using $1/\sqrt{s} = \Lambda^2/P^+$. It is possible to produce such step function cutoffs with a similarity transformation,[2] but this leads to pathologies at higher order. More importantly, I have been forced to introduce *a second cutoff*,

$$xp^+ > \epsilon P^+\;, \qquad (36)$$

because there is a logarithmic divergence in the loop longitudinal momentum integration even with the gaussian cutoff in place. This second cutoff must be taken to zero and no new counterterms can be added to the hamiltonian, so all divergences must cancel before it is taken to zero.

We have no choice about whether this divergent operator is in the hamiltonian if we use coupling coherence. We can only choose between putting it in $\mathcal{H}_0^\Lambda$ or in $\mathcal{V}^\Lambda$. I make different choices in QED and QCD, and the arguments are based on physics.

The *divergent* electron 'mass' is a complete lie. We encounter a term proportional to $e_\Lambda^2\Lambda^2\ln(1/\epsilon)/P^+$ when the scale is Λ; however, we can reduce this scale as far as we please in perturbation theory. Photons are massless, so the electron continues to dress itself with small-x photons to arbitrarily small Λ. Since I believe that this divergent self-energy is exactly canceled by mixing with small-x photons, and that this mixing is perturbative in QED, I simply put it in $\mathcal{V}^\Lambda$.

There are two time-ordered diagrams involving photon exchange between an electron with initial momentum p_1 and final momentum p_2, and a positron

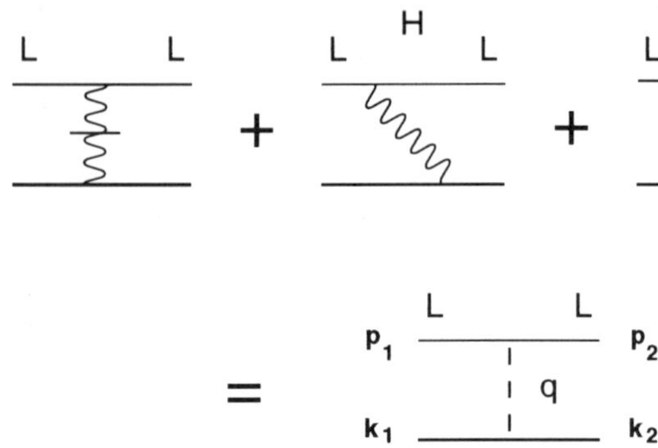

Figure 5. Effective two-body interaction between low energy constituents resulting from: (i) the canonical instantaneous exchange interaction, and (ii) the elimination of direct coupling between low energy two-body states and high energy states containing an additional gauge particle.

with initial momentum k_1 and final momentum k_2. These are shown in Fig. 5, along with the instantaneous exchange diagram. I refer the reader to longer articles where details are given [2,9] and concentrate here on the essential results.

Photon exchange above the cutoff produces an effective interaction that cancels the instantaneous photon exchange operator in the canonical hamiltonian, replacing it with a Coulomb interaction and magnetic interactions that are partially responsible for fine structure. Instantaneous photon exchange below the cutoff remains, and as is discussed in the section on QCD it produces a long-range confining interaction. In QED this long range interaction is exactly cancelled by further exchange of massless low energy photons in bound state perturbation theory. In QCD the long range interaction also acts on gluons, blocking such a cancellation and producing the essential difference between QED and QCD in this approach.

This means that we can concentrate on photon exchange above the cutoff to leading order, which still leaves us with a complicated operator in light-front coordinates. In order to present an analytic analysis I make assumptions that are justified *a posteriori*. First I assume that the electron and positron momenta can be arbitrarily large, but that in low-lying states their relative momenta satisfy

$$|\mathbf{p}_\perp - \mathbf{k}_\perp| \sim \alpha m \; , \tag{37}$$

$$|p^+ - k^+| \sim \alpha(p^+ + k^+) \; . \tag{38}$$

This allows us to use power counting to evaluate the perturbative strength of operators for small coupling.

Given these order of magnitude estimates for momenta, we can drastically simplify all of the operators in the hamiltonian. At this point we can complete the zeroth order analysis of positronium using the state,

$$|\Psi(P)\rangle = \sum_{\sigma\lambda} \int \frac{dp^+ d^2p_\perp}{16\pi^3 p^+} \frac{dk^+ d^2k_\perp}{16\pi^3 k^+} \sqrt{p^+k^+} 16\pi^3 \delta^3(P-p-k)$$
$$\phi(p,\sigma;k,\lambda) b^\dagger(p,\sigma) d^\dagger(k,\lambda)|0\rangle \,, \tag{39}$$

where $\phi(p,\sigma;k,\lambda)$ is the wave function for the relative motion of the electron and positron, with the center-of-mass momentum being P. We need to choose the longitudinal momentum appearing in the cutoff, and I use the natural scale P^+.

If we want to find a cutoff for which the ground state is dominated by the electron-positron component of the wave function, we need the cutoff to remove the important part of the electron-positron-photon phase space. Since the exchanged photon energy is typically $\mathcal{O}(\alpha m^2)$, we need:

$$\Lambda^2 < \alpha m^2 \,. \tag{40}$$

On the other hand, we cannot allow the cutoff to remove the region of the electron-positron phase space from which the wave function receives most of its strength. This requires

$$\Lambda^2 > \alpha^2 m^2 \,. \tag{41}$$

For cutoffs that satisfy $\alpha m^2 > \Lambda^2 > \alpha^2 m^2$, the bound state equation can be simplified to:

$$-E\phi(\mathbf{k}_1) = \frac{\mathbf{k}_1^2}{m}\phi(\mathbf{k}_1) - \alpha \int \frac{d^3k_2}{(2\pi)^3} \frac{1}{\left(\mathbf{k}_1^2 - \mathbf{k}_2^2\right)} \phi(\mathbf{k}_2) \,, \tag{42}$$

where a simple change of variables has replaced longitudinal momentum fractions with a z-component of momentum, making the system's nonrelativistic dynamics manifest. The cutoffs drop out to leading order, leaving us with the familiar nonrelativistic Schrödinger equation for positronium in momentum space. The solution is

$$\phi(\mathbf{k}) = \frac{\mathcal{N}}{\left(\mathbf{k}^2 + mE\right)^2} \,, \tag{43}$$

$$E = \frac{1}{4}\alpha^2 m \,. \tag{44}$$

$\mathcal{N}$ is a normalization constant. This is the Bohr energy for the ground state of positronium, and it is obvious that the entire nonrelativistic spectrum is reproduced to leading order.

Beyond this leading order result the calculations become much more interesting, and in any hamiltonian formulation they rapidly become complicated. The leading correction to the binding energy is $\mathcal{O}(\alpha^4)$, and producing these corrections is a much more serious test of the renormalization procedure. We have shown that the fine structure of positronium is correctly reproduced when the first- and second-order corrections from bound state perturbation theory are added.[9] This is a formidable calculation, because the exact Coulomb bound and scattering states appear in second-order bound state perturbation theory.

A complete calculation of the Lamb shift in hydrogen would also require a fourth-order similarity calculation of the hamiltonian; however, the dominant contribution to the Lamb shift that was first computed by Bethe [34] can be computed using a hamiltonian determined to $\mathcal{O}(\alpha)$.[10] In our calculation a Bloch transformation was used rather than a similarity transformation because the Bloch transformation is simpler and small energy denominator problems can be avoided in analytical QED calculations.

The primary obstacle to using our light-front strategy for precision QED calculations is algebraic complexity. We have successfully used QED as a testing ground for this strategy, but these calculations can be done much more conveniently using other methods. The theory for which we believe our methods are best suited is QCD.

3.2 Light-front QCD

We only require the QCD hamiltonian determined to $\mathcal{O}(\alpha)$ to discuss a simple confinement mechanism that appears naturally in light-front QCD and to complete reasonable zeroth order calculations for heavy quark bound states. To this order the QCD hamiltonian in the quark-antiquark sector is almost identical to the QED hamiltonian in the electron-positron sector. Of course the QCD hamiltonian differs significantly from the QED hamiltonian in other sectors, and this is essential for justifying my choice of $\mathcal{H}_0^\Lambda$ for non-perturbative calculations.

The basic strategy for doing a sequence of (hopefully) increasingly accurate QCD bound state calculations is almost identical to the strategy for doing QED calculations. Find an expansion for H^Λ in powers of the QCD coupling constant to a finite order. Divide the hamiltonian into a non-perturbative part, $\mathcal{H}_0^\Lambda$, and a perturbative part, $\mathcal{V}^\Lambda$. The division is based on the physical argument that adding a parton in an intermediate state should require more energy than indicated by the free hamiltonian, and that as a result these states 'freeze out' as the cutoff approaches Λ_{QCD}. When this happens the evolution of the hamiltonian as the cutoff is lowered further changes qualitatively, and operators that were consistently canceled over an infinite number of scales also freeze, so that their effects in the few parton sectors can be studied directly. A one-body operator and a two-body operator arise in this fashion, and serve to confine both quarks and gluons.

The simple confinement mechanism I outline is certainly not the final story, but it may be the seed for the full confinement mechanism. One of the most serious problems we face when looking for non-perturbative effects such as confinement is that the search itself depends on the effect. A candidate mechanism must be found and then shown to produce itself self-consistently as the cutoff is lowered towards Λ_{QCD}.

Once we find a candidate confinement mechanism, it is possible to study heavy quark bound states with little modification of the QED strategy. Of course the results in QCD differ from those in QED because of the new choice of $\mathcal{H}_0^\Lambda$, and in higher orders because of the gluon interactions.

When we compute the QCD hamiltonian to $\mathcal{O}(\alpha)$, several significant new features appear. First are the familiar gluon interactions. In addition to the many gluon interactions found in the canonical hamiltonian, there are modifications to the instantaneous gluon exchange interactions, just as there were modifications to the electron-positron interaction. For example, a Coulomb interaction automatically arises at short distances. In addition the gluon self-energy differs drastically from the photon self-energy.

The photon develops a self-energy because it mixes with electron-positron pairs, and this self energy is $\mathcal{O}(\alpha\Lambda^2/P^+)$. When the cutoff is lowered below $4m^2$, this mass term dies exponentially because it is no longer possible to produce electron-positron pairs. For all cutoffs the small bare photon self-energy is exactly canceled by mixing with pairs below the cutoff. I do not go through the calculation, but because the gluon also mixes with gluon pairs in QCD, the gluon self-energy acquires an infrared divergence, just as the electron did in QED. In QCD both the quark and gluon self-energies are proportional to $\alpha\Lambda^2\ln(1/\epsilon)/P^+$, where ϵ is the secondary cutoff on parton longitudinal momenta introduced in the last section. This means that even when the primary cutoff Λ^2 is finite, the energy of a single quark or a single gluon is infinite, because we are supposed to let $\epsilon \rightarrow 0$.

In QED I argued that the bare electron self-energy is a complete lie, because the bare electron mixes with photons carrying arbitrarily small longitudinal momenta to cancel this bare self-energy and produce a finite mass physical electron. However, in QCD there is no reason to believe that this perturbative mixing continues to arbitrarily small cutoffs. There are *no* massless gluons in the world. In this case, the free QCD hamiltonian is a complete lie and cannot be trusted at low energies.

On the other hand, coupling coherence gives us no choice about the quark and gluon self-energies as computed in perturbation theory. The question is not whether large self-energies appear in the hamiltonian. The question is whether these self-energies are canceled by mixing with low energy multi-gluon states. As the cutoff approaches Λ_{QCD}, I speculate that these cancellations cease to occur because perturbation theory breaks down and a mass gap between states with and without extra gluons appears.

But if the quark and gluon self-energies diverge, and the divergences cannot be canceled by mixing between sectors with an increasingly large number of partons, how is it possible to obtain finite mass hadrons? The parton-parton interaction also diverges, and the infrared divergence in the two-body interaction *exactly cancels* the infrared divergence in the one-body operator for color singlet states.

Of course, the cancellation of infrared divergences is not enough to obtain confinement. The cancellation is exact regardless of the relative motion of the partons in a color singlet state, and confinement requires a residual interaction. The $\mathcal{O}(\alpha)$ QCD hamiltonian contains a logarithmic potential in both longitudinal and transverse directions. There is no rigorous demonstration that the confining interaction is linear, and a logarithmic potential is of interest phenomenologically for heavy quark bound states.[35] I would be delighted if a better light-front calculation produces a linear potential, but this may not be necessary even for successful light hadron calculations.

The calculation of how the quark self-energy changes when a similarity transformation lowers the cutoff on energy transfer is almost identical to the electron self-energy calculation. We find the one-body operator required by coupling coherence,

$$\Sigma^{\Lambda}_{coh}(p) = \frac{g^2 C_F}{8\pi^2 p^+}\left\{2y\Lambda^2 \ln\left(\frac{y^2\Lambda^2}{(y\Lambda^2+m^2)\epsilon}\right) - \frac{3}{2}y\Lambda^2 + \frac{1}{2}\frac{ym^2\Lambda^2}{y\Lambda^2+m^2} \right.$$
$$\left. + 3m^2 \ln\left(\frac{m^2}{y\Lambda^2+m^2}\right)\right\} + \mathcal{O}(\epsilon/y) , \qquad (45)$$

where $C_F = (N^2-1)/(2N)$ for a SU(N) gauge theory.

The calculation of the quark-antiquark interaction required by coupling coherence is also nearly identical to the QED calculation. Just as in QED the coupling coherent interaction induced by gluon exchange above the cutoff partially cancels instantaneous gluon exchange. For the discussion of confinement the part of V_{coh} that remains is not important, because it produces the short range part of the Coulomb interaction. However, the part of the instantaneous interaction that is not canceled is

$$\tilde{V}^{\Lambda}_{instant} = -8g_\Lambda^2 C_F \sqrt{p_1^+ p_2^+ k_1^+ k_2^+} \left(\frac{1}{q^+}\right)^2 \delta_{\sigma_1\sigma_2}\delta_{\lambda_1\lambda_2}$$
$$\times\ \theta\big(|p_1^+ - p_2^+| - \epsilon P^+\big)$$
$$\times\ \exp\Bigg[-2s(p_1^- - p_2^- - q^-)(k_2^- - k_1^- - q^-)\Bigg] . \tag{46}$$

Note that this interaction contains a cutoff that projects onto exchange energies below the cutoff, because the interaction has been screened by gluon exchange above the cutoffs. This interaction can become important at long distances, if parton exchange below the cutoff is dynamically suppressed. In QED I argued that this singular long range interaction is exactly canceled by photon exchange below the cutoff, because such exchange is not suppressed no matter how low the cutoff becomes. Photons are massless and experience no significant interactions, so they are exchanged to arbitrarily low energies as effectively free photons. This cannot be the case for gluons.

For the discussion of confinement, place the most singular parts of the quark self-energy and the quark-antiquark interaction in $\mathcal{H}_0^\Lambda$. To see that all infrared divergences cancel and that the residual long range interaction is logarithmic, study the matrix element of these operators for a quark-antiquark state,

$$|\Psi(P)\rangle = \sum_{\sigma\lambda}\sum_{rs}\int \frac{dp^+ d^2p_\perp}{16\pi^3 p^+}\frac{dk^+ d^2k_\perp}{16\pi^3 k^+}\sqrt{p^+k^+}16\pi^3\delta^3(P-p-k)$$
$$\phi(p,\sigma,r;k,\lambda,s)b^{r\dagger}(p,\sigma)d^{s\dagger}(k,\lambda)|0\rangle , \tag{47}$$

where r and s are color indices and ϕ is a color singlet.

When the expectation value of the hamiltonian is taken using this state there are divergences as $\epsilon \to 0$ in both the expectation value of the self-energy and the surviving piece of instantaneous gluon exchange. These divergences cancel exactly for any color-singlet state. The cancellation resembles what happens in the Schwinger model. If the state is a color octet the divergences are both positive and cannot cancel. Since the cancellation occurs in the matrix element, we can let $\epsilon \to 0$ before diagonalizing $\mathcal{H}_0^\Lambda$.

The fact that the divergences cancel exactly does not indicate that confinement occurs. This requires the residual interactions to diverge at large distances. It is easily shown that for large longitudinal separations the interaction becomes:

$$V(x^-) = \frac{g_\Lambda^2 C_F \Lambda^2}{2\pi^2} \ln\left(|x^-|\right) . \tag{48}$$

At large transverse separations it becomes:

$$V(x_\perp) = \frac{g_\Lambda^2 C_F \Lambda^2}{\pi^2} \ln(|\mathbf{x}_\perp|) . \tag{49}$$

The strength of the long-range logarithmic potential is not spherically symmetrical in these coordinates, with the potential being larger in the transverse than in the longitudinal direction. Of course, there is no reason to demand that the potential be rotationally symmetric in these coordinates, because rotations are dynamical and are supposed to alter the Fock space composition of states in addition to rotating the state in a given Fock space sector.

Had we computed the quark-gluon or gluon-gluon interaction, we would find essentially the same residual long range two-body interaction in every Fock space sector, although the strengths would differ because different color operators appear. In QCD gluons have a divergent self-energy and experience divergent long range interactions with other partons if we use coupling coherence. In this sense, the assumption that gluon exchange below some cutoff is suppressed is consistent with the hamiltonian that results from this assumption. To show that gluon exchange is suppressed when $\Lambda \rightarrow \Lambda_{QCD}$, rather than some other scale (*i.e.*, zero as in QED), a non-perturbative calculation of gluon exchange is required. This same confinement mechanism appears using the similarity transformation developed by Walhout.[29]

I provide only a brief summary of our heavy quark bound state calculations,[20,21] and refer the reader to the original articles for details. We follow the strategy that has been successfully applied to QED, with modifications suggested by the fact that gluons experience a confining interaction.

For heavy quark bound states [21] we can simplify the hamiltonian by making a nonrelativistic reduction and solving a Schrödinger equation. We must then choose values for Λ, α, and M. These should be chosen differently for bottomonium and charmonium. The cutoff for which the constituent approximation works well depends on the constituent mass, as in QED where it is obviously different for positronium and muonium. In order to fit the ground state and first two excited states of charmonium, we use $\Lambda = 2.5 GeV$, $\alpha = 0.53$, $M_c = 1.6 GeV$. In order to fit these states in bottomonium we use $\Lambda = 4.9 GeV$, $\alpha = 0.4$, and $M_b = 4.8 GeV$. Violations of rotational invariance from the remaining parts of the potential are only about 10%, and we expect corrections from higher Fock state components to be at least of this magnitude for the couplings we use.

These calculations show that the approach is reasonable, but they are not yet very convincing. There are a host of additional calculations that must be done before the success of this approach can be judged.

Acknowledgments

I wish to thank the organizers of the APCTP-RCNP Joint International School on Physics of Hadrons and QCD for putting together an excellent meeting. I would also like to acknowledge many useful discussions with Brent Allen, Stan Głazek, Roger Kylin, Rick Mohr, Jim Steele, and Ken Wilson. This program has been advanced by many theorists, including Edsel Ammons, Martina Brisudova, Koji Harada, Avaroth Harindranath, Billy Jones, Yizhang Mo, Dave Robertson, Tim Walhout, and Wei-Min Zhang. This work was supported by National Science Foundation grant PHY-9800964.

References

1. K. G. Wilson, T. S. Walhout, A. Harindranath, W.-M. Zhang, R. J. Perry and St. D. Głazek, Phys. Rev. D **49**, 6720 (1994).
2. R. J. Perry, in the Proceedings of *Hadrons '94*, V. Herscovitz and C. Vasconcellos, eds. (World Scientific, Singapore, 1995), and revised version hep-th/9411037.
3. S. D. Głazek and K.G. Wilson, Phys. Rev. D **48**, 5863 (1993);
   S. D. Głazek and K.G. Wilson, Phys. Rev. D **49**, 4214 (1994).
4. F. Wegner, Ann. Physik **3**, 77 (1994).
5. S. D. Głazek and K.G. Wilson, Phys. Rev. D **57**, 3558 (1998).
6. R. J. Perry and K. G. Wilson, Nucl. Phys. B **403**, 587 (1993).
7. R. J. Perry, Ann. Phys. **232**, 116 (1994).
8. B. H. Allen and R. J. Perry, Phys. Rev. D **58**, 125017 (1998), hep-th/9804136;
   R. D. Kylin, B. H. Allen, and R. J. Perry, hep-th/9812080 (1998).
9. B. D. Jones, R. J. Perry and St. D. Głazek, Phys. Rev. D **55**, 6561 (1997).
10. B. D. Jones and R. J. Perry, Phys. Rev. D **55**, 7715 (1997).
11. Sérgio Szpigel and Robert J. Perry, in these Proceedings (1999).
12. P.A.M. Dirac, Rev. Mod. Phys. **21**, 392 (1949).
13. M. Burkardt, Phys. Rev. D **44**, 4628 (1993).
14. M. Burkardt, Nuclear Summer School NUSS 97, hep-ph/9709421.

15. S.J. Chang, R.G. Root, and T.M. Yan, Phys. Rev. D **7**, 1133 (1973).
16. D. Mustaki, hep-ph/9404206.
17. K. G. Wilson and M. Brisudová, in Proceedings of the International Workshop on *Light-cone QCD*, S. Głazek, ed. (World Scientific, Singapore, 1995).
18. K. G. Wilson, Rev. Mod. Phys. **47**, 773 (1975).
19. M. Brisudová, S. Szpigel and R. J. Perry, Phys. Lett. B **421**, 334 (1998).
20. M. Brisudová and R. J. Perry, Phys. Rev. D **54**, 1831 (1996).
21. M. Brisudová, R. J. Perry and K. G. Wilson, Phys. Rev. Lett. **78**, 1227 (1997).
22. B. H. Allen and R. J. Perry, work in progress.
23. J. Schwinger, Phys. Rev. **125**, 397 (1962), Phys. Rev. **128**, 2425 (1962).
24. J. Lowenstein and A. Swieca, Ann. Phys. (N.Y.) **68**, 172 (1971).
25. H. Bergknoff, Nucl. Phys. B **122**, 215 (1977).
26. K. G. Wilson, Phys. Rev. **140**, B445 (1965).
27. K. G. Wilson, Phys. Rev. D **2**, 1438 (1970).
28. K. G. Wilson and J. B. Kogut, Phys. Rep. **12C**, 75 (1974).
29. T. S. Walhout, hep-th/9806097.
30. R. Oehme, K. Sibold, and W. Zimmerman, Phys. Lett. B **147**, 115 (1984); R. Oehme and W. Zimmerman, Comm. Math. Phys. **97**, 569 (1985); W. Zimmerman, Comm. Math. Phys. **97**, 211 (1985); J. Kubo, K. Sibold, and W. Zimmerman, Nuc. Phys. B **259**, 331 (1985); R. Oehme, Prog. Theor. Phys. Supp. **86**, 215 (1986).
31. S. J. Brodsky and G. P. Lepage, in *Perturbative Quantum Chromodynamics*, A. H. Mueller, ed. (World Scientific, Singapore, 1989).
32. W. M. Zhang and A. Harindranath, Phys. Rev. D **48**, 4868; 4881; 4903 (1993).
33. R. J. Perry, in Proceedings of the NATO Advanced Study Institute *Confinement, Duality, and Non-Perturbative Aspects of QCD*, Newton Institute, Cambridge, England, June, 1997 (Plenum, London, 1998).
34. H. A. Bethe, Phys. Rev. **72**, 339 (1947).
35. C. Quigg and J.L. Rosner, Phys. Lett. **71B**, 153 (1977).

CHIRAL SYMMETRY IN NUCLEAR PHYSICS [a]

Mannque RHO
Service de Physique Théorique, CE Saclay
91191 Gif-sur-Yvette, France [b]
and
School of Physics, Korea Institute for Advanced Study
Seoul 130-012, Korea
E-mail: rho@spht.saclay.cea.fr

How chiral symmetry – which is a basic ingredient of quantum chromodynamics (QCD) for light-quark hadrons – enters and plays an eminent role in nuclear physics is discussed. This is done in two steps. In the first step, I introduce the notion of the Cheshire Cat Principle which describes how a strong-interaction phenomenon can be described in various different languages. In the second step, I treat two cases of widely different density regimes. The first case deals with formulating nuclear physics of *dilute* nuclear systems with a focus on both bound and scattering processes in two-nucleon systems in terms of effective field theories (EFT), e.g., in the framework of chiral perturbation theory. The second case requires formulating effective field theories for *dense* hadronic matter such as heavy nuclei and nuclear matter under normal as well as extreme conditions expected to be encountered in compact stars and in heavy-ion collisions at relativistic energy. The first case is rigorously formulated but the second relies on several assumptions that appear to be reasonable but need ultimately to be justified. The Cheshire Cat Principle is proposed to play an essential role in lending support to the latter development.

1 Introduction

Chiral symmetry, now known to be associated with the fact that the up and down quarks are very light, has been playing an important role in nuclear physics since a long time, since even before the advent of QCD [1,2]. In this lecture, I will describe some of the more recent developments in this area in which I have participated.

One of the most notable results obtained in nuclear physics that are consequences of chiral symmetry is the exchange current contribution to nuclear responses to electroweak currents. Although exchange currents are clearly es-

[a]Lectures given at "The APCTP-RCNP Joint International School on Physics of Hadrons and QCD," Osaka, Japan, 12-13 October 1998 and at "The 1998 YITP Workshop on QCD and Hadron Physics," Kyoto, Japan, 14-16 October 1998

[b]Permanent address

tablished by theory and experiment, the role of chiral symmetry – specifically in terms of the pion – is not well appreciated in nuclear physics community. I believe that this issue marks the turning point from the "classical" nuclear physics to the next generation of hadronic physics probing hadronic matter under extreme density and/or temperature conditions.

It was pointed out in 1971 by Chemtob and Rho[3] that the meson-exchange current, a long-standing problem in nuclear physics, could be efficiently organized in such a way that the main contribution be calculated by low-energy theorems based on chiral symmetry. Riska and Brown [4], by exploiting the Chemtob-Rho organization, succeeded to explain quantitatively the radiative np capture process $n + p \to d + \gamma$. In 1978, Kubodera, Delorme and Rho[5] proposed what is now called the "chiral filter hypothesis" which states that whenever soft-pion exchanges are *un-suppressed* by kinematics or selection rules, they should provide the dominant model-independent contribution with the remaining corrections markedly suppressed while whenever they are *suppressed* for the same reason, then short-distance effects can and do enter at the same level as higher order corrections, rendering systematic calculations difficult if not unfeasible. The predictions that followed from this hypothesis were that the soft-pion effects should be prominent in (1) nuclear electromagnetic M1 transitions and (2) weak axial-charge transitions. Both these predictions were unambiguously confirmed by experiments. The former was confirmed in the large momentum transfer electrodisintegration of the deuteron $e+d \to n+p+e$ performed in 1980's [6] and the latter in a series of experiments and analyses of Warburton in early 1990's [7]. These predictions which were made without direct recourse to low-energy effective field theories of QCD were more recently given a justification in terms of chiral perturbation theory in 1991 [8] [#1] and more rigorously treated in the context of effective field theories using a cut-off regularization as described below [9].

What I believe to be the most crucial aspect in understanding nuclear phenomena in the context of QCD is that a low-energy process can be described very accurately in terms of variables that represent physical (color singlet) objects, that is, the hadron degrees of freedom that are measured in the laboratory although the appropriate QCD variables are quarks and gluons. The way this can be understood from the point of view of QCD is in terms of what is now called "Cheshire Cat Principle" (CCP) [2] which in a nutshell can be

[#1] In the context of chiral perturbation theory, it is natural that the pion-exchange currents dominate whenever they are unsuppressed over short-ranged terms although in the potential both the pion-exchange and contact interactions without derivatives figure on the same footing. The chiral filter phenomena in the *currents* can be understood by a simple chiral counting as shown in this paper.

summarized as an (albeit approximate) equivalence between the description in terms of hadronic variables and that in terms of QCD variables for processes that take place in long-wavelength kinematic regimes. In certain cases, the CCP can be exact. For instance, in two space-time dimensions, the bosonization technique allows one to precisely describe fermionic excitations in terms of bosonic and vice versa and in the presence of supersymmetry, there can be exact "duality," a concept which is becoming very popular in particle physics community [#2]. In nonsupersymmetric and dimension-four world, there is no known transformation that can give an exact CCP but there is no proof that such a transformation does not exist. I will show a recent case where in QCD at four dimensions a fairly good CCP can be seen in nucleon structure. Another intriguing case, though somewhat different in nature, is discussed in [11].

At very low energy and low density, nuclear physics can be accurately described in terms of low-mass hadronic variables with a cutoff set at the chiral scale given by the vector meson mass, ~ 1 GeV. I will illustrate this in terms of effective field theories that involve only the nucleon fields and pionic fields. In fact we will see that even the pion can be integrated out by setting an effective cutoff at the pion mass and yet a quantitative agreement can be obtained in the nucleon-only theory. I will focus on two-nucleon systems which can be described in terms of a Lagrangian whose parameters are defined in matter-free space.

I will next jump over the intermediate density regime to one for which no systematic chiral perturbation theory approach is available. This is the case with nuclear matter where new scales enter due to the presence of a Fermi sea. For this case, two effective field theories will figure; one, chiral Lagrangian field theory in a dense background and the other, Fermi-liquid fixed-point theory. These two by themselves are superb effective field theories. I will argue that in going to dense hadronic matter beyond nuclear matter density, the two effective field theories could be married once Brown-Rho scaling[12] is introduced. The arguments used here are more conjectural than rigorous but I will show a number of evidence that the marriage is a successful one.

2 Cheshire Cat Principle (CCP)

When the large N_C property of QCD which implies the skyrmion picture for baryons and the bag structure of QCD with asymptotic freedom are properly combined, we have the chiral bag [2]. When the topological structure of the skyrmion is incorporated into the chiral bag, we obtain the (approximate)

[#2]In modern string theory [10], this is summarized by the two-fold duality, "elementality $\leftrightarrow$ composite" and "classical $\leftrightarrow$ quantum."

Cheshire Cat structure. It has been shown [2,13] that at least at low energies, practically all physical observables, with few exceptions, obey the Cheshire Cat Principle, in some cases exactly (such as the baryon charge thanks to topological invariance) and some approximately (those properties not connected to topology). The reason why the Cheshire Cat Principle is operative is probably connected to the existence of something deep in the form of a web of dualities in gauge theories.

Among the few cases that appeared to deviate from the Cheshire Cat Principle, a most recent and intriguing one is the flavor singlet axial charge (FSAC) of the proton, sometimes mistakenly identified as "proton spin." A straightforward calculation of the FSAC in the chiral bag would indicate that the FSAC is small at small bag radius and grows to order unity at large bag radius which would mean that the Cheshire Cat is grossly violated. The purpose of this part of the lecture is to show that if certain vacuum properties of the bag related to anomalies are properly taken into account, the CC can be recovered [14]. Here breaking of chiral symmetry (i.e., chiral $U_A(1)$) due to quantum effects will be involved.

2.1 Chiral Bag

The most convenient formalism to discuss the CCP is the chiral bag model (CBM) which is a model that interpolates between the "bagged" quark description known as MIT bag and the topological model, skyrmion, obtainable in the large N_C limit of QCD. This model consists of the bag of radius R within which the QCD degrees of freedom, quarks and gluons, live surrounded by the cloud of pseudo-Goldstone bosons (pions, kaons etc.) and, if needed for short-distance physics, massive mesons with communication between the inside and the outside mediated by suitable boundary conditions. The power of this model is that it can combine both long-distance (outside) and short-distance (inside) physics controlled by the boundary terms. Thus it can describe long-wavelength processes as well as short-wavelength processes within a single model which is difficult for the skyrmion or the MIT bag separately. For convenience, one usually uses – as we shall do for this lecture – a spherical geometry for the bag with a sharp boundary but in principle one could use an arbitrarily shaped bag with a fuzzy surface. Because of the leakage of various charges across the boundary, it should not really matter what shape and how sharp the boundary should be in. The CCP in (octet or triplet) flavor space has been extensively discussed, e.g., in [2], so I shall not go into that. Let me just focus on the flavor singlet $U_A(1)$ channel which is of particular interest because of the chiral anomaly associated with the massive η' degree of freedom. The model we consider has the η' field (which we shall simply call η) outside[#3]. This η field couples on the surface to the quarks living inside. The coupling condition follows from the equation of motion for the quark field and takes the form

$$in^\mu \gamma_\mu \psi = e^{i\frac{\eta}{f}\gamma_5}\psi(R) \tag{1}$$

where f is a mass-dimension-1 constant related to the η decay constant and n^μ is a unit normal to the surface. This condition assures at the classical level that the baryon current is conserved [#4]

When the $U_A(1)$ field is present, it can also couple to the gluons living inside. That gives a boundary condition that involves the η field and the gluon field. To deduce this condition, consider the normal components of the FSAC on both sides of the boundary. Coming from outside, one has

$$fn^\mu \partial_\mu \eta(R). \tag{2}$$

This is well-defined at the surface since it is a nice local operator. Now coming from inside, we have

$$\frac{1}{2} n^\mu \bar{\psi}\gamma_\mu \gamma_5 \psi(R). \tag{3}$$

This however is not well-defined since the bilinear in fermion fields at the same space-time point is singular. Regularizing it by point-splitting a la Schwinger reveals that the regularized form of (3) has an additional term involving the Chern-Simons current $K^\mu = \epsilon^{\mu\nu\alpha\beta}(G_\nu^a G_{\alpha\beta}^a - \frac{2}{3} f^{abc} g G_\nu^a G_\alpha^b G_\beta^c)$ given in terms of the color gauge field G_μ^a,

$$\frac{1}{2}\hat{n}\cdot(\bar{\psi}\vec{\gamma}\gamma_5\psi)_{NO} + \frac{N_F g^2}{16\pi^2}\hat{n}\cdot K \tag{4}$$

where the subscript "NO" stands for normal ordering. Now the Chern-Simons current on the surface is not gauge-invariant, so (4) is not gauge-invariant. This violation of gauge invariance, which is not acceptable, can be understood as "color anomaly." The color anomaly arises as a consequence of the celebrated chiral anomaly associated with the $U_A(1)$ symmetry which is broken due to

---

[#3] The pion field outside plays an important role in inducing the leakage of the baryon charge. This phenomenon is very well understood and will be taken into account in the result although we will not discuss it here.

[#4] We know however that the hedgehog pions induce a violation of the baryon current but the η field does not participate in this violation since there is no topological component associated with this pseudoscalar meson.

quantum fluctuations in the vacuum. The flavor singlet axial current inside has the anomaly and hence the color leaks out as a consequence. To prevent the leakage of the color, the simplest (though perhaps not the only) way is to put a boundary condition that "soaks up" this leaking color charge. It has been shown [2] that the following surface counter term does the job:

$$\mathcal{L}_{CT} = i\frac{g^2}{32\pi^2}\oint_{\Sigma} d\beta K^{\mu} n_{\mu} (\mathrm{Tr}\ln U^{\dagger} - \mathrm{Tr}\ln U) \tag{5}$$

where N_F is the number of flavors (here taken to be =3), β is a point on a surface Σ, n^{μ} is the outward normal to the bag surface, U is the $U(N_F)$ matrix-valued field written as $U = e^{i\pi/f}e^{i\eta/f}$. Note that (5) manifestly breaks color gauge invariance (both large and small, the latter due to the bag), so the action of the chiral bag model with this term is not gauge invariant at the classical level but as shown in [15], when quantum fluctuations are calculated, there appears an induced anomaly term on the surface which exactly cancels this term. Thus gauge invariance is restored at the quantum level. Presence of the boundary term (5) implies that the color electric and magnetic fields must satisfy boundary conditions that have coupling to the η field,

$$\hat{n}\cdot\vec{E}^a = -\frac{N_F g^2}{8\pi^2 f}\hat{n}\cdot\vec{B}^a\eta, \tag{6}$$

$$\hat{n}\times\vec{B}^a = \frac{N_F g^2}{8\pi^2 f}\hat{n}\times\vec{E}^a\eta. \tag{7}$$

These together with the "matter" boundary conditions (1) and the axial current continuity

$$\frac{1}{2}\hat{n}\cdot(\bar{\psi}\vec{\gamma}\gamma_5\psi)_{NO}(R) = f n^{\mu}\partial_{\mu}\eta(R) \tag{8}$$

define the surface boundary conditions of our CBM to be satisfied by the usual equations of motion of the quarks and gluons inside the bag and of the massive η field outside of the bag.

2.2 Flavor-Singlet Axial Charge (FSAC)

We now study the flavor singlet axial charge of the proton. We define the axial current as

$$A^{\mu} = A^{\mu}_B\Theta_B + A^{\mu}_M\Theta_M. \tag{9}$$

Since we will be dealing only with the flavor-singlet axial current, we will omit the flavor index in the current. We shall use the short-hand notations $\Theta_B = \theta(R-r)$ and $\Theta_M = \theta(r-R)$ with R being the radius of the bag. We demand that the $U_A(1)$ anomaly be given in this model by

$$\partial_\mu A^\mu = \frac{\alpha_s N_F}{2\pi} \sum_a \vec{E}^a \cdot \vec{B}^a \Theta_B + f m_\eta^2 \eta \Theta_M. \tag{10}$$

Our task is to construct the FSAC in the chiral bag model that is gauge-invariant and consistent with this anomaly equation. Our basic assumption is that in the nonperturbative sector outside of the bag, the only relevant $U_A(1)$ degree of freedom is the massive η' field. This assumption allows us to write

$$A_M^\mu = A_\eta^\mu = f \partial^\mu \eta \tag{11}$$

with the divergence

$$\partial_\mu A_\eta^\mu = f m_\eta^2 \eta. \tag{12}$$

Now the question is: what is the gauge-invariant and regularized A_B^μ such that the anomaly (10) is satisfied? To address this question, we rewrite the current (9) as

$$A^\mu = A_{B_Q}^\mu + A_{B_G}^\mu + A_\eta^\mu \tag{13}$$

such that

$$\partial_\mu (A_{B_Q}^\mu + A_\eta^\mu) = f m_\eta^2 \eta \Theta_M, \tag{14}$$

$$\partial_\mu A_{B_G}^\mu = \frac{\alpha_s N_F}{2\pi} \sum_a \vec{E}^a \cdot \vec{B}^a \Theta_B. \tag{15}$$

The subindices Q and G imply that these currents are written in terms of quark and gluon fields respectively. In writing (14), we have ignored the up and down quark masses.

The matrix element of the "matter" currents satisfying (14) is quite simple. The quark current inside the bag is

$$A_{B_Q}^\mu = \bar{\Psi} \gamma^\mu \gamma_5 \Psi \tag{16}$$

where Ψ should be understood to be the *bagged* quark field. The proton matrix element

$$a_{B_Q}^0 \equiv \langle p| \int_B d^3 r \bar{\Psi} \gamma_3 \gamma_5 \Psi |p\rangle \tag{17}$$

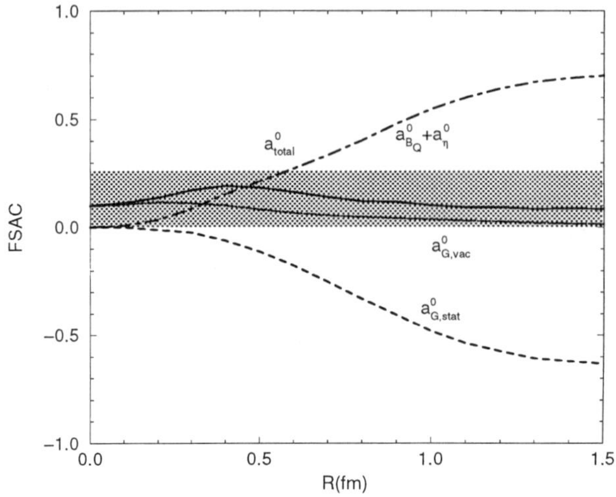

Figure 1: Various contributions to the flavor singlet axial current of the proton as a function of bag radius and comparison with the experiment: (a) "matter" (quark plus η) contribution ($a^0_{B_Q} + a^0_\eta$), (b) the contribution of the static gluons due to quark source ($a^0_{G,stat}$) , (c) the gluon vacuum contribution ($a^0_{G,vac}$), and (d) their sum (a^0_{total}). The shaded area corresponds to the range implied by experiments.

can be computed readily taking into account of the leakage of the baryon charge due to the hedgehog pion we know how to compute. As for the η contribution, one sees that the boundary condition (8) relates it directly to the quark contribution. A simple calculation gives

$$a^0_\eta \equiv \frac{1+y_\eta}{2(1+y_\eta)+y_\eta^2}\langle p|\int_B d^3r\bar{\Psi}\gamma_3\gamma_5\Psi|p\rangle \tag{18}$$

with $y_\eta = m_\eta R$. The first crucial observation at this point is that the matter contribution, the sum of (17) and (18), will vanish as the radius is shrunk to zero and will increase monotonically as the bag is increased. This can be seen in Fig.1. This feature by itself would show a violent breakdown of the CCP.

We now turn to the gluon contribution which is a lot more difficult to calculate accurately. Only an approximate calculation exists at the moment but I believe it is qualitatively correct.

To see what is involved, we first write down the relevant matrix element which follows from the relation (15),

$$a_G^0 = \langle p| - \frac{N_F \alpha_s}{\pi} \int_B d^3 r x_3 \vec{E}^a \cdot \vec{B}^a |p\rangle. \tag{19}$$

It is not obvious but this matrix element can be shown to be gauge-invariant. Since we are dealing with a cavity system *and* the color anomaly which requires coupling to the η field in the gluon boundary conditions (6) and (7), we have a double Casimir problem here. Only when the Casimir contributions are properly taken into account can we recover a fully consistent gauge-invariant result. Doing this calculation in full generality is not going to be an easy matter [#5], so we shall proceed less rigorously.

The gluon contribution can be roughly divided into two parts: One is the (presumably dominant) effect coming from the quark and η sources and the other is the contribution coming from a vacuum change due to the η coupling. The first involves the valence quarks and static gluon fields. The calculation for this was already made in previous papers (references are given in [14]) and only a slight extension including the η field in the boundary condition yields the desired result. The result shown in Fig.1 as $a^0_{G,stat}$ provides us with the second important information and that is that the dominant gluon contribution comes with a negative sign, so canceling nearly completely the matter contribution. It seems reasonable as a first approximation to conclude that $a^0_{B_Q} + a^0_\eta + a^0_{G,stat} \approx 0$. What is left must therefore be the Casimir contribution.

The simplest way to compute the Casimir effect is to assume CCP and check that the assumption is verified a posteriori. In [15], it is shown that the CCP allows us to write – via the color boundary conditions (6) and (7) – the following local operator

$$\vec{E}^a \cdot \vec{B}^a(x) \approx -\frac{N_F g^2}{8\pi^2} \frac{\eta(R)}{f} \frac{1}{2} G^2(x). \tag{20}$$

Here only the term up to the first order in η is retained in the right-hand side. Applied to the CBM, the couplings are to be understood as the average bag couplings and the gluon fields are to be expressed in the *cavity vacuum* through a mode expansion. In fact, by comparing the expression for the η' mass derived in [15] using Eq.(20) with that obtained by Novikov et al [16] in a QCD sum-rule method, we note that the matrix element of the G^2 in (20) should be evaluated in the *absence* of light quarks. This means, in the bag model, the *cavity vacuum*. The corresponding Casimir calculation involves a mode sum regularized, say, by the heat-kernel method. There is some subtlety involved

[#5] Such a calculation is in progress at the time of this writing.

in this calculation but we will not go into details here (see [14] for details). The results are summarized in Fig.1 compared with the presently accepted range of the experimental values obtained in deep inelastic lepton scattering from the nucleon. We see that the CCP is recovered via the Casimir effect.

To summarize, the CCP holds fairly well in *all* low-energy properties of baryon dynamics. More tantalizingly, there is a hint for a one-to-one mapping between quark-gluon matter at high density and hadron matter at low density [11]. We will later argue that the CCP holds in normal nuclear matter as well as in high density matter before the phase transition.

3 Effective Field Theories (EFT) For Nuclei

3.1 The Essence of EFT

The idea of effective quantum field theory[17,18,19] is extremely simple. At low energy where non-perturbative effects of QCD dominate, the relevant degrees of freedom are not quarks and gluons but hadrons. We do not have to know how the quark-gluon variables get transformed into hadron variables. We may simply assume the CCP described above as the mechanism for the transformation. Let the relevant hadronic degrees of freedom be represented by the generic field Φ. Separate the field into two parts; the low-energy part Φ_L in which we are interested and the high-energy part Φ_H which does not interest us directly. We delineate the two parts at the momentum scale characterized by a cutoff, say, Λ_1. In the generating functional Z that we want to compute, integrate out the Φ_H field and define an effective action $S^{eff}_{\Lambda_1}[\Phi_L] = \frac{1}{i}\ln\left(\int[d\Phi_H]e^{iS[\Phi_L,\Phi_H]}\right)$. Then what we need to compute is

$$Z = \int_{\Lambda_1}[d\Phi_L]e^{iS^{eff}_{\Lambda_1}[\Phi_L]}. \tag{21}$$

Given the right degrees of freedom effective below the cutoff Λ_1, we expand the effective action as $S^{eff}_{\Lambda_1}[\Phi_L] = \sum_i^\infty C_i Q_i[\Phi_L]$, where $Q_i[\Phi_L]$ are local field operators allowed by the symmetries of the problem, and C_i's are constants, which are required to satisfy the "naturalness condition". The $Q_i[\Phi_L]$'s are ordered in such a manner that, for low-energy processes, the importance of the $Q_i[\Phi_L]$ term diminishes as i increases. The contribution of the Φ_H degrees of freedom integrated out from the action is not simply discarded but instead gets lodged in the coefficients C_i as well as in higher-dimensional operators. The strategy of effective field theories is to truncate the series at a manageable finite order, and obtain desired accuracy to capture the essence of the physics involved. The separation of Φ into Φ_L and Φ_H involves two steps: The first is to "decimate" the degrees of freedom, that is, to choose the fields Φ_L relevant for the physics to be incorporated explicitly into the Lagrangian, and the second is a regularization procedure which regulates the high-momentum contribution. Let m be the lightest mass of the degrees of freedom that are integrated out. Clearly the most physically transparent regularization is to introduce a momentum cutoff with $\Lambda_1 \sim m$. A different cutoff introduces different coefficients C_i for the given set of fields. The idea is then to pick a value of the cutoff that is low enough to avoid unnecessary complications but high enough to avoid throwing away relevant physics.

This defines an effective field theory. The rest is just prescription and work. Technical details may differ, some more convenient and more elegant than others, but when done well, they should all give equally good results.

In the full (untruncated) theory, the location of the cutoff is entirely arbitrary, so we could have chosen any values of the cutoff other than Λ_1 without changing physics. The cut-off dependence of $Q_i[\Phi_L]$ is compensated by the corresponding changes in C_i through the renormalization group equation. In such a theory, which is rarely available in four dimensions, physical observables should be strictly independent of where the cutoff is set. This of course will not be the case when the series is truncated. However, for a suitably truncated effective theory to be predictive, the observables should be more or less insensitive to where the cutoff is put. Should there be strong cutoff dependence, it would signal that there is something amiss in the scheme. For instance, it could be a signal for the presence of new degrees of freedom or of "new physics" that must be incorporated and/or for the necessity of including higher-order terms.

An important aspect of effective field theory is that the result should not depend upon what sort of regularization one uses. In principle, therefore, one could equally well use the dimensional regularization or the cut-off regularization or something else as long as one makes sure that all relevant symmetries are properly implemented. Some of these issues in the context of the dimensional regularization and its variants were addressed recently in[20]. Here we will use the cut-off regularization for the reason that the procedure is simple and physically transparent[21] when limited to the order we shall adopt, namely, to the next-to-leading order (NLO). For some of the problems treated here such as neutron-proton (np) scattering, one could also use in a predictive manner a modified dimensional regularization in which certain power divergences are removed (i.e., PDS scheme)[22].

3.2 An EFT At Work

We shall apply the above strategy to two-nucleon systems at very low energy. Suppose we look at processes that involve momentum $p \ll m_\pi \approx$ 140 MeV. We can approach the problem in two ways. Since the momentum involved is much less than the pion mass, we can simply integrate out the pion and treat the nucleon field only. This means that the physics of the pion will appear in the coefficients of the terms we write down in the Lagrangian that we wish to solve. Next we will restore the pion and put it explicitly into the calculation. The coefficients of the terms in the Lagrangian will now be different from the case when the pion was integrated out. We will compare the two cases and learn how "new physics" in the form of the pion emerges [23,24]. As mentioned, if done correctly, it should not matter which regularization one uses. Here we shall use the cut-off regularization which is now known to be the simplest and trouble-free regularization in the market. There have been lots of papers written in connection with the *np* scattering at low energy discussing the merits and demerits of various regularization schemes. Readers interested in what the hot debates are all about could find them in [25]. The issue is quite interesting theoretically but at the end of the day, it is fair to say that using the cut-off regularization is still the physically most transparent (although perhaps not the most elegant) way to do the calculation. I won't go into this matter here which is in some sense academic. I could also spend all my lecture time discussing scattering only but I won't do this either. Instead I will treat *all* two-nucleon problems at the same time, including *np* and *pp* channels, bound states, scattering states as well as electroweak transitions involving the deuteron and unbound states. Work along this line with a cutoff regularization was done previously by Ordóñez et al [26].

- **Counting rule**

Since we are going to use the cut-off regularization, Weinberg's original chiral counting rule is perfectly fine. This will deviate from the PDS scheme in treating the pion when pions figure in the theory. As I shall mention later, Weinberg's counting scheme has a predictive power in the form of chiral filter which the PDS scheme lacks. More on this later.

Start with the case where the pion is present. Let Q be the typical momentum scale of the process or the pion mass, which is regarded as "small" compared to the chiral scale $\Lambda_\chi \sim 1$ GeV. The counting rule given by Weinberg[27] is that, a Feynman graph comprised of A nucleons, N_E external (electroweak) fields, L *irreducible* loops and C-separated pieces is order of $\mathcal{O}(Q^\nu)$ with

$$\text{ChPT:} \quad \nu = 2L + 2(C-1) + 2 - (A + N_E) + \sum_i \bar{\nu}_i, \tag{22}$$

$$\bar{\nu}_i \equiv d_i + \frac{n_i}{2} + e_i - 2 \tag{23}$$

where we have characterized each vertex i by d_i the number of derivatives and/or the power of the pion mass, n_i that of nucleon fields and e_i that of external (electroweak) fields. The quantities $\bar{\nu}_i$ and C are defined so that $\bar{\nu}_i \geq 0$ even in the presence of external fields [8] and $(C-1) = 0$ for connected diagrams.

Now if the pion is integrated out, there are no irreducible loops and the corresponding counting rule can be obtained simply by putting the number of loops L equal to zero,

$$\text{EFT:} \quad \nu = 2(C-1) + 2 - (A + N_E) + \sum_i \bar{\nu}_i \tag{24}$$

with the same definition for $\bar{\nu}_i$ given in eq.(23). But the meaning of d_i in eq.(23) is changed: Since the pion field is integrated out, the d_i stands here only for the number of derivatives, and not the power of the pion mass. Then, up to the next-to-leading order (NLO), the resulting irreducible vertex or potential consists of contact interactions and two spatial derivatives thereof.

Electromagnetic interactions have different counting rules and should be treated separately. Since the Coulomb interaction (between protons) is given as $\frac{\alpha}{\mathbf{q}^2}$ (where $\alpha \simeq 1/137$ is the fine-structure constant and $\mathbf{q}$ is the momentum transferred), it is of relevance in an extremely small momentum region while it becomes irrelevant in other regions due to the smallness of α. Since we will be interested also in the proton fusion at threshold, we will explicitly include the Coulomb potential in the *pp* sector.

- **Putting the cutoff**

To the order we consider here for the irreducible graphs, that is, to NLO, all we need is to put the cutoff on the irreducible vertex (call potential from now on). Given a potential in the momentum space $\mathcal{V}$ which we are to iterate to infinite order in the *reducible* channel to get the bound or quasi-bound state, the regularized potential in coordinate space is taken as

$$V(\mathbf{r}) \equiv \int \frac{d^3\mathbf{q}}{(2\pi)^3}\, e^{i\mathbf{q}\cdot\mathbf{r}}\, S_\Lambda(\mathbf{q}^2)\, \mathcal{V}(\mathbf{q}), \tag{25}$$

where $S_\Lambda(\mathbf{q}^2)$ is the regulator with a cutoff Λ. For our purpose it is convenient to take the Gaussian regulator

$$S_\Lambda(\mathbf{q}^2) = \exp\left(-\frac{\mathbf{q}^2}{2\Lambda^2}\right). \tag{26}$$

Summing the infinite diagrams with the potential is equivalent to solving Schrödinger equation with the potential $V(\mathbf{r})$. Since in our scheme we should not let Λ become too large, the Schrödinger equation is well-defined.

- **Potential without the pion**

In the absence of the pion, the potential to NLO is simply

$$\mathcal{V}_{no-\pi}(\mathbf{q}) = \frac{4\pi}{M}\left[C_0 + (C_2\delta^{ij} + D_2\sigma^{ij})q^i q^j\right] + Z_1 Z_2 \frac{\alpha}{\mathbf{q}^2} \tag{27}$$

with

$$\sigma^{ij} = \frac{3}{\sqrt{8}}\left(\frac{\sigma_1^i\sigma_2^j + \sigma_1^j\sigma_2^i}{2} - \frac{\delta^{ij}}{3}\sigma_1\cdot\sigma_2\right), \tag{28}$$

where $\mathbf{q}$ is the momentum transferred, $M \simeq 940$ MeV the nucleon mass, and $Z_1Z_2 = 1$ for the pp channel and zero otherwise. For the proton fusion process, we will also consider $\mathcal{O}(\alpha^2)$ corrections, i.e., the vacuum polarization (VP) potential and the two-photon-exchange (C2) potential; as in [28], these will be treated perturbatively. The parameters C's and D_2 are defined for each channel. The D_2 term, however, is effective only for the spin-triplet channel. Thus, there are two parameters (C_0 and C_2) for each of the pp and np $^1S_0$ channels, and three (C_0, C_2 and D_2) for the $^3S_1$ channel. These parameters will be determined from scattering and bound-state experimental data.

- **Potential with the pion**

To the order considered, the contribution of the pion is simple and parameter-free. There are no loops in the irreducible channel, so all we need is the potential

$$\mathcal{V}(\mathbf{q}) = -\tau_1\cdot\tau_2\frac{g_A^2}{4f_\pi^2}\frac{\sigma_1\cdot\mathbf{q}\,\sigma_2\cdot\mathbf{q}}{\mathbf{q}^2 + m_\pi^2} + \mathcal{V}_{no-\pi}(\mathbf{q}) \tag{29}$$

where $g_A \simeq 1.26$ is the axial-vector coupling constant and $f_\pi \simeq 93$ MeV is the pion decay constant.

• **Renormalization**

Since the cutoff is finite, there are no divergences, so for a given Λ, the renormalization amounts to determining the constants that appear in the potential. All the constants are fixed in our procedure by the scattering lengths and effective ranges for *np* and *pp* in ${}^1S_0$ channel and the binding energy of the deuteron (B_d), the deuteron D/S ratio (η_d) and the wave function normalization factor (A_s), all of which are given very accurately by experiments. For the *pp* channel, one has to take into account the Coulomb and other electromagnetic effects. This can be done in various ways with little ambiguity [29].

• **Observables**

The following observables will be *predicted* and compared with experiments.

1. Phase shifts for *np* ${}^1S_0$ channel.
2. The charge radius r_d, quadrupole moment Q_d, D-state probability p_D and magnetic moment μ_d of the deuteron.
3. The radiative *np* capture $n + p \to d + \gamma$.
4. The solar fusion process $p + p \to d + e^+ + \nu_e$.

3.3 The Power of EFT: Comparison with Experiments

We will now show that the theory works impressively. We will also see the power of the chiral filter in our scheme. The deuteron properties and electroweak matrix elements will be treated *with* and *without* the pion. The scattering phase shifts will be shown later.

• **Results without the pion**

Without the pion, we expect that the cutoff should be around the pion mass $\sim$ 140 MeV. The results are listed in Table 1. They are remarkably stable against the cutoff as long as it is near or above the pion mass. The agreement with experiment is also quite impressive.

• **Results with the pion**

With the pion figuring explicitly, we expect the cutoff to be at the next mass scale $\gtrsim 2m_\pi \sim 300$ MeV. The results are given in Table 2. The results are even more spectacular.

Table 1: The next-to-leading order (NLO) results *without* the pion field. The static properties of the deuteron, the M1 and GT amplitudes are listed for various choices of the cutoff Λ. The low-energy input parameters are the scattering lengths and effective ranges for the np and pp $^1S_0$ channels, and B_d, A_s and η_d for the deuteron channel.

Λ (MeV)	140	150	175	200	225	250	Exp.	$v_{18}{}^{30}$
r_d (fm)	1.973	1.972	1.974	1.978	1.983	1.987	1.966(7)	1.967
Q_d (fm^2)	0.259	0.268	0.287	0.302	0.312	0.319	0.286	0.270
P_D (%)	2.32	2.83	4.34	6.14	8.09	9.90	–	5.76
μ_d	0.867	0.864	0.855	0.845	0.834	0.823	0.8574	0.847
$\mathcal{M}_{\rm M1}$ (fm)	3.995	3.989	3.973	3.955	3.936	3.918	–	3.979
$\mathcal{M}_{\rm GT}$ (fm)	4.887	4.881	4.864	4.846	4.827	4.810	–	4.859

Table 2: The NLO results *with* pion field. See the caption of Table 1.

Λ (MeV)	200	250	300	350	400	500	Exp.	$v_{18}{}^{30}$
r_d (fm)	1.963	1.965	1.966	1.967	1.968	1.968	1.966(7)	1.967
Q_d (fm^2)	0.261	0.268	0.272	0.273	0.274	0.274	0.286	0.270
P_D (%)	3.16	4.11	4.77	5.16	5.35	5.39	–	5.76
μ_d	0.862	0.856	0.853	0.850	0.849	0.849	0.857	0.847
$\mathcal{M}_{\rm M1}$ (fm)	3.987	3.976	3.968	3.963	3.958	3.952	–	3.979
$\mathcal{M}_{\rm GT}$ (fm)	4.884	4.874	4.867	4.862	4.859	4.854	–	4.859

• Meson-exchange current corrections

In order to complete the calculations for the np capture and the proton fusion process, the exchange-current corrections should be calculated.

For the M1 transition for the np capture, the leading correction to the dominant single-particle matrix element comes at the next order from one-soft-pion exchange. This is in accordance with the chiral filter argument. A detailed bench-mark calculation in ChPT [23] that includes loop corrections (NNLO relative to the leading M1 matrix element) finds 9.29% correction to the cross section for the radiative np capture getting $\sigma^{th} = 334 \pm 3$ mb in agreement with the experiment $\sigma^{exp} = 334.2 \pm 0.5$ mb, the 1% uncertainty in the theoretical result being due to short-distance physics incompletely accounted for in chiral perturbation theory.

The meson-exchange correction to the Gamow-Teller matrix is a different story since it is *not* protected by the chiral filter. In order to assure an accuracy, one would have to go to $\mathcal{O}(Q^n)$ with $n \geq 3$. So far, calculations have been done up to $\mathcal{O}(Q^3)$ but there is no reason to believe that terms of order with $n \geq 4$ are negligible compared with what has been calculated. The predicted S factor for the solar burning process [28] comes out to be $S^{ChPT} = 4.05 \times 10^{-25}$ MeV − barn with a possible uncertainty of $\sim$ 2% due to higher-order corrections in the meson-exchange currents that have not been evaluated yet. There is no direct check of this prediction. It is however quite close to what has been obtained by the usual potential model calculation (with realistic potentials), $S^{pot} = 4.00 \times 10^{-25}(1 \pm 0.007^{+0.020}_{-0.011})$ MeV − barn. As is well known now, this leads to the celebrated solar neutrino problem within the Standard Model. The neutrino mass resolution to the problem is a hot topic lying in a different domain of physics.

• np $^1S_0$ phase shifts

Because of the large scattering length involved in this channel, naive dimensionally regularized EFTs do not work in a natural way. Much "sweat and blood' have been shed on the issue of how to rescue the dimensional regularization scheme from demise. The final rescue was found in the PDS scheme [22] with a slight modification of the counting rule together with the demotion of the pion to a sub-dominant role. All this technical problem is avoided in a straightforward way in the cut-off regularization scheme. It is now generally accepted [25] that the PDS scheme is equivalent to the cut-off scheme if the chiral counting is done properly to the same order.

The calculation with a cutoff to NLO in the irreducible vertex (or potential) without any additional fiddling is fully consistent [#6]. Figure 2 shows that the phase shifts come out very well, fairly independently of the cutoff, up to $p \sim 100$ MeV if the pion is present and to $p \sim 70$ MeV if the pion is absent. Again the presence of the pion as an explicit degree of freedom markedly improves the result.

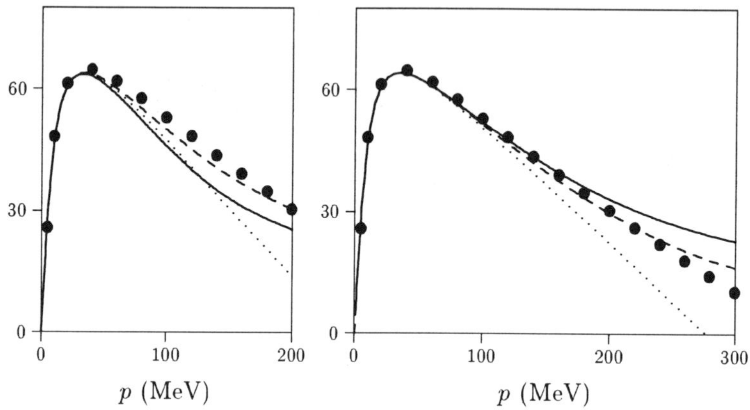

Figure 2: The $np\ ^1S_0$ phase shift (degrees) vs. the center-of-mass (CM) momentum p. We show the predicted results both without (left) and with (right) the pion field. For each case, our NLO results are given for two extreme values of the cutoff: The solid curve represents the lower limit, $\Lambda = 100$ MeV (without the pion) and 200 MeV (with the pion), and the dotted curve the upper limit, $\Lambda = 300$ MeV (without the pion) and 500 MeV (with the pion). The "experimental points" (obtained from the Nijmegen multi-energy analysis) are given by the solid circles. Our theory with $\Lambda = 150$ MeV (without the pion) and $\Lambda = 250$ (with the pion) are shown by the dashed line to show that the theory is in almost perfect agreement with the data up to $p \lesssim 200$ MeV.

- **The goodness of EFT: Insensitivity to Λ**

Although there is little dependence on the value of Λ, the results are not entirely independent of it. This is expected since the series is truncated in the way the cutoff procedure goes. The reason is simply that the (chiral) counting in the scheme with the cutoff is not fully consistent, with the error residing in higher orders. That is, a genuine Λ independence requires including certain

[#6]Furthermore it is so simple that it makes one wonder what all this fuss about the PDS and other schemes is about.

terms of the next order and we are not including them here, so they represent the error committed. That the results are quite insensitive to the cutoff value indicates however that the error committed is small enough.

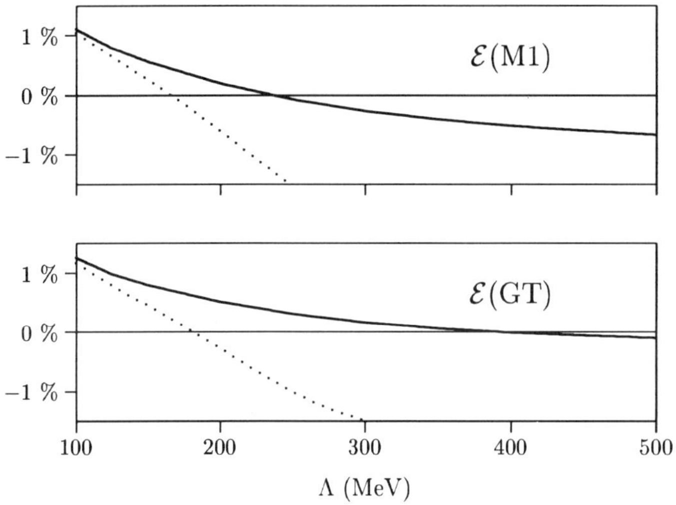

Figure 3: $\mathcal{E}$(M1) (upper) and $\mathcal{E}$(GT) (lower) vs. the cutoff Λ. The solid curves represents the NLO results with pions and the dotted curves without pions.

Let us quantitatively examine whether and how our theory is a truly *effective* effective theory. We do this using the electroweak processes. This test has not yet been made by other workers in the field. For this we should demand not only that the theory agree with experiments but also that the result not depend sensitively on the cutoff. We test the latter by looking at the following quantities:

$$\mathcal{E}(\mathrm{M1}) \equiv \frac{\mathcal{M}^{\mathrm{th}}_{\mathrm{M1}} - \mathcal{M}^{v18}_{\mathrm{M1}}}{\mathcal{M}^{v18}_{\mathrm{M1}}}, \quad \mathcal{E}(\mathrm{GT}) \equiv \frac{\mathcal{M}^{\mathrm{th}}_{\mathrm{GT}} - \mathcal{M}^{v18}_{\mathrm{GT}}}{\mathcal{M}^{v18}_{\mathrm{GT}}}, \tag{30}$$

where $\mathcal{M}^{\mathrm{th}}_{\mathrm{M1}}$ and $\mathcal{M}^{v18}_{\mathrm{M1}}$ denote, respectively, the M1 transition matrix element of our NLO calculation and that of the Argonne v_{18} potential [30], and similarly for $\mathcal{E}$(GT). Here we are taking the Argonne potential as "experiment" since it is fitted to experiments. Since these quantities are not the entire story that can be directly compared with experiments (due to the exchange currents), this is the best we can do to NLO. We see in Fig.3 that our criterion is very well satisfied. This is the more so when the pion is incorporated. *We see in a quantitative way how accounting for "relevant degrees of freedom" in effective theories improves the result as well as the consistency of the theory.*

• The chiral filter

I mentioned above that the advantage over other consistent schemes (such as PDS) of the Weinberg counting combined with the cut-off regularization we adopted is that the chiral filter mechanism allows one to make certain *predictions, not post-dictions* that are not easily accessible to other schemes. The soft pion plays a particularly prominent role in nuclear processes and this predominance is visible in the scheme. Thus the importance of the soft-pion exchange in nuclear isovector M1 transitions and in nuclear axial-charge transitions was *predicted* before experiments came to confirm it. A further forecast we can make is that when the pion-exchange is suppressed for whatever reason, chiral perturbation expansion will lose much, if not all, of its predictive power and one would be forced to adopt a different strategy than the usual one. Indeed this is exemplified by the inability of low-order chiral perturbation calculations to explain the experimental cross section for the $p+p \to p+p+\pi^0$ process. In other schemes, the particular role of pions could be hidden, though not absent, and one would have difficulty in "seeing" it and hence in making predictions that hinge on it.

As a specific example, let me take the radiative *np* capture. The precise agreement between the chiral perturbative calculation and the experiment for the capture cross section obtained in [23] is the bench-mark in nuclear physics and the result was understood due to the clear identification of the important role of the pion exchange. In the recent calculation of the same process in the PDS scheme [31], the authors emphasize the fact that the pion-exchange graph in their scheme is *only one of several NLO terms in the expansion, is scale- and scheme-dependent and hence has no particular significance.* Since the experimental cross section is used to fix one unknown NLO dimension-6 counter term, this calculation is not a prediction. Indeed the counter term so extracted turns out to be as large as the terms explicitly calculated, and there is nothing which says that higher order (nonanalytic) terms and counter terms will be negligible in the scheme.

I should hasten to say that there is nothing basically wrong in this PDS calculation. It is just that it does not seem as predictive. My personal opinion on this matter is this. The PDS scheme is perhaps more consistent in the spirit of EFT than the scheme of ref.[23]. However by strict adherence to the consistency of the counting, it misses an important piece of physics, namely, the short-range correlation that is present in nuclear wave functions.

So what is this short-range correlation (SRC)? As far as I can see, this is not understood within the framework of EFT. It is an effect of short-distance physics that is "integrated out" from the EFT, so in principle if the expansion of the EFT makes sense, the SRC effect presumably is lodged in the parameters of the EFT. In this scheme, then, the question is: How far does one have to go in the expansion to capture the essence of SRC physics? In the nuclear physics community, the SRC appears in the wave function as a sort of "black box" determined by looking at experiments. In [23], the SRC so invoked is found to suppress *all* contact terms *in the current* [#7]. The result shows that this is consistent with the chiral filter, the precise agreement providing a support for this reasoning: The error incurred is found to be negligible. It seems possible to justify this SRC mechanism by resorting, along the line developed by Lepage[21], to the OPE in the wave function to factorize the short-distance (asymptotically free) part of the QCD interaction inaccessible to chiral perturbation theory. By delegating the short-distance physics to the wave function as we did, we "enhance" the terms which are allowed by chiral filter and "suppress" those which are not. This is why there is no predictive power for our ChPT when soft-pion terms are suppressed. On the other hand, the PDS scheme may allow a "systematic calculation," but there is no reason to expect that it would work well.

4 Effective Field Theories For Dense Matter

The strategy found to be so successful at low energy for two-nucleon systems is being extended to three-body systems[32] but it would require a quantum jump to get to nuclear matter and denser matter. This means that there cannot be a "straightforward" application of chiral perturbation theory in the immediate future. All we can do for the moment is to resort to some educated guesses and check simplifying assumptions we make against nature to see whether our guesses pass the test. We eliminate wrong guesses and adopt the right ones by trial and error.[#8] I will describe in the rest of this lecture how we move forward in this direction and meet with some success. This part is based on a series of papers written recently[33,34,35,36].

The basic idea is to combine two superbly effective field theories available

---

[#7] Note that contact terms in the irreducible diagrams contributing to the potential are not suppressed by the same mechanism as explained in [8]. This is the modern understanding of the chiral filter conjecture.

[#8] My point is that if we sit and worry about the enormous complications brought in by many-body dynamics under extreme conditions, we will get nowhere. The analogy would be a centipede who sits and worries how it is going to march forward with so many legs potentially in the way.

in the market via a simple scaling assumption which is known in the literature as "BR scaling." The two EFTs that I will invoke are chiral Lagrangian field theory we now understand well and Landau Fermi liquid theory for dense matter with a Fermi sea which can be formulated as an effective field theory [18]. We know how the first works in dilute systems and how the latter works in strongly correlated fermionic systems. Our aim is to bridge the two in a way that can be applied to dense hadronic matter.

4.1 Nuclear Matter as a Chiral Fermi Liquid

The first question we have to address is: What is nuclear matter in chiral Lagrangian field theory? One might address this question in terms of a skyrmion with a large winding number A (see [2] for references) but so far very little has been learned from this. It seems more profitable to start with a chiral Lagrangian with nucleons and mesons incorporated, calculate the effective action in a suitable chiral expansion and look for a (nontopological) soliton with a given baryon number A. Such a solitonic solution would then describe the ground state of a system with baryon number A which could be closely resembling a heavy nucleus in its ground state. Lynn [37] has argued, using a somewhat oversimplified model, that such a system is a chiral liquid that gives a realistic description of a droplet of nuclear matter. Although there is no convincing derivation of the chiral liquid solution, the reasoning is very plausible and I am going to assume that this conjecture is correct. The next crucial step is to identify this chiral liquid structure with the mean field solution of a phenomenologically successful effective chiral action such as that written by Furnstahl et al [38] in terms of an expansion based on "naturalness condition." The effective chiral Lagrangian theory of Furnstahl et al for nuclear matter is a generalization of Walecka model [39], so we arrive at the first conclusion that the chiral liquid structure can be identified with the mean field solution of a Walecka-type model.

4.2 Effective Chiral Lagrangian with BR Scaling

I will next argue that when BR scaling is implemented into an effective chiral Lagrangian, the mean-field theory of this Lagrangian corresponds to Walecka mean field theory *extended* a la Furnstahl et al.

- **BR scaling**

It is highly likely that when baryon matter is present, the vacuum itself on which excitations take place is modified. This is so since the quantity that characterizes the vacuum, the quark condensate $\langle\bar{q}q\rangle$, must change in the presence of matter as indicated by such models as Nambu-Jona-Lasinio model (see[2]). We may consider writing an effective Lagrangian in a space that contains a matter of density ρ directly in terms of the parameters defined in that space. The parameters then must change as the density changes. One can think of the parameters depending on some field variables that change according to the change of the condensate and hence indirectly according to density. We will then assume that the effective Lagrangian defined in that space characterized by a particular field value (or density) be dictated by the symmetries of QCD. In [12], assuming the relevant symmetries to be the chiral symmetry and the scale symmetry of QCD phrased in the form of large N_c theory, i.e., skyrmion, the mass parameters and the (pion decay) coupling constant of the effective theory are found to scale as

$$\frac{M_N^\star}{m_N} \approx \frac{m_\omega^\star}{m_\omega} \approx \frac{m_\rho^\star}{m_\rho} \approx \frac{m_\phi^\star}{m_\phi} \approx \frac{f_\pi^\star}{f_\pi} \equiv \Phi. \tag{31}$$

Here the $\star$ represents in-medium quantity, $M_N^\star$ a scaling nucleon mass unaffected by the pion cloud (the "Landau effective mass" of the nucleon $m_N^\star$ will be defined later), ω and ρ are respectively the isoscalar vector meson and the isovector vector meson degrees of freedom and ϕ an isoscalar scalar meson field which is a chiral singlet effective in nuclear matter at a mass ~ 500 MeV.

• Chiral symmetry and Walecka model

It has often been asked why Walecka model [39] works. In fact, Cohen et al [40] gave a QCD sum rule justification of the large vector and scalar mean fields found in the Walecka-type models. Contrary to what some people have claimed, the success of the model is *not* due to any of the following: (1) renormalizability, (2) relativity, (3) accuracy of the mean field at large density.

The theory is an effective theory and *not* a fundamental theory, so neither does renormalizability have any raison d'être nor should it be valid at asymptotic density at which the model – unless fundamentally changed – should be irrelevant. Since heavy-baryon (nonrelativistic) chiral perturbation treatment (as given below) successfully reproduces the result of Walecka-type models, relativity is clearly not an essential ingredient for the success either.

So why does Walecka model work?

I suggest the reason is that it is consistent with chiral symmetry, as explained in [41,42]. To see this, suppose we write an effective Lagrangian in heavy-fermion formalism with *all* except nucleon fields integrated out. The Lagrangian will consist of the usual nucleon bilinear term and multi-Fermion interactions, $(\bar{N}\cdots N)^n$ with $n \geq 2$ with various covariants and derivatives appearing so that chiral symmetry is preserved. In particular the four-Fermi interactions of the form

$$\bar{N}\Gamma^\alpha N\bar{N}\Lambda_\alpha N \tag{32}$$

where $\Gamma^\alpha = \Lambda^\alpha = 1$ and $\Gamma^\alpha = \Lambda^\alpha = \gamma^\alpha$ have respectively scalar meson ("σ") and isoscalar vector meson ("ω") quantum numbers which would, when suitably treated in mean field, give rise to the same effects as the σ and ω fields in a Walecka-type theory. It is clear from this argument that the scalar is not the sigma of the linear sigma model. It is a chiral singlet and hence safe from the disaster in nuclear matter that the σ of the linear sigma model produces.

The most reasonable way of interpreting the scalar is that it is a "quarkonium" component of the dilaton that figures in the trace anomaly [43]. This is therefore a chiral singlet. In free space, it must be massive $\sim$ 700 MeV but go down to $\sim$ 500 MeV at nuclear matter density. This scalar must ultimately join the pions at the chiral phase transition to make up the quartet of the $O(4)$ chiral symmetry. This could occur via Weinberg's mended symmetry [44,43]. Thus as density increases toward the chiral symmetry restoration point, the mass of the scalar must drop. BR scaling (31) prescribes how the mass could drop.

The vector meson ω is also a chiral singlet. According to BR scaling, its mass must drop as a function of density. How does this happen? The way to see that the mass dropping is "universal" as in (31) is that it is essentially a scalar tadpole acting on all hadrons in the way that the Nambu-Jona-Lasinio model would describe, namely through the dynamically generated mass of a constituent quark making up the hadrons. It is the higher dimension multi-Fermi operators present in the effective Lagrangian that would generate the tadpole mechanism.

- **The chiral liquid as Landau Fermi liquid**

If we assume that we have a chiral liquid defined with the Fermi momentum k_F, then excitations around the liquid ground state can be treated again as an effective field theory as described beautifully in [18]. This will give essentially Landau Fermi liquid theory, with two fixed points, one the effective mass of the nucleon $m_N^\star$ which we will identify as Landau effective mass, and the other, Fermi liquid interactions $\mathcal{F}$ between quasiparticles.

The next important argument we need can be borrowed from Matsui [45] who showed that Walecka theory for nuclear matter can be mapped to Landau Fermi liquid theory with the vector-meson-mediated interaction related to the Landau parameter F_1. Since this Landau parameter governs also the Landau effective mass of the nucleon, the scaling of the nucleon mass must be related *mainly* to the vector degree of freedom #9 . With BR scaling this means that the scaling of all hadrons must be governed by the Fermi-liquid fixed-point parameter F_1.

Our conclusion then is: chiral Lagrangian with BR scaling, via its identification with Walecka mean field theory, can be mapped to Landau Fermi liquid fixed point theory. For details of the reasoning, I refer to [33,34,35,36].

4.3 Predictions at Normal Nuclear Matter Density

Here we shall establish a crucial relation between BR scaling and the Landau parameter F_1 and give some predictions based on the relation.

• Nuclear matter

We learned from the chain of arguments developed above that in order to describe nuclear matter, all we need to do is to write the simplest form of Walecka model with masses and coupling constants affixed with a star and let the masses scale according to (31). The Lagrangian is

$$\begin{aligned} \mathcal{L} &= \bar{N}[\gamma_\mu(i\partial^\mu - g_v^\star(\rho)\omega^\mu) - M^\star(\rho) + h^\star\phi]N \\ &+\frac{1}{2}[(\partial\phi)^2 - m_s^{\star 2}(\rho)\phi^2] - \frac{1}{4}F_\omega^2 + \frac{1}{2}m_\omega^{\star 2}(\rho)\omega^2 \end{aligned} \tag{33}$$

where N is the nucleon field, ω_μ the isoscalar vector field and ϕ an isoscalar scalar field. Pseudoscalar fields can be suitably incorporated as done later but they do not figure in the symmetric nuclear matter that we are concerned with here. The scaling behavior of the constants g_v and h is left arbitrary. It was shown in [34] that in the mean field, this Lagrangian with the BR scaling (31), a suitable scaling of the constant g_v and no scaling of h gives a surprisingly good description of the ground state with a compression modulus well within the accepted value $200 \sim 300$ MeV. The result is given in Fig.4. The scaling $\Phi(\rho_0) = 0.78$ needed here is forced on us, as described below, by QCD sum-rule calculations and orbital gyromagnetic ratios in heavy nuclei. Contrary to the worry expressed by some people, when properly interpreted, the density dependence of the parameters of the Lagrangian does not spoil any thermodynamic consistency of the theory [35].

#9 Note that this differs from the usual Walecka model where the nucleon mass shift comes from the VEV of the ϕ field.

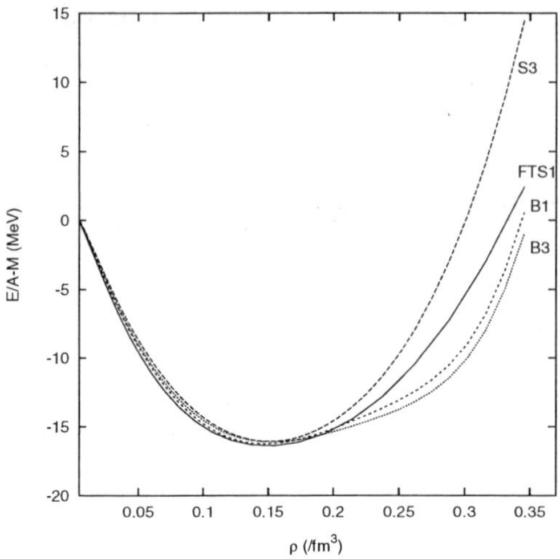

Figure 4: $E/A - M$ vs. ρ for this model ($S3$) compared with the elaborate model (labeled as $FTS1$) of Furnstahl et al [38]. $B1$ and $B3$ include higher polynomial terms in the Lagrangian (33) that represent fluctuations beyond the mean field.

• BR scaling Φ and Landau parameter F_1

From Galilean invariance (or Lorentz invariance) follows the Landau effective mass of the nucleon

$$\frac{m_N^\star}{m_N} = 1 + F_1/3 = \left(1 - \tilde{F}_1/3\right)^{-1} \tag{34}$$

where $\tilde{F} = (m_N/m_N^\star)F$. On the other hand, the chiral Lagrangian with BR scaling gives in tree order

$$\frac{m_N^\star}{m_N} = \Phi(1 + F_1^\pi/3) = \left(\Phi^{-1} - \tilde{F}_1^\pi/3\right)^{-1} \tag{35}$$

where F_1^π is the contribution from the pion to the Landau F_1 parameter. Now comparing (34) and (35), we see that

$$\tilde{F}_1 - \tilde{F}_1^\pi = 3(1 - \Phi^{-1}). \tag{36}$$

It seems reasonable to assume that the contribution to F_1 from the ω channel saturates the rest. Then we have

$$\tilde{F}_1^\omega = 3(1 - \Phi^{-1}). \tag{37}$$

This is a crucial relation between BR scaling and the ω degree of freedom. It shows that the nucleon effective mass is primarily governed by the ω degree of freedom as mentioned above.

Calculation of the response of a nucleon sitting on top of the Fermi sea to electromagnetic current gives the gyromagnetic ratio that agrees precisely with the Landau-Migdal formula [46]

$$g_l = \frac{1+\tau_3}{2} + \delta g_l \tag{38}$$

with

$$\delta g_l = \frac{1}{6}(\tilde{F}_1' - \tilde{F}_1)\tau_3 \tag{39}$$

where F_1' is the isovector counterpart to F_1. The chiral Lagrangian with BR scaling gives in tree order

$$\delta g_l = \frac{4}{9}\left[\Phi^{-1} - 1 - \tilde{F}_1^{\pi}\right]\tau_3. \tag{40}$$

Since the F_1^{π} is completely known by chiral symmetry, given a δg_l from experiment, one can determine Φ at nuclear matter density. We can turn the problem around and calculate δg_l, given $\Phi(\rho_0)$. We shall do the latter by noting that $\Phi(\rho_0)$ can be calculated from Gell-Mann-Oakes-Renner formula for the pion in medium or from QCD sum-rule calculations of the vector meson mass in medium and using BR scaling (31). The two ways give the same answer. Take the QCD sum-rule value [47]

$$\Phi(\rho_0) = 0.78 \pm 0.08. \tag{41}$$

With this value and the known F_1^{π} at $\rho = \rho_0$ (for which the Landau mass comes out to be $m_N^\star(\rho_0)/m_N \approx 0.70$), we get

$$\delta g_l(\rho_0) = 0.227\tau_3. \tag{42}$$

This agrees precisely with the experimental value obtained for the proton from giant dipole resonances in heavy nuclei

$$\delta g_l^{(exp)} = 0.23 \pm 0.03. \tag{43}$$

- **Fluctuations around nuclear matter ground state**

Pseudo-Goldstone octet bosons fields can be readily restored in the Lagrangian (33) in a way consistent with chiral symmetry. Their fluctuations will describe their properties in the medium *defined* by the scaling parameter Φ. For example, kaonic fluctuations would have the form

$$\mathcal{L}^{eff} = \frac{-6i}{8f_\pi^{\star 2}}\overline{K}\partial_t K N^\dagger N + \frac{\Sigma_{KN}}{f_\pi^{\star 2}}\overline{K}K \tag{44}$$

where Σ_{KN} is the KN sigma term. In tree order (i.e., mean field), $\overline{N}N \approx N^\dagger N \approx \rho$. The potential felt by the kaon in the background of nuclear matter is given by

$$V_{K^\pm} = \pm\frac{3}{8f_\pi^{\star 2}}\rho, \tag{45}$$

$$S_{K^\pm} = -\frac{\Sigma_{KN}}{f_\pi^{\star 2}}\rho_s \tag{46}$$

where $\rho = \langle N^\dagger N\rangle \approx \rho_s = \langle \overline{N}N\rangle$. At nuclear matter density, we can identify these results as one-third of the corresponding potentials for nucleons, so we can write

$$V_{K^\pm} \approx \pm\frac{1}{3}V_N \tag{47}$$

and

$$S_{K^\pm} \approx \frac{1}{3}S_N. \tag{48}$$

Phenomenology in Walecka-type mean-field theory gives $(S_N - V_N) \lesssim -600$ MeV for $\rho = \rho_0$. This leads to the prediction that at nuclear matter density

$$S_{K^-} + V_{K^-} \lesssim -200 \text{ MeV}. \tag{49}$$

This is consistent with the result coming from the analysis in K-mesic atoms $(S_{K^-} + V_{K^-})^{K-atom} \sim -200$MeV. This order of the mass shift in the kaon mass is also seen in heavy-ion experiments and found to lead to an interesting consequence on the formation of neutron stars and light-mass black holes. This issue is discussed by Li, Lee and Brown [48].

- **Strongly enhanced axial-charge transitions in heavy nuclei**

As mentioned at the beginning, Warburton's experiments and analyses on weak axial-charge transitions [7] confirmed one of the chiral filter predictions. There was also a surprise there which excited quite a few theorists. Warburton found that not only was there a strong mesonic contribution to the axial-charge matrix element but even more interestingly, the enhancement over the single-particle matrix element in heavy nuclei (such as lead) was 100 %. It turns out that this strong enhancement in dense nuclei can be simply explained by BR scaling as proposed by Kubodera and Rho [49].

Consider, following Warburton, the quantity ϵ_{MEC} which is the ratio of the *measured* axial-charge matrix element over the *theoretical* single-particle matrix element. Warburton's result for the lead region was that

$$\epsilon_{MEC}^{exp} \approx 1.9 \sim 2.0. \tag{50}$$

Now using our chiral Lagrangian with BR scaling in the tree order, we get the very simple formula [36]

$$\epsilon_{MEC}^{th} \approx \Phi^{-1}(1+\mathcal{R}) \tag{51}$$

where $\mathcal{R}$ is the ratio of the (soft-pion dominated) meson-exchange current matrix element over the single-particle matrix element which for $\rho = \rho_0$ comes out to be $0.56 \sim 0.61$ (this is a quite reliable range protected by the chiral filter mechanism [36,50]). For the lead region in question, we may take $\rho \approx \rho_0$. We find for the range involved for $\mathcal{R}$,

$$\epsilon_{MEC}^{th}(\rho_0) = 2.0 \sim 2.1. \tag{52}$$

In my opinion, this is as strong a confirmation of the notion of BR scaling as the orbital gyromagnetic ratio discussed above.

4.4 Predictions For Higher Density $\rho > \rho_0$

We have so far seen that used at the mean field level, the chiral effective Lagrangian with BR scaling can reasonably well describe nuclear properties at nuclear matter density. The question is: Can one extend the treatment to higher densities at which various phase changes such as kaon condensation, color superconductivity, chiral restoration etc. could take place? In [48], it was argued that kaon condensation could take place at $\rho \lesssim 3\ \rho_0$ within the framework. What about the others? For this, one must go beyond the mean field with the effective Lagrangian but so far there is no systematic formulation of going higher order with the Lagrangian with BR scaling.

The mean field arguments are expected to work provided quasiparticle notions apply to the hadrons relevant to the process. For instance, BR scaling has been successfully applied to the CERES dilepton processes [51] which involve densities of $2 \sim 3\ \rho_0$ and temperatures of $\gtrsim 100$ MeV. Here the principal mechanism that is invoked is the mass shift of the ρ meson with the width playing a secondary role. The reasoning here is that when one fluctuates around the background at a given density with scaling masses and other parameters, most of the strong correlations are included in those effective parameters and the residual interactions on top of the given background are *weak* so that a weakly-interacting local field description is valid. Thus it may be that viewed from BR scaling, widths do not destroy the quasiparticle picture.

On the other hand, if one were to build correlations starting from the matter-free background, one is in a strong-coupling regime and it seems inevitable that interactions broaden the width of the hadrons involved, in particular, of the ρ meson. It is found that the medium-broadened width of the vector meson can also explain reasonably well the CERES data [52].

Is there any connection between the mass-shifted vector meson defined at a density $\rho \neq 0$ and the broadened-width vector meson built on the $\rho = 0$ background? My conjecture is that there is a dual description of the same process in terms of different languages, similar to the CCP discussed in the first lecture. Here the languages dual to each other are the "partonic" picture of ref. [51] (with BR-scaling hadrons) and the "hadronic" picture of ref. [52] (with many-body interactions). This is somewhat like the quark-hadron duality one sees in heavy-light meson decay processes [53]. How this dual structure can work in the case of the CERES dilepton process is explained in a recent paper co-authored by the two schools – BR scaling and dynamical width [54].

Acknowledgments

This lecture is based on a series of work I have done over the years with my collaborators, in particular, Gerry Brown, Marc Chemtob, Jean Delorme, Bengt Friman, Kuniharu Kubodera, Kurt Langfeld, Chang-Hwan Lee, Dong-Pil Min, Byung-Yoon Park, Tae-Sun Park, Chaejun Song, Vicente Vento, Ismail Zahed. This lecture was prepared at Korea Institute for Advanced Study (KIAS) and I am grateful to Chung Wook Kim, Director of KIAS, for his support and encouragement.

References

1. M. Rho and G.E. Brown, "The role of chiral invariance in nuclei," *Comments on Nucl. and Part. Phys.* **10**, 201 (1980)
2. For a comprehensive review, see M.A. Nowak, M. Rho and I. Zahed, *Chiral Nuclear Dynamics* (World Scientific Pub., Singapore, 1996); M. Rho, "Cheshire Cat hadrons," *Phys. Repts.* **240**, 1 (1994)

3. M. Chemtob and M. Rho, "Meson-exchange currents in nuclear weak and electromagnetic interactions," *Nucl. Phys.***A 163**, 1 (1971)
4. D.O. Riska and G.E. Brown, "Meson exchange effects in $n + p \to d + \gamma$," *Phys. Lett.***B 38**, 193 (1972)
5. K. Kubodera, J. Delorme and M. Rho, "Axial currents in nuclei," *Phys. Rev. Lett.* **40**, 755 (1978)
6. See for review B. Frois and J.-F. Mathiot, *Comments on Nucl. and Part. Phys.* **18**, 307 (1989)
7. E.K. Warburton, "Mesonic enhancement of the weak axial-vector current evaluated from β-decay in the lead region," *Phys. Rev. Lett.* **66**, 1823 (1991); E.K. Warburton and I.S. Towner, *Phys. Lett.***B 294**, 1 (1992)
8. M. Rho, "Exchange currents from chiral Lagrangians," *Phys. Rev. Lett.* **66**, 1275 (1991)
9. T.-S. Park, K. Kubodera, D.-P. Min and M. Rho, "The power of effective field theories for nuclei," nucl-th/9807054
10. For a recent colloquium-type review, see A. Sen, "Developments in superstring theory," hep-th/9810356
11. T. Schäfer and F. Wilczek, "Continuity of quark and hadron matter," hep-ph/9811473
12. G.E. Brown and M. Rho, "Scaling effective Lagrangians in dense medium," *Phys. Rev. Lett.* **66**, 2720 (1991)
13. A. Hosaka and H. Toki, "Chiral bag model for the nucleon," *Phys. Rept.* **277**, 65 (1996)
14. H.-J. Lee, B.-Y. Park, D.-P. Min, M. Rho and V. Vento, "The proton spin in the chiral bag model: Casimir contributions and Cheshire Cat Principle," hep-ph/9810539
15. H.B. Nielsen, M. Rho, A. Wirzba and I. Zahed, "Color anomaly in a hybrid bag model," *Phys. Lett.***B 269**, 389 (1991); "The tale of the η' from the Cheshire Cat principle," *Phys. Lett.***B 281**, 345 (1992)
16. V.A. Novikov, M.A. Shifman, A.I. Vainshtein and V.I. Zakharov, "Are all hadrons alike?" *Nucl. Phys.***B 191**, 301 (1981)
17. See, e.g., S. Weinberg, *The Quantum Theory of Fields II* (Cambridge Press, 1996); *Nature* **386**, 234 (1997)
18. J. Polchinski, in *Recent Directions in Particle Theory* eds. by J.Harvey and J. Polchinski (World Scientific, Singapore, 1994); R. Shankar, *Rev. Mod. Phys.* **66**, 129 (1994).
19. G.P. Lepage, "What is renormalization?", TASI-89 Summer School Lecture.
20. J. Gegelia, "Are pions perturbative in effective field theory?", nucl-th/9806028.

21. G.P. Lepage, "How to renormalize the Schrödinger equation", nucl-th/9706029.
22. D.B. Kaplan, M.J. Savage and M.B. Wise, "A new expansion for nucleon-nucleon interaction," *Phys. Lett.***B 424**, 390 (1998); "Two-nucleon systems from effective field theory," *Nucl. Phys.***B 534**, 329 (1998); "A perturbative calculation of the electromagnetic form factors of the deuteron," nucl-th/9804032.
23. T.-S. Park, D.-P. Min and M. Rho, "Radiative n-p capture in chiral Lagrangians," *Phys. Rev. Lett.* **74**, 4153 (1995); *Nucl. Phys.* **A 596**, 515 (1996)
24. T.-S. Park, K. Kubodera, D.-P. Min and M. Rho, "Effective field theory for low-energy two-nucleon systems", hep-ph/9711463, *Phys. Rev.***C 58**, R637 (1998)
25. T.D. Cohen, "Systematic power counting in cutoff effective field theories for nucleon-nucleon interactions and the equivalence with PDS," nucle-th/9808006; T. Mehen and I.W. Stewart, "Renormalization schemes and the range of two-nucleon effective field theory," nucl-th/9809095; D.R. Phillips, S.R. Beane and M.C. Birse, "Scheming in dimensional regularization," hep-th/9810049
26. C. Ordóñez, L. Ray and U. van Kolck, "The two-nucleon potential from chiral Lagrangians," *Phys. Rev.***C 53**, 2086 (1996) and references given therein.
27. S. Weinberg, "Nuclear forces from chiral Lagrangians," *Phys. Lett.* **B 251**, 288 (1990); *Nucl. Phys.* **B 363**, 3 (1991); *Phys. Lett.* **B 295**, 114 (1992)
28. T.-S. Park, K. Kubodera, D.-P. Min and M. Rho, "The solar proton burning process revisited in chiral perturbation theory," *Ap.J* **507**, 443 (1998)
29. Most recent treatment of this matter in EFT is given by X. Kong and F. Ravndal, "Proton-proton scattering lengths from effective field theories," nucl-th/9811076
30. R.B. Wiringa, V.G.J. Stoks and R. Schiavilla, "Accurate nucleon-nucleon potential with charge-independence breaking," *Phys. Rev.* **C 51**, 38 (1995) and references therein.
31. M.J. Savage, K.A. Scaldeferri and M.B. Wise, "$n+p \to d+\gamma$ in effective field theory," nucl-th/9811029
32. P.F. Bedaque, H.-W. Hammer and U. van Kolck, "Renormalization of the three-body system with short-range interactions," nucl-th/9809025; "The three-body system with short-range interactions," nucl-th/9811046
33. B. Friman and M. Rho, "From chiral Lagrangians to Landau Fermi-liquid

theory of nuclear matter," *Nucl. Phys.*A **606**, 303 (1996)

34. Chaejun Song, G.E. Brown, D.-P. Min and M. Rho, "Fluctuations in 'Brown-Rho' scaled chiral Lagrangians," *Phys. Rev.*C **56**, 2244 (1997)
35. Chaejun Song, D.-P. Min and M. Rho, "Thermodynamic properties of effective chiral Lagrangians with BR scaling," *Phys. Lett.*B **424**, 226 (1998)
36. B. Friman, M. Rho and Chaejun Song, "Scaling of chiral Lagrangians and Landau Fermi liquid theory for dense hadronic matter," nucl-th/9809088
37. B.W. Lynn, "Chiral $SU(2)_L \times SU(2)_R$ liquids: a theory of heavy nuclei and neutron stars," *Nucl. Phys.*B **402**, 281 (1993)
38. R.J. Furnstahl, H.-B. Tang and S.D. Serot, "Vacuum contribution in a chiral effective Lagrangian for nuclei," *Phys. Rev.*C **52**, 1368 (1995); R.J. Furnstahl, S.D. Serot and H.-B. Tang, NPA **615**, 441 (1997)
39. B.D. Serot and J.D. Walecka, *Adv. Nucl. Phys.* **16**, 1 (1988); *Int. Jour. Mod. Phys.* **E6**, 515 (1997)
40. T.D. Cohen, R.J. Furnstahl and K. Griegel, "From QCD sum rules to relativistic nuclear physics," *Phys. Rev. Lett.* **67**, 961 (1991)
41. G.E. Brown and M. Rho, "From chiral mean field to Walecka mean field and kaon condensation," *Nucl. Phys.*A **596**, 503 (1996)
42. G. Gelmini and B. Ritzi, "Chiral effective Lagrangian description of bulk nuclear matter," *Phys. Lett.*B **357**, 431 (1995)
43. S. Beane and U. van Kolck, "The dilated chiral quark model," *Phys. Lett.*B **328**, 137 (1994)
44. S. Weinberg, "Unbreaking symmetries," in *Salamfestschrift*, Trieste Proceedings 1993
45. T. Matsui, "Fermi-liquid properties of nuclear matter in a relativistic mean-field theory," *Nucl. Phys.*A **370**, 365 (1981)
46. A.B. Migdal, *Theory of finite Fermi systems and applications to finite nuclei* (Interscience, London, 1967)
47. T. Hatsuda and S.H. Lee, "QCD sum rules for vector mesons in the nuclear medium," *Phys. Rev.*C **46**, R34 (1992); X. Jin and D.B. Leinweber, *Phys. Rev.*C **52**, 3344 (1995)
48. G.Q. Li, C.-H. Lee and G.E. Brown, "Kaon production in heavy-ion collisions and maximum mass of neutron stars," *Phys. Rev. Lett.* **79**, 5214 (1997); "Kaons in dense matter, kaon production in heavy-ion collisions and kaopn condensation in neutron stars," *Nucl. Phys.*A **625**, 372 (1997)
49. K. Kubodera and M. Rho, "Axial-charge transitions in heavy nuclei and in-medium effective chiral Lagrangians," *Phys. Rev. Lett.* **67**, 3479 (1991)
50. T.-S. Park, D.-P. Min and M. Rho, "Chiral dynamics and heavy-fermion

formalism in nuclei: Exchange axial currents," *Phys. Repts.* **233**, 341 (1993)
51. G.Q. Li, C.M. Ko and G.E. Brown, *Phys. Rev. Lett.* **75**, 4007 (1995)
52. W. Cassing, E.L. Bratkovskaya, R. Rapp and J. Wambach, "Probing the rho spectral function in hot and dense nuclear matter by dileptons," *Phys. Rev.*C **57**, 916 (1998)
53. B. Grinstein and R.F. Lebed, "Explicit quark-hadron duality in heavy-light meson decays in the 't Hooft model," *Phys. Rev.*D **57**, 1366 (1998)
54. G.E. Brown, C.Q. Li, R. Rapp, M. Rho and J. Wambach, "Medium dependence of the vector-meson mass: dynamical and/or Brown-Rho scaling?" nucl-th/9806026, *Acta Physica Polonica* **B29**, 2309 (1998)

QCD WITH DOMAIN WALL QUARKS

T. BLUM
RIKEN BNL Research Center, Upton, NY, 11973-5000, USA
E-mail: tblum@bnl.gov

I present a pedagogical review of domain wall fermions and review relevant numerical results in lattice QCD. Topics include the spectrum, weak interaction matrix elements (including new preliminary results for $B_K(\mu = 2GeV)$, the $n_f = 2$ finite temperature phase transition, and topology and zero modes.

1 Introduction

Recently, progress has been made in numerical simulations using lattice fermions that are chirally symmetric up to soft breaking terms which can be made arbitrarily small. The origin of these fermions is due to Kaplan who constructed free lattice chiral fermions by adding an *odd* infinite dimension to the usual *even* dimensional spacetime[1]. The mass of the fermion is given the shape of a domain wall in this extra dimension(Fig. (1)). Chiral zero modes arise naturally on the wall where the mass is zero. If the extra dimension is periodic, an anti-domain wall appears at the other end of the lattice with a zero mode of the opposite chirality. These two chiral zero modes are coupled to the same four dimensional($4d$) gauge field to simulate a *vector* gauge theory, like QCD. The chiral symmetry of the vector theory is manifest. Since the left and right handed fermions have been separated globally in the extra dimension, they may be rotated independently just as in the continuum.

After Kaplan's original work, Shamir developed the *boundary variant of domain wall fermions*(DWF)[2]. The crucial observation was that the dynamics of the extra dimension which lead to low energy bound states on the mass defect originate from only the right-half-s-plane (for the shape of the defect shown in Fig. (1)), so one can just throw away the left-half-s-plane. The zero modes then exist on the boundaries of the extra dimension. This has the nice feature of reducing the computational burden by a factor of two. Also, the introduction of the $4d$ quark mass is much simpler: the two boundaries are coupled with anti-periodic boundary conditions, but with a strength equal to m. m turns out to be proportional to the quark mass in the low energy effective theory living on the boundaries. m explicitly couples the right-handed mode on one boundary to the left-handed mode on the other boundary which is the same as the quark mass term in the continuum lagrangian. When the extra dimension has a *finite* number of sites, N_s, the zero modes on each

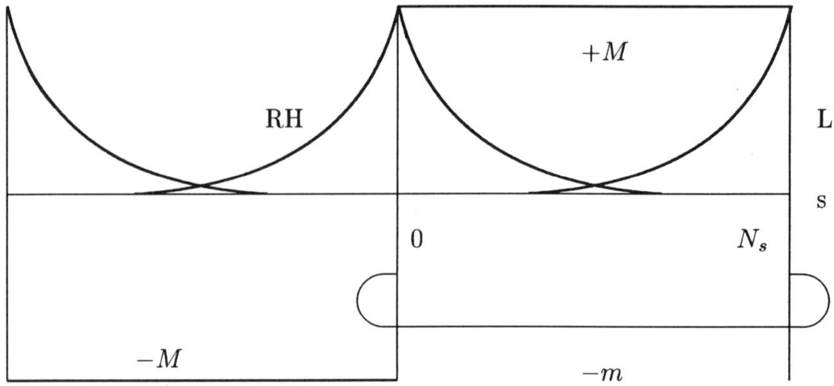

Figure 1. Scematic of domain wall fermions. The mass $M(s)$ of the five dimensional fermion is given the shape of a domain wall. Four dimensional chiral zero modes are bound to the defect. Throwing away the left half plane yields the boundary fermion variant of DWF. The zero modes on the remaining boundaries are coupled with a strength $-m(m > 0)$ which is proporional to the $4d$ quark mass when $N_s \to \infty$.

boundary have an exponentially small overlap(see Fig. (1)) which generates an intrinsic quark mass. This effect can be made arbitrarily small by increasing the size of the extra dimension. So, the chiral limit of DWF is $N_s \to \infty$ and then $m \to 0$. This similar to the case in ordinary four dimensional theories where the chiral limit is $V \to \infty$ and then $m_q \to 0$. The boundary fermion variant, which will henceforth also be called DWF, is discussed in more detail in section 2.

The above realization of chiral zero modes was already known to occur in the continuum, for instance, by Callan and Harvey who studied topological defects embedded in higher dimensions[3]. There and in Kaplan's original study one may suppose that the extra 5th dimension is physical: our 4d spacetime is bound to the wall at low energies, but at high enough energies the 5th dimension is accessible. It is interesting to note such ideas have recently been proposed as a natural mechanism for the spontaneous breaking of supersymmetry[4]. Similar constructions relate 11 dimensional supergravity to 10 dimensional string theory. Loosely speaking, domain walls as described here are low energy D-branes[5]. These higher dimensional theories may describe quantum gravity which propagates in all dimensions whereas the Standard Model particles are bound to a lower $4d$ spacetime[6]. So, while DWF may only be a nice trick for maintaining chiral symmetry on the lattice, the reason they work may be because that is the way our world really works.

Shortly after Kaplan's original proposal and also motivated by an independent proposal by Frolov and Slavnov[7], Naryanan and Neuberger transformed the extra dimension into an infinite flavor space of heavy regulator fields[8]. Using a transfer matrix formalism, the chiral determinant is given as the overlap of the ground states of two Hamiltonians(hence the name, the overlap formulation, or the overlap, for short). The overlap and Kaplan's DWF are equivalent in the particular case where the extra dimension is infinite. The overlap provides an elegant framework for studying the Index theorem on the lattice. It is the starting point for a new lattice Dirac operator that also maintains chiral symmetry on the lattice[9]. An extensive review of the overlap is given in Ref. [10].

The original proposal by Kaplan was intended as a lattice chiral gauge theory. However, it was soon recognized that the discretization would be useful for lattice QCD[2] since conventional methods explicitly break chiral symmetry. Recent numerical studies have shown the actual extent of the extra dimension can be modest, 10-20 sites for simulations of quenched QCD, and still preserve the chiral symmetry to a high degree of accuracy[11,12,13]. Further, DWF are accurate to order $O(a^2)$ since no (mass) dimension five operators that are also chirally symmetric exist to cancel $O(a)$ errors[14,12]. Indeed, the initial numerical results indicate improved scaling. This scaling (if it holds up after further scrutiny) offsets the cost of the extra dimension.

For a finite extra dimension with N_s sites, the leading discretization errors are expected to be $O(a)\exp(-CN_s)$ where C is the rate of exponential fall off in the extra dimension. This has been verified at one loop in perturbation theory for QCD[15]. In particular, the authors of Ref. [15] calculate the one loop fermion self-energy and find that the mass of the light mode remains exponentially small. Even after accounting for exponentially small terms in the free propagator that were neglected in Ref. [15], Ref. [16] also concludes that the exponential suppression stays intact after one loop radiative corrections.

2 Boundary Fermion Variant

In this section I closely follow the original work of Shamir[2]. The action for DWF is essentially a standard $5d$ Wilson fermion action:

$$S_{dwf} = -\sum_{x,y,s,s'} \bar{\psi}\,(\not{D}_{x,y}\delta_{s,s'} + \not{D}_{s,s'}\delta_{x,y})\,\psi,$$

$$\not{D}_{x,y} = \frac{1}{2}\sum_{\mu}(\,(1+\gamma_\mu)U_{x,\mu}\delta_{x+\hat{\mu},y}$$

$$
\begin{aligned}
&+ (1-\gamma_\mu)U^\dagger_{y,\mu}\delta_{x-\hat\mu,y}\ ) + (M-4)\delta_{x,y}, \\
\not{D}_{s,s'} = \ & \begin{matrix} P_R\delta_{1,s'} - mP_L\delta_{N_s-1,s'} - \delta_{0,s'} \\ P_R\delta_{s+1,s'} + P_L\delta_{s-1,s'} - \delta_{s,s'} \\ -mP_R\delta_{0,s'} + P_L\delta_{N_s-2,s'} - \delta_{N_s-1,s'} \end{matrix} \quad ,
\end{aligned}
\tag{1}
$$

where M is the full $5d$ Dirac mass, s and s' denote the extra dimension, and $P_{R,L} = (1 \pm \gamma_5)/2$. There are important differences with the ordinary Wilson action, however. The links in the s direction are set to unity, the relative sign between the Wilson term and the 5d mass term is opposite to the usual convention, and, as mentioned above, the boundaries in the s direction are coupled with a strength m. The boundary conditions are anti-periodic since the Kaplan model is periodic over $2N_s$. Finally, the extra hopping terms have γ_5's in place of γ_μ's; there is no chirality operator in odd dimensions.

In the free theory for $N_s = \infty$ and $p_\mu, m \to 0$ it is straightforward to show that the mass of the light quark is $m_q = mM(2-M)$ and that the singular part of the propagator (on one wall) is

$$G_R(s,s',p) = P_R\frac{M(2-M)}{\not{p} + m_q}(1-M)^{s+s'}. \tag{2}$$

A similar result holds for the other wall. Aside from the extra factor $M(2-M)(1-M)^{s+s'}$, this is same form found in the continuum. The free propagator is exponentially suppressed in the extra dimension with a rate that depends on M. For $M = 1$ the zero momentum mode does not penetrate the extra dimension at all. Note, the doublers have been removed for the choice $0 < M < 2$. As M is increased in increments of two, up to ten, the zero modes disappear from the spectrum and new ones appear, depending on which corners of the Brillouin zone, according to the factor $b(p) = 1 - M + \sum_\mu(1-\cos(p_\mu))$, contribute(see [2] for details). After $M = 10$, no zero modes exist. At tree level, the critical value where the light quark first appears is $M_c = 0$. In Ref. [17], the leading order corrections to the quark mass for finite N_s were given,

$$m_q = M(2-M)(m + (1-M+O(p^2))^{N_s}). \tag{3}$$

The overlap of the two modes in the extra dimension has generated an exponentially small quark mass. This mass can be made arbitrarily small compared to m by increasing N_s.

In the interacting theory M is additively renormalized, just like ordinary Wilson fermions, $M \to \widetilde{M} = M - M_c$. At one loop, the dominant effect comes from the tadpole diagram which is diagonal in s[15]; hence the above free field results are unchanged up to a simple shift in M. If one uses a reasonable choice for the coupling constant, then the magnitude of the tadpole contribution agrees nicely with the original nonperturbative estimate (near quenched $6/g^2 \approx 6.0$) for the optimal value of M which minimizes the overlap of the two light modes[11,12]. A similar shift occurs for the dynamical simulations, as well[18]. Eq. (3) also suggests why we do not expect the usual "exceptional" configurations for DWF for *even* N_s: the contribution to m_q is always positive.

Four dimensional quark fields are constructed from the five dimensional fields by taking their chiral projections on the boundaries.

$$
\begin{aligned}
q_x &= \frac{(1+\gamma_5)}{2}\psi_{x,0} + \frac{(1-\gamma_5)}{2}\psi_{x,N_s-1}, \\
\bar{q}_x &= \bar{\psi}_{x,N_s-1}\frac{(1+\gamma_5)}{2} + \bar{\psi}_{x,0}\frac{(1-\gamma_5)}{2}.
\end{aligned} \tag{4}
$$

These definitions are the simplest choice for interpolating operators that create and destroy quarks. Other definitions are possible; *i.e.* one may average over some width around each boundary or use an exponential ansatz to overlap mainly with the lightest mode. However, since the exponent is not known for the interacting case and depends on the gauge coupling, this is not simple to implement.

Operators constructed from the definitions in Eq. (4) satisfy a set of axial Ward identities[19].

$$
\Delta_\mu \left\langle A^a_\mu(x) O(y) \right\rangle = 2m \left\langle J^a_5(x) O(y) \right\rangle + i \left\langle \delta^a_A O(y) \right\rangle + 2 \left\langle J^a_{5q}(x) O(y) \right\rangle, \tag{5}
$$

where $A^a_\mu(x)$ is the $4d$ conserved axial vector current constructed by summing the conserved $5d$ current over the 5th dimension.

$$
\begin{aligned}
A^a_\mu(x) = \sum_s \mathrm{sign}(N_s/2 - 1 - s) &\left(\bar{\psi}(x,s)\frac{1+\gamma_\mu}{2}\lambda^a \psi(x+\hat{\mu},s) \right. \\
&\left. - \, , \bar{\psi}(x+\hat{\mu},s)\frac{1-\gamma_\mu}{2}\lambda^a \psi(x,s) \right),
\end{aligned} \tag{6}
$$

The pseudoscalar densities are constructed from fields at the boundaries (J_5) and in the middle (J_{5q}) of the extra dimension.

$$
\begin{aligned}
J^a_5 &- \bar{\psi}(x, N_s - 1)\frac{1+\gamma_5}{2}\lambda^a \psi(x,0) - \bar{\psi}(x,0)\frac{1-\gamma_5}{2}\lambda^a \psi(x, N_s - 1), \\
J^a_{5q} &= \bar{\psi}(x, \frac{N_s}{2} - 1)\frac{1+\gamma_5}{2}\lambda^a \psi(x, \frac{N_s}{2}) - \bar{\psi}(x, \frac{N_s}{2})\frac{1-\gamma_5}{2}\lambda^a \psi(x, \frac{N_s}{2} - 1)
\end{aligned} \tag{7}
$$

The identities are derived in the usual way, by demanding invariance of expectation values under infinitesimal transformations. Here, we take advantage of the fact that the left and right handed modes are globally separated in the fifth dimension in order to rotate them independently. The first two terms on the r.h.s of Eq. (5) appear in the continuum while the last term is "anomalous" and results from the non-invariance of the finite N_s action under the axial transformation. For the flavor nonsinglet case it vanishes in the limit $N_s \to \infty$ before the continuum limit is taken[19]. Thus, DWF fermions have the full axial symmetry of the continuum at *any* lattice spacing. However, this does not rule out the possibility that the doublers reappear at strong coupling in order to trivially satisfy Eq. (5). The axial ward identities have also been verified explicitly at one loop in QCD[20]. For the singlet case, the last term gives rise to the usual axial anomaly[21].

Using the quark fields defined in Eq. (4), the pseudoscalar density can be written in its familiar form, $J_5 = \bar{q}\gamma_5 q$.

3 Numerical Results

A thorough review of numerical results for QCD using DWF is given in Ref. [22]. Below I present a sample from that review as well as a few new preliminary results concerning weak matrix elements.

3.1 PCAC Ward identity, zero modes, topology, and finite N_s effects

The anomalous term in Eq. (5) for the PCAC Ward identity ($O = J_5$) can be used to estimate the effect of finite N_s in simulations. Taking the ratio of Eq. (5) with the pseudoscalar correlator leads to $2m + 2\langle J_{5q}(x) J_5(y)\rangle / \langle J_5(x) J_5(y)\rangle$ on the r.h.s., so, as long as the second term is small compared to the explicit quark mass m, one expects good chiral behavior.

Results for the relative anomalous contribution are shown in Fig. (2). For $M = 1.7$ all the points are at least an order of magnitude less than the value of $m(0.05)$ used in the simulations, except the point at $N_s = 10$, $6/g^2 = 5.85$. As shown below, these simulations display good chiral behavior whereas the pion mass, for example, at $6/g^2 = 5.85$ and $N_s = 10$ has a small but nonzero value when extrapolated to $m = 0$. Note also that the effects decrease significantly as the coupling is weakened while the rate appears to increase. At $6/g^2 = 5.85$, $M = 1$, the anomalous contribution is large, and we do not expect to see light modes. The above behavior is expected because of the additive renormalization of M which is closely related to ordinary Wilson fermions (see Ref. [15,23] for a detailed explanation).

The Argonne group has been calculating the low lying spectrum of the

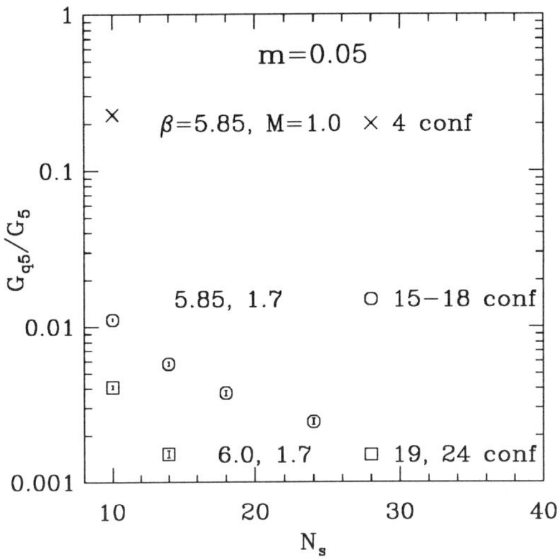

Figure 2. The relative anomalous contribution to the PCAC Ward identity.

Hermitean DWF operator, $R\gamma_5 D_{DWF}$, in order to study the behavior of flavor singlet correlation functions and the restoration of the $U_A(1)$ symmetry above the chiral symmetry phase transition[24]. A sample of their results is shown in Fig. (3) for quenched $6/g^2 = 6.2$. Similar results hold at $6/g^2 = 6.0$. Note that $N_s \leq 10$ in the figure. The lowest eigenvalue is clearly exponential in N_s while the next lowest is N_s independent. For configurations with a net number of instantons, they find the same number of exponentially small eigenvalues, in accord with the Atiyah-Singer Index theorem. They explicitly calculate the topological charge $Q_{top} \equiv m \int d^4x \bar{q}_x \gamma_5 q_x$ and find that it is integer valued within small errors, an impressive result considering the values of N_s used. The deviation of Q_{top} near the origin is a finite N_s effect. Strictly speaking, the Index theorem for DWF is only exact in the limit $N_s \to \infty$. Away from the continuum DWF(and the overlap) do not have a unique index; it depends on M(see Ref. [25], for example). These results are strong evidence for fermionic zero modes arising from gauge field topology.

The Columbia group has seen dramatic evidence of unsupressed fermionic zero modes in quenched and smooth background gauge field configurations by calculating the chiral condensate with DWF[26,18](Fig. (4)). This behavior had not been seen with conventional discretizations because of lattice artifacts, but is expected in the contiuum. Since the number of zero modes grows like

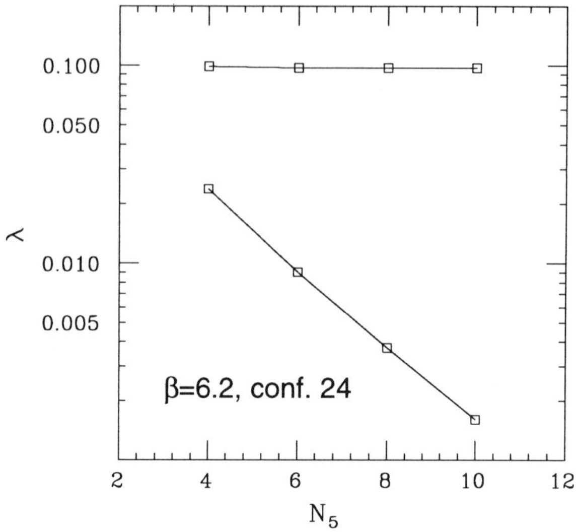

Figure 3. The lowest two eigenvalues of the hermitian DWF operator vs. N_s on a configuration with $Q_{top} = 1$. The plot is from the Argonne group[24].

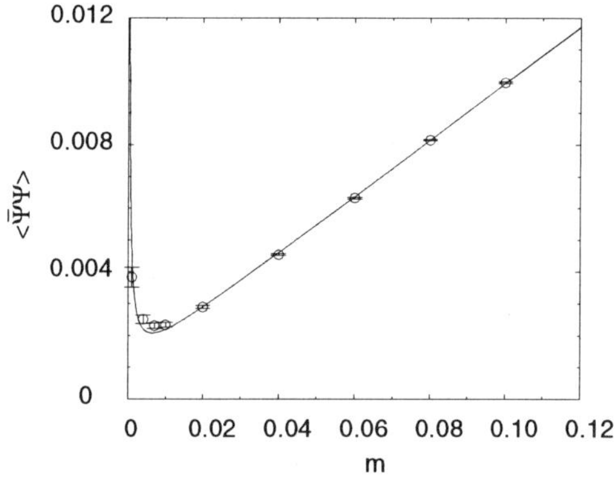

Figure 4. $\langle\bar{\psi}\psi\rangle$ at $6/g^2 = 5.85$ on an $8^3 \times 32$ lattice. It diverges as $m \to 0$ due to the presence of zero modes which are not supressed in the quenched approximation. The result is from the Columbia collaboration[18].

$V^{1/2}$, the effect in $\langle\bar{\psi}\psi\rangle$ should vanish as $V \to \infty$. They see a significant decrease in the coefficient of the m^{-1} term on a $16^3 \times 32$ *vs.* $8^3 \times 32$ lattice[18].

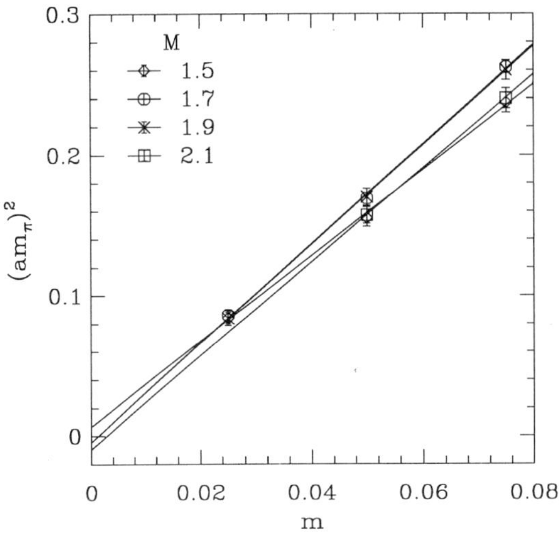

Figure 5. The pion mass squared as a function of m. $N_s = 14$, $6/g^2 = 6.0$.

3.2 Light hadron spectrum

In Fig. (5) the pion mass squared is plotted as a function of m for $1.5 \leq M \leq 2.1$ at $6/g^2 = 6.0$ on a $16^3 \times 32$ lattice. Each curve linearly extrapolates to zero, in accord with lowest order chiral perturbation theory. Fig. (6) shows the same data versus M. m_π^2 has the characteristic quadratic shape expected from free field theory but with a maximum that is shifted from its tree level value. From the figure the shift is approximately 0.8, in good agreement with the estimates of M_c from perturbation theory and 4d Wilson simulations[15,23]. The lines correspond to fits to Eq. (3) with M_c fixed to the value corresponding to κ_c for Wilson fermions(see Ref. [23] for details).

The Columbia group has calculated the light hadron spectrum at quenched $6/g^2 = 5.7$, $8^3 \times 32xN_s$[18]. As mentioned above, the chiral symmetry breaking effects due to finite N_s are larger at stronger coupling, so they have used $N_s = 48$ to reach the large N_s limit. Even so, the pion mass is not zero in the limit $m = 0$, but corresponds to roughly 230 MeV, which is likely an ordinary finite volume effect. Preliminary results suggest this is the case[27]. The quark masses used in their study are smaller than those in Fig. (5) which explains why they see these effects. Similar finite volume effects were clearly found in studies of the 2d vector Schwinger model[28].

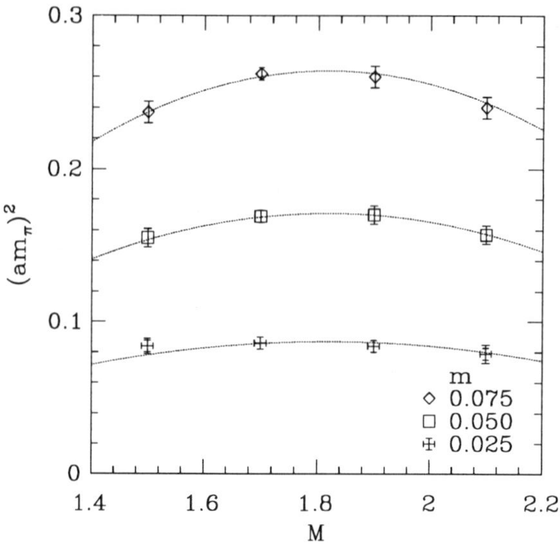

Figure 6. The pion mass squared as a function of the domain wall height, M. The plot is from Ref. [23]. The lines correspond to fits to Eq. (3) using $\widetilde{M}$ and the nonperturbative value of M_c.

3.3 Weak interaction matrix elements

The study of weak interaction matrix elements was one of the original motivations for using DWF for QCD simulations[11,12]. This is because the matrix elements of many important weak operators between pseudoscalar states vanish linearly with the quark mass, which is a direct consequence of chiral symmetry[29]. Ordinary Wilson quarks explicitly break chiral symmetry, so the naive lattice transcription of these operators does not vanish in the chiral limit. This is a well known, long standing problem. In principle, the solution is to fine tune the continuum-like operators with others that transform under different chiral representations to restore the correct behavior in the chiral limit. Alternatively, one may use Kogut-Susskind(KS) quarks which do exhibit the correct behavior since they maintain an exact abelian axial symmetry on the lattice. This remnant of the continuum is enough to ensure the proper behavior. However, the price of KS quarks is the breaking of the continuum flavor symmetries which induces significant lattice artifacts. As we shall see, DWF seem to have smaller discretization errors than the KS variety.

The vanishing of weak matrix elements in the chiral limit is a stringent

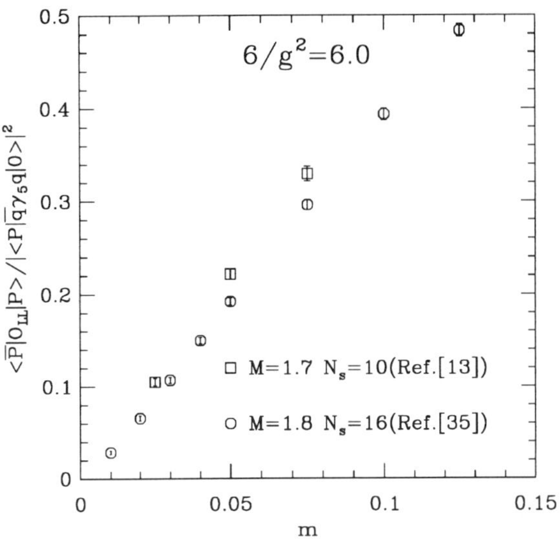

Figure 7. The matrix element of O_{LL} between degenerate pseudoscalar states. It vanishes in the chiral limit, a stringent test of the chiral symmetry properties of DWF.

test of the chiral symmetry of DWF since even with operator (un)mixing this is a difficult achievement (see Refs.[30,31] for recent results with nonperturbative and perturbative mixing, respectively). Fig. (7) shows the matrix element of the left-left weak operator ($O_{LL} = (\bar{s}\gamma_\mu(1-\gamma_5)d)^2$) that describes $K^0 - \bar{K}^0$ mixing. It vanishes as $m \to 0$, as predicted by chiral perturbation theory[29].

The matrix element of O_{LL} is conventionally normalized to its value in the vacuum saturation approximation which yields the B parameter. The kaon B parameter, B_K, is an important phenomenological parameter used to extract part of the CKM matrix from experimental measurements, which in turn constrains the Standard Model. The quenched DWF result is shown for several lattice spacings in Fig. (8). The KS result of Ref. [32] is shown for comparison. Both gauge variant and invariant operators are used. After including perturbative corrections which restore gauge invariance, the results are fit simultaneously and constrained to extrapolate to the same value in the continuum limit. Also shown is the (smaller) result of Ref. [33] which uses perturbatively improved gauge invariant operators. Since O_{LL} is generated by the operator product expansion, it is scale dependent, and therefore the result at each coupling must be run to a common scale before the continuum limit is taken. In the continuum, the vacuum saturated value of $\langle K^0|O_{LL}|\bar{K}^0\rangle$ is proportional to the square of the axial current matrix element which does not get renormalized since the axial current is partially conserved. As explained above, DWF have a partially conserved axial current, but this is not the current that is needed for the vacuum saturation of O_{LL} on the lattice since we have chosen to construct O_{LL} from the quark operators defined in Eq. (4). Therefore, the denominator in B_K also receives a (finite) multiplicative renormalization.

In Fig. (8) we show the bare (DWF) and renormalized (KS and DWF) values of B_K. The renormalized values are given in the NDR scheme at 2 GeV. The renormalization for the KS case is perturbative while that for DWF is a preliminary nonperturbative result using the method of the Rome group[34] which appears quite promising. The renormalized DWF operator is about ten percent less than the bare operator at $6/g^2 = 5.85$ and 6.0. We do not yet have results for 6.3. The present error on the renoramlization factors is about five percent and is not reflected in Fig. (8). Thus this should be considered a first rough estimate. New results at $6/g^2 = 5.85$ and 6.0 and the nonperturbative renoramlization factors are part of a new collaborative and on-going project to calculate the spectrum and weak interaction matrix elements using DWF[35].

We see from Fig. (8) that the DWF results are systematically low compared to the KS case, but finite volume effects and quark mass dependence need to be checked thoroughly before concluding there is a significant difference. There are two different volumes at 5.85 (12^3 and 16^3) which are consistent with each other and a 24^3 volume at 6.0 is consistent with 16^3. However, the quark mass dependence is not well understood on the larger volumes since only two quark masses are currently available on each. Furthermore, B_K exhibits a strong dependence on the quark mass in the physically interesting region. This dependence is consistent with expectations from chiral perturbation theory[36] and is seen for KS quarks as well, but could be mimicked by finite volume effects.

The indication from Fig. (8) is that discretization errors are significantly smaller in the DWF case. Needless to say, it is a high priority to reduce the statistical errors to see to what extent this remains true. It is important to point out that the lattice volumes of the last two KS points in Fig. (8) are $56^3 \times 96$ and $48^3 \times 96$ and correspond to physical volumes of 2.3fm^3 and 2.8fm^3, respectively. If the improved scaling of DWF holds up, then the added cost of the extra dimension will be significantly offset by the smaller $4d$ lattice volumes allowed by larger lattice spacings.

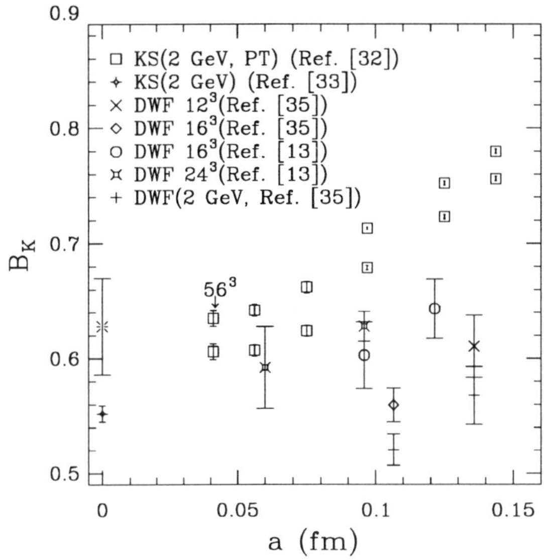

Figure 8. The kaon B parameter. Renormalized operators are in the NDR scheme. The KS results use perturbative renormalization factors while the DWF results use nonperturbative ones.

3.4 Dynamical simulations and non-zero temperature

The Columbia group has begun the first simulations using dynamical DWF to study the two flavor QCD chiral symmetry restoration phase transition at non-zero temperature[18]. These simulations are obviously quite demanding.

Using DWF, the two flavor chiral phase transition in QCD appears qualitatively similar to the KS result. In Fig. (9) there is a smooth crossover in both the Polyakov loop and $\langle\bar{\psi}\psi\rangle$ starting at $6/g^2 \approx 5.2$ ($N_\tau = 4$). In the limit $m \to 0$, below the transition $\langle\bar{\psi}\psi\rangle$ is spontaneously broken while above the transition chiral symmetry is restored (Fig. (9)). It should be stressed that for the first time the correct number of light pions is being simulated in dynamical studies of the $n_f = 2$ QCD chiral phase transition, which may be crucial to the dynamics. The Columbia group has also found that the scalar partners of the pions are not much heavier than the pions themselves which also has significant implications for the dynamics of the phase transition.

In general, dynamical simulations require a Pauli-Villars(PV) subtraction of the heavy states of the extra dimension. The action for these $N_s - 1$ states dominates the total action in the limit $N_s \to \infty$[8]. The form of the subtraction is not unique; a second order operator is suggested in Ref. [2] while in Ref. [17,18] a

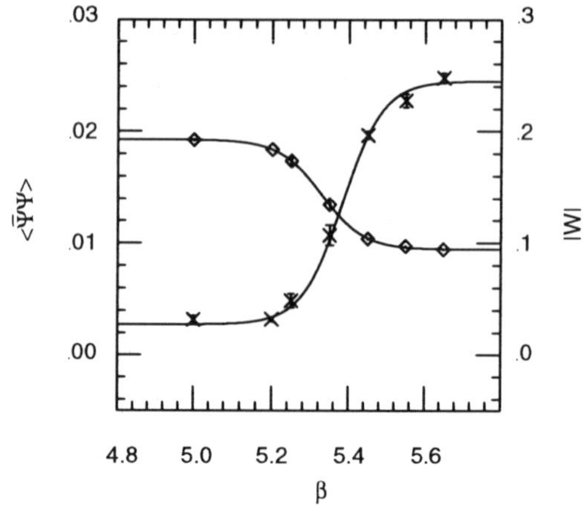

Figure 9. $\langle\bar{\psi}\psi\rangle$(diamonds), Polyakov loop(crosses) for $n_f = 2$ from the Columbia group[18].

first order operator is used. The latter version is vulnerable to "exceptional" configurations[22]. On the other hand, the authors of Ref. [18] have seen no evidence for these modes in their simulations. A method for suppressing such configurations has been suggested in Ref. [37].

4 Conclusions

We have seen that DWF with a modest number of sites in the extra dimension maintain their good chiral properties, which are exact in the limit $N_s \to \infty$. The results show promising behavior for weak matrix element phenomenology, dynamical simulations of the chiral symmetry restoration phase transition, and the relation between topology and fermionic zero modes. Indications of improved scaling with DWF need to be verified since this may significantly offset the cost of the added dimension.

I am grateful for many helpful discussions with my lattice colleagues at Brookhaven National Lab and Columbia University. Numerical simulations were done on the NERSC T3E and RIKEN BNL QCDSP.

References

1. D. Kaplan, Phys. Lett. B **288**, 342 (1992).
2. Y. Shamir, Nucl. Phys. B **406**, 90 (1993).
3. C.G. Callan and J.A. Harvey, Nucl. Phys. B **250**, 427 (1984).
4. G. Dvali and M. Shifman, Nucl. Phys. **B504**, 127 (1997).

5. J. Polchinski, TASI lectures on D-Branes, hep-ph/9611050.
6. R. Sundrum, hep-ph/9807348, and references therein.
7. S.A. Frolov and A.A. Slavnov, Saclay preprint SPhT/92-051 (1992).
8. R. Narayanan and H. Neuberger, Phys. Lett. B **302**, 62 (1993).
9. H. Neuberger, Phys. Lett. B **417**, 141 (1998); H. Neuberger, hep-lat/9807009 and references therein.
10. R. Narayanan and H. Neuberger, Nucl. Phys. B **443**, 305 (1995).
11. T. Blum and A. Soni, Phys. Rev. D **56**, 174 (1997).
12. T. Blum and A. Soni, Phys. Rev. Lett. **79**, 3595 (1997).
13. T. Blum and A. Soni, hep-lat/9712004.
14. Y. Kikukawa, *et al.* , Phys. Lett. B **399**, 105 (1997).
15. S. Aoki and Y. Taniguchi, hep-lat/9711004.
16. Y. Kikukawa, H. Neuberger, and A. Yamada, Nucl. Phys. B **526**, 572 (1998); H. Neuberger, Phys. Rev. **D57**, 5417 (1998).
17. P. Vranas, Phys. Rev. **D57** 1415 (1998).
18. P. Chen, *et al.* , hep-lat/(9812011, 9811026, 9811013, 9809159) .
19. Y. Shamir and V. Furman, Nucl. Phys. B **439**, 54 (1995).
20. S. Aoki and Y. Taniguchi, hep-lat/9811007.
21. Y. Shamir, Nucl. Phys. B **417**, 167 (1994).
22. T. Blum, plenary review, Lattice 98, hep-lat/9810017.
23. T. Blum, A. Soni, and M. Wingate, hep-lat/9809065; *ibid* in preparation.
24. J. Lagae, hep-lat/9809134.
25. R. Narayanan and P. Vranas, Nucl. Phys. B **506**, 373 (1997).
26. P. Chen, *et al.* , hep-lat/9807029.
27. The Columbia lattice group, private communication.
28. A. Jaster, hep-lat/9605011.
29. C. Bernard, *et al.* , Phys. Rev. **D32** 2343 (1985).
30. S. Aoki, *et al.* , Nucl. Phys. B **60A** (Proc. Suppl.) 67, (1998).
31. L. Lellouch and C. J. Lin, hep-lat/980914.
32. S. Aoki, *et al.* , Phys. Rev. Lett. **80** 5271, (1998).
33. G. Kilcup and D. Pekurovsky, Nucl. Phys. B **63** (Proc. Suppl.), 293 (1998).
34. M. Ciuchini, *et al.* , Nucl. Phys. B **415**, 403 (1994).
35. Calculations are being carried out on the RIKEN BNL supercomputer by the BNL lattice group: M. Creutz, C. Dawson, A. Soni, the Columbia lattice group: P. Chen, N. Christ, G. Fleming, A. Kaehler, C. Malureanu, R. Mawhinney, G. Siegert, P. Vranas, Y. Zhestkov, and the RIKEN BNL lattice group: T. Blum, S. Sasaki, and M. Wingate.
36. S. Sharpe, Phys. Rev. **D46**, 3146 (1992).
37. Y. Shamir, hep-lat/9807012.

HADRON STRUCTURES AND PERTURBATIVE QCD [a]

YUJI KOIKE
Department of Physics, Niigata University
Ikarashi, Niigata 950-2181, Japan
E-mail: koike@nt.sc.niigata-u.ac.jp

In the first part of this talk, I will summarize recent developments in the study of the chiral-odd spin-dependent parton distributions $h_1(x, Q^2)$ and $h_L(x, Q^2)$ of the nucleon, in particular, (i) Next-to-leading order Q^2 evolution of $h_1(x, Q^2)$ and (ii) Leading order Q^2 evolution of the twist-3 distribution $h_L(x, Q^2)$ and the universal simplification of the Q^2 evolution of all the twist-3 distributions in the large N_c limit. The second part of this talk will be devoted to a systematic analysis on the light-cone distribution amplitudes of vector mesons (ρ, ω, ϕ, K^* etc) relevant for exclusive processes producing these mesons. In particular, twist-3 distribution amplitudes are discussed in detail.

1 Introduction

High energy processes can be classified into two categories, inclusive and exclusive processes. Quark-gluon substructures of hadrons involved in these processes reveal themselves as a form of parton distribution functions in the inclusive processes, and light-cone distribution amplitudes in the exclusive proceeses. Understanding on both quantities constitutes a crutial step for the QCD description of the high energy processes. In this talk, I will summarize our recent studies on the quark distribtution functions in the nucleon and the light-cone distribution amplitudes for the light vector mesons.

Spin dependent parton distribution functions for the nucleon measured by the polarized beams and targets represent "spin distributions" carried by quarks and gluons inside the nucleon. They are functions of Bjorken's x which represent parton's momentum fraction in the nucleon and a scale Q^2 at which they are measured. Untill now, most data on the nucleon's distribution functions have been obtained through the lepton-nucleon deep inelastic scattering (DIS). The chiral-odd distributions, $h_{1,L}(x, Q^2)$, are the new type of distribution functions which have not been measured so far: Due to the chiral-odd nature, they decouple from the inclusive DIS. They can, however, be measured by the nucleon-nucleon polarized Drell-Yan process and semi-inclusive DIS which detect particular hadrons in the final state. They will hopefully be measured by planned experiments using polarized accelerators at BNL, DESY,

---

[a] Invited talk presented at "RCNP International School of Physics of Hadrons and QCD", October 12-13, 1998, Osaka, Japan. To be published in the proceedings.

CERN and SLAC etc[1]. In particular, RHIC at BNL is expected to provide first data on these distributions.

In the study of these distribution functions, perturbative QCD plays an important role in predicting their Q^2-dependence: Given a distribution function, say $h_1(x, Q_0^2)$, at one scale Q_0^2, perturbative QCD predicts the shape of $h_1(x, Q^2)$ at an arbitrary scale Q^2. This Q^2 evolution is necessary not only in extracting low energy hadron properties from high energy experimental data but also in testing the x-dependence predicted by a non-perturbative QCD technique or a model with the high energy data. In the first part of this talk, I will summarize our recent studies on the Q^2-dependence of $h_{1,L}(x, Q^2)$.

Light-cone distribution amplitudes (wave functions) for the vector mesons (ρ,ω,ϕ, and K^*) appear in various exclusive processes producing these vector mesons in final states, such as B decay, $B \to \ell\nu V$, $B \to \ell^+\ell^- V$, $B \to \gamma V$, and electro-production, $e + N \to N' + V$. (Study on the wave functions for pseudoscalr mesons is less involved, and has been done by many works.) Analysis on the wave functions is indispensable to test applicability of perturbative QCD to exclusive proceeses. In particular, test of the standard model through the rare B decay requires the knowledge on these wave functions. In the second part of this talk, we present a complete classification of the two-particle (quark-antiquark) wave functions for the vector mesons based on twist, chirality and spin. This can be done in parallel with that for the nucleon's parton distribution functions. In particular, for the twist-3 wave functions, we identify the contribution from the three-particle (quark-gluon-antiquark) twist-3 distribution amplitudes, using QCD equation of motion. The renormalization and the model building for the twist-3 wave functions can be/should be done starting from these exact relations, which is discussed by Tanaka in the workshop.

2 Distribution Function of the Nucleon in Inclusive Processes

2.1 Chiral-Odd Distributions $h_{1,L}(x, Q^2)$

Inclusive hard processes can be generally analyzed in the framework of the QCD factorization theorem[2]. This theorem generalizes the idea of the Bjorken-Feynman's "parton model" and allows us to include QCD correction in a systematic way. Here I restrict myself to the hard processes with the nucleon target, such as deep-inelastic lepton-nucleon scattering (DIS, $l + p \to l' + X$), Drell-Yan ($p + p' \to l^+l^- + X$), semi-inclusive DIS ($l + p \to l' + h + X$). According to the above theorem, the cross section (or the nucleon structure function) for these processes can be factorized into a "soft part" and a "hard part": The soft part represents the parton (quark or gluon) distribution in the nucleon and the hard part describes the short distance cross section between the parton and the external hard probe which is calculable within perturbation theory. For example, a nucleon structure function in DIS can be written as the imaginary part of the virtual photon-nucleon forward Compton scattering amplitude. (Fig. 1 (b)) According to the above theorem, in the Bjorken limit, i.e. $Q^2, \nu = P \cdot q \to \infty$ with $x = Q^2/2\nu =$ finite, ($Q^2 = -q^2$ is the virtuality of the space-like photon, P is the nucleon's four momentum), the structure function can be written as

$$W(x, Q^2) = \sum_a \int_x^1 \frac{dy}{y} H^a(\frac{x}{y}, \frac{Q^2}{\mu^2}, \alpha_s(\mu^2))\Phi^a(y, \mu^2), \tag{1}$$

where Φ^a represents a distribution of parton a in the nucleon and H^a describes the short distance cross section of the parton a with the virtual photon. μ^2 is the factorization scale. In Fig. 1(b), Φ^a is identified by the dotted line. (Fig.1 (a)). Similarly to DIS, the cross section for the nucleon-nucleon Drell-Yan process can also be written in a factorized form at $s = (P_A + P_B)^2, Q^2 \to \infty$ with a fixed Q^2/s ($P_{A,B}$ are the momenta of the two nucleons, Q is the momentum of the virtual photon):

$$d\sigma \sim \sum_{a,b} \int_{x_a}^1 dy_a \int_{x_b}^1 dy_b H^{ab} \left[\frac{x_a}{y_a}, \frac{x_b}{y_b}, Q^2; \frac{Q^2}{\mu^2}, \alpha_s(\mu^2)\right] \Phi^a(y_a, \mu^2)\Phi^b(y_b, \mu^2) \tag{2}$$

where the two parton distributions, Φ^a and Φ^b, for the beam and the target appear as was shown by dotted lines in Fig. 1(c).

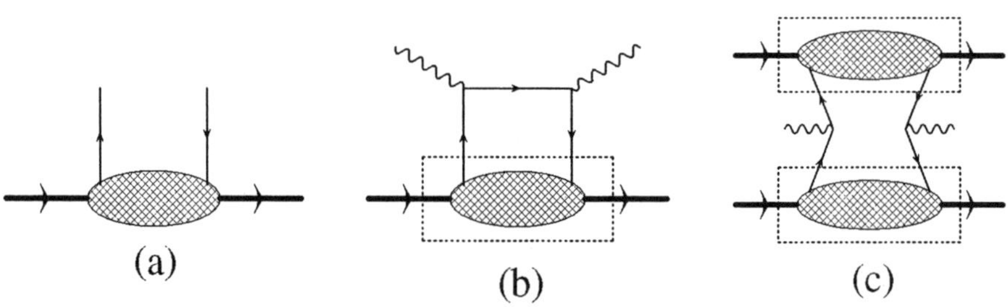

Figure 1: (a) Quark distribution function. (b) Nucleon structure function in DIS. (c) Cross section for the nucleon-nucleon Drell-Yan process.

As is seen from Figs. 1(b),(c), the parton distribution can be regarded as a parton-nucleon forward scattering amplitude shown in Fig. 1 (a) which appear in several different hard processes. In particular, the quark distribution in the nucleon moving in the $+\hat{e}_3$ direction can be written as the light-cone Fourier transform of the quark correlation function in the nucleon:[3]

$$\Phi^a(x,\mu^2) = P^+ \int_{-\infty}^{\infty} \frac{dz^-}{2\pi} e^{ixP\cdot z} \langle PS|\bar{\psi}^a(0)\Gamma\psi^a(z)|_\mu|PS\rangle, \tag{3}$$

where $|PS\rangle$ denotes the nucleon (mass M) state with momentum P^μ and spin S^μ, and ψ^a is the quark field with flavor a. In (3), we have suppressed for simplicity the gauge link operator which ensures the gauge invariance and $|_\mu$ indicates the operator is renormalized at the scale μ^2. A four vector a^μ is decomposed into two light-cone components $a^\pm = \frac{1}{\sqrt{2}}(a^0 \pm a^3)$ and the transverse component $\vec{a}_\perp$. In (3), $z^+ = 0$, $\vec{z}_\perp = \vec{0}$, and $z^2 = 0$. Γ generically represents γ-matrices, $\Gamma = \gamma_\mu, \gamma_\mu\gamma_5, \sigma_{\mu\nu}, 1$. $\Phi^a(x,\mu^2)$ measures the distribution of the parton a to carry the momentum $k^+ = xP^+$ in the nucleon, which is independent from particular hard proceeses.

If one puts $\Gamma = \gamma_\mu, \gamma_\mu\gamma_5$, the chirality of $\bar{\psi}$ and ψ becomes the same, namely it defines the distributions. Likewise, putting $\Gamma = \sigma_{\mu\nu}, 1$ defines the chiral-odd disributions. For the case of the deep-inelastic scattering (Fig. 1 (b)), the quark line emanating from the target nucleon comes back to the original nucleon after passing through the hard interactions. Since the perturbative interaction in the standard model preserves the chirality except a tiny quark mass effect, the chirality of the two quark lines entering the nucleon in Fig. 1(b) is the same. Hence the DIS can probe only the quark distributions. On the other hand, in the Drell-Yan process (Fig. 1 (c)), there is no correlation in chirality between two quark lines entering each nucleon. Therefore the Drell-Yan process probes both chiral even and odd distributions.

The chiral-odd distributions $h_1^a(x,\mu^2)$, $h_L^a(x,\mu^2)$ in our interest are defined by putting $\Gamma = \sigma_{\mu\nu}i\gamma_5$ in (3):[4]

$$\begin{aligned} \int \frac{d\lambda}{2\pi} e^{i\lambda x} \langle PS|\bar{\psi}^a(0)\sigma_{\mu\nu}i\gamma_5\psi^a(\lambda n)|_\mu|PS\rangle \\ = 2[h_1^a(x,\mu^2)(S_{\perp\mu}p_\nu - S_{\perp\nu}p_\mu)/M \\ +h_L^a(x,\mu^2)M(p_\mu n_\nu - p_\nu n_\mu)(S\cdot n) \\ +h_3^a(x,\mu^2)M(S_{\perp\mu}n_\nu - S_{\perp\nu}n_\mu)] \end{aligned} \tag{4}$$

where we introduced two light-like vectors p, n ($p^2 = n^2 = 0$) by the relation $P^\mu = p^\mu + \frac{M^2}{2}n^\mu$, $p\cdot n = 1$, $p^- = n^+ = 0$. If we write $P^+ = \mathcal{P}$, $p = \frac{\mathcal{P}}{\sqrt{2}}(1,0,0,1)$, $n = \frac{1}{\sqrt{2}\mathcal{P}}(1,0,0,-1)$. $\mathcal{P}$ is a parameter which specifies the Lorentz frame of the system: $\mathcal{P} \to \infty$ corresponds to the infinite momentum frame, and $\mathcal{P} \to M/\sqrt{2}$ the rest frame of the nucleon. $S_\perp^\mu$ is the transverse component of S^μ defined by $S^\mu = (S\cdot n)p^\mu + (S\cdot p)n^\mu + S_\perp^\mu$. One can show that Φ^a defined in (3) has a support $-1 < x < 1$. If one replaces the quark field ψ in (3) by its charge conjugation field $C\bar{\psi}^T$, it defines the anti-quark distribution $\bar{\Phi}^a$. In particular $h^a_{1,L,3}(x,\mu^2)$ in (4) are related to their anti-quark distribution by $h^a_{1,L,3}(-x,\mu^2) = -\bar{h}^a_{1,L,3}(x,\mu^2)$.

Φ^a appears in a physical cross section in the form of the convolution with a short distance cross section in a parton level as is shown in (1) and (2). The cross section can be expanded in powers of $\frac{1}{\sqrt{Q^2}}$ as

$$\sigma(Q^2) \sim A(\ln Q^2) + \frac{M}{\sqrt{Q^2}}B(\ln Q^2) + \frac{M^2}{Q^2}C(\ln Q^2) + \cdots, \tag{5}$$

where each coefficient A, B, C receives logarithmic Q^2-dependence due to the QCD radiative correction. In order to see how $h_{1,L,3}$ can contribute in the expansion (5), it is convenient to move into the infinite momentum frame ($\mathcal{P} \sim Q \to \infty$). In this limit the coefficient of $h_{1,L,3}$ in (4) behaves, respectively, as $O(Q)$, $O(1)$, $O(1/Q)$. Therefore if h_1 contributes to the A term in (5), h_L can contribute at most to the B-term, and h_3 can contribute at most to the C-term. In general, when a distribution function contributes to hard processes at most in the order of $\left(\frac{1}{\sqrt{Q^2}}\right)^{\tau-2}$, the distribution is called twist-τ. Therefore h_1, h_L, h_3 in (4) is, respectively, twist-2, -3 and -4.

Twist-2 distribution h_1 can be measured through the transversely polarized Drell-Yan[5,6,4,7], semi-inclusive deep inelastic scatterings which detect pion[8], polarized baryons[6,9,10], correlated two pions[11].

From the discussion above, one sees that it is generally difficult to isolate experimentally higher twist ($\tau \geq 3$) distributions in hard proceeses, since they are hidden by the leading twist-2 contribution (A term in (5)). However, this is not the case for h_L and g_T. In particular spin asymmetries, they contribute to the B-term in the absence of A-term: g_T can be measured in the transversely polarized DIS [12], and h_L appears in the longitudinal versus transverse spin asymmetry in the polarized nucleon-nucleon Drell-Yan process[4]. Therefore the Q^2-evolution of g_T and h_L can be a new test of perturbative QCD beyond the twist-2 level.

Insertion of other γ-matrices in (3) defines other distributions. In Table 1, we show the classification of the quark distributions up to twist-3.[4] There f_1, $g_{1,T}$, e is defined, respectively, by $\Gamma = \gamma_\mu, \gamma_\mu\gamma_5, 1$ in (3). A similar classification can also be extended to the gluon distributions[14]. The distribution f_1 contributes to the spin averaged structure functions $F_{1,2}(x,Q^2)$ familiar in DIS. The helicity distribution g_1 contributes to the $G_1(x,Q^2)$ structure function measured in the longitudinally polarized DIS. By now there has been much

spin	average	longitudinal	transverse
twist-2	f_1	g_1	$\underline{h_1}$
twist-3	$\underline{e}$	$\underline{h_L}$	g_T

Table 1: Clasification of the quark distributions based on spin, twist and chirality. Underlined distributions are chiral-odd. Others are .

accumulation of experimental data on f_1 and g_1, and the data on g_1 triggered lots of theoretical discussion on the "origin of the nucleon spin"[1]. The first nonzero data on g_2 ($= g_T - g_1$) was also reported in Ref. [13].

2.2 Next-to-leading order (NLO) Q^2-evolution of $h_1(x, Q^2)$

As we saw in the previous section, h_1 is the third and the final twist-2 quark distribution. It has a simple parton model interpretation as can be seen by the Fourier expansion of ψ in (4). It measures the probability in the transversely polarized nucleon to find a quark polarized parallel to the nucleon spin minus the probability to find it oppositely polarized. Here the transverse polarization refers to the eigenstate of the transverse Pauli-Lubański operator $\gamma_5 \not{S}_\perp$. If one replaces the transverse polarization by the longitudinal one, it becomes the helicity distribution g_1. For nonrelativistic quarks, $h_1(x, \mu^2) = g_1(x, \mu^2)$. A model calculation suggests, h_1 is the same order as g_1.[4,15,16]

The Q^2-evolution of h_1 is described by the usual DGLAP evolution equation[17]. Because of its chiral-odd nature it does not mix with gluon distributions. Therefore the Q^2-dependence of h_1 is described by the same equation both for singlet and nonsinglet distributions. For f_1 and g_1, the NLO Q^2 evolution was derived long time ago[18,19,20,21] and has been frequently used for the analysis of experiments[22,23]. The leading order (LO) Q^2-evolution for h_1 has been known for some time[6]. In the recent literature, the next-to-leading order (NLO) Q^2-evolution has also been completed by two papers[24,25]: Vogelsang[24] presented the light-cone gauge calculation for the two-loop splitting function of h_1 in the formalism originally used for f_1[20]. We[25] carried out the Feynman gauge calculation of the two-loop anomalous dimension following the method of Ref. [18] for f_1. The results of these calculations in the $\overline{\text{MS}}$ scheme agreed completely. In the following, I briefly discuss the characteristic feature of the NLO Q^2 evolution of h_1 following Refs.[25,26].

Analysis of (4) gives the connection between the n-th moment of h_1 and a tower of twist-2 operators:

$$\mathcal{M}_n[h_1(\mu^2)] \equiv \int_{-1}^{1} dx\, x^n h_1(x, \mu^2) = \frac{-1}{2M} \langle PS_\perp | O_n^\perp(\mu^2) | PS_\perp \rangle,$$

$$O_n^{\perp} = S_{\perp\nu}\bar{\psi}\sigma^{\nu\alpha}n_\alpha i\gamma_5(in\cdot D)^n\psi, \tag{6}$$

where $S_\perp$ stands for the transverse polarization and $O_n^\perp(\mu^2)$ indicates the operator $O_n^\perp$ is renormalizaed at the scale μ^2. The contraction with n^μ and $S_\perp^\mu$ (recall $S_\perp\cdot n = 0$, $n^2 = 0$) in (6) projects out the relevant twist-2 contribution from the composite operator. ("Twist" for local composite operators is defined as dimension minus spin.) By solving the renormalization group equation for $O_n^\perp$, one gets the NLO Q^2 dependence of $\mathcal{M}_n[h_1(\mu^2)]$ as

$$\frac{\mathcal{M}_n[h_1(Q^2)]}{\mathcal{M}_n[h_1(\mu^2)]} = \left[\frac{\alpha_s(Q^2)}{\alpha_s(\mu^2)}\right]^{\frac{\gamma_n^{(0)}}{2\beta_0}}\left[1+\frac{\alpha_s(Q^2)-\alpha_s(\mu^2)}{4\pi}\frac{\beta_1}{\beta_0}\left(\frac{\gamma_n^{(1)}}{2\beta_1}-\frac{\gamma_n^{(0)}}{2\beta_0}\right)\right], \tag{7}$$

where $\alpha_s(Q^2)$ is the NLO QCD running coupling constant given by

$$\frac{\alpha_s(Q^2)}{4\pi} = \frac{1}{\beta_0\ln(Q^2/\Lambda^2)}\left[1-\frac{\beta_1\ln\ln(Q^2/\Lambda^2)}{\beta_0^2\ln(Q^2/\Lambda^2)}\right], \tag{8}$$

with the one-loop and two-loop coefficients of the β-function $\beta_0 = 11-2/3N_f$ and $\beta_1 = 102-38/3N_f$ (N_f is the number of quark flavor) and the QCD scale parameter Λ. $\gamma_n^{(0)}$ and $\gamma_n^{(1)}$ are the one-loop and two-loop coefficients of the anomalous dimension γ_n for $O_n^\nu S_{\perp\nu}$ defined as

$$\gamma_n = \frac{\alpha_s}{4\pi}\gamma_n^{(0)} + \left(\frac{\alpha_s}{4\pi}\right)^2\gamma_n^{(1)}. \tag{9}$$

If one sets $\beta_1 \to 0$ and $\gamma_n^{(1)} \to 0$ in (7), the leading order (LO) Q^2 evolution is obtained. $\gamma_n^{(0)}$ and $\gamma_n^{(1)}$ are obtained, respectively, by calculating the one-loop and two-loop corrections to the two-point Green function which imbeds $O_n^\nu S_{\perp\nu}$. To obtain $\gamma_n^{(1)}$, calculation of 18 two-loop diagrams is required in the Feynman gauge. Since the expression for $\gamma_n^{(1)}$ is quite complicated, we refer the readers to Refs. [24,25] for them.

In order to get a rough idea about the NLO Q^2 dependence of h_1, we plotted in Fig. 2 $\gamma_n^{h(1)}$ ($\gamma_n^{(1)}$ for h_1) in comparison with $\gamma_n^{fg(1)}$ ($\gamma_n^{(1)}$ for the nonsinglet f_1 and g_1) for $N_f = 3, 5$. One sees from Fig. 2 $\gamma_n^{h(1)} > \gamma_n^{fg(1)}$ especially at small n. This suggests that the NLO Q^2 evolution of h_1 is quite different from that of f_1 and g_1 in the small x region. The relation $\gamma_n^{h(1)} > \gamma_n^{fg(1)}$ is in parallel with and even more conspicuous than the LO anomalous dimensions which read

$$\gamma_n^{h(0)} = 2C_F \left(1 + 4\sum_{j=2}^{n+1} \frac{1}{j}\right),$$

$$\gamma_n^{fg(0)} = 2C_F \left(1 - \frac{2}{(n+1)(n+2)} + 4\sum_{j=2}^{n+1} \frac{1}{j}\right). \tag{10}$$

To illustrate the generic feature of the Q^2 evolution, we have applied the obtained Q^2 evolution to a reference distribution for g_1 and h_1. As a reference distribution, we take GRSV g_1 distribution[23] and assume $h_1(x, \mu^2) = g_1(x, \mu^2)$ at a low energy input scale ($\mu^2 = 0.23$ GeV^2 for LO and $\mu^2 = 0.34$ GeV^2 for NLO evolution) as is suggested by a nucleon model[4,15]. We then evolve them to $Q^2 = 20$ GeV^2 and see how much deviation is produced between them. The result is shown in Fig. 3. As is expected from the anomalous dimension, the drastic difference in the Q^2 evolution between h_1 and g_1 is observed in the small x region, and this tendency is more significant for the NLO evolution.[27,28,26] (Although g_1 for u-quark mixes with the gluon distribution, the same tendency in the difference from h_1 is observed for the nonsinglet distribution.)

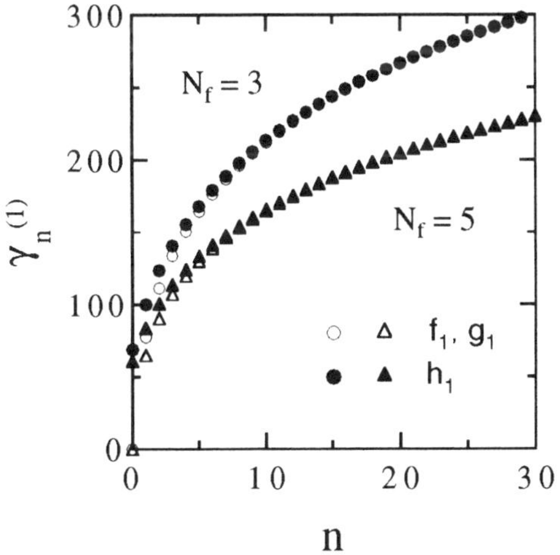

Figure 2: The NLO anomalous dimension $\gamma_n^{h(1)}$ in comparison with $\gamma_n^{fg(1)}$. This figure is taken from Ref. [25].

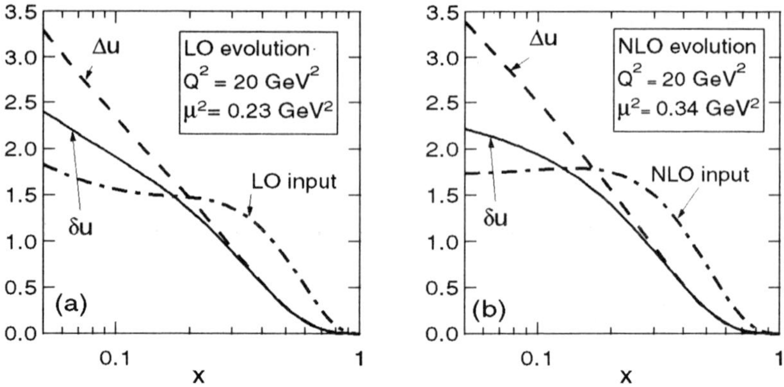

Figure 3: (a) The LO Q^2 evolution of h_1 (denoted by δu) and g_1 (denoted by Δu) for the u-quark. (b) The NLO Q^2 evolution of h_1 and g_1 for the u-quark. This figure is taken from Ref. [26]

In Ref. [30], the Regge asymptotics of h_1 was studied and the small-x behavior was predicted to be $h_1(x) \sim$ constant $(x \to 0)$. On the other hand, the rightmost singularity of $\gamma_n^{h(0)}$ and $\gamma_n^{h(1)}$ are, respectively, located at $n = -2$ and $n = -1$ in the complex n plane, which, respectively, corresponds to $h_1(x) \sim x$ and $h_1(x) \sim$ constant as $x \to 0$. Therefore inclusion of the NLO effect in the DGLAP asymptotics gives consistent behavior at $x \to 0$ as the Regge asymptotics. This is in contrast to the (nonsinglet) f_1 and g_1 distributions, whose LO and NLO DGLAP asymptotics are the same.

One of the interesting applications of the obtained NLO Q^2 evolution of h_1 is the preservation of the Soffer's inequality,[31] $2|h_1^a(x,Q^2)| \leq f_1^a(x,Q^2) + g_1^a(x,Q^2)$. Although the validity of this inequality hinges on schemes beyond LO[32], the NLO Q^2 evolution maintains the inequality at $Q^2 > Q_0^2$ if it is satisfied at some (low) scale Q_0^2 in suitably defined factorization schemes such as $\overline{\text{MS}}$ and Drell-Yan factorization schemes.[33,24].

As was discussed in Sec. 2, a physical cross section is a convolution of a parton distribution and a short distance cross section. (See (1) and (2)) For the double transverse spin asymmetry (A_{TT}) in the Drell-Yan process, the NLO short distance cross section has been calculated in Ref. [34] in the $\overline{\text{MS}}$ scheme. The analysis on A_{TT} combined with the NLO tranversity distribution predicts modest but not negligible NLO effect.[35]

2.3 Q^2-evolution of $h_L(x, Q^2)$ and its $N_c \to \infty$ limit

In general, higher twist ($\tau \geq 3$) distributions represent quark-gluon correlation in the nucleon. Using the QCD equation of motion (see (33) later), one obtains from (4) the following relation ($m_q = 0$)[36]:

$$h_L(x,\mu^2) = 2x \int_x^1 \frac{dy}{y^2} h_1(y,\mu^2) + \widetilde{h}_L(x,\mu^2), \tag{11}$$

$$\widetilde{h}_L(x,\mu^2) = \frac{iP^+}{M} \int_{-\infty}^{\infty} \frac{dz^-}{2\pi} e^{-2ixP\cdot z} \int_0^1 udu \int_{-u}^{u} tdt$$
$$\times \langle PS_{\|}|\bar{\psi}(uz) i\gamma_5 \sigma_{\mu\alpha} g G_\nu{}^\alpha(tz) z^\mu z^\nu \psi(-uz)|PS_{\|}\rangle, \tag{12}$$

where $z^2 = 0$, $z^+ = 0$ and $S_{\|}$ stands for the longitudinal polarization for the nucleon ($S^\mu = S_{\|}^\mu = p^\mu - \frac{M^2}{2} n^\mu$). This equation means that h_L consists of the twist-2 contribution and $\widetilde{h}_L$ which represents quark-gluon correlation in the nucleon. We call the latter contribution "purely twist-3" contribution. (Expansion of (12) produces twist-3 local operators. See (15) below.) Equation (11) reminds us of the Wandzura-Wilczek relation[37] for g_T:

$$g_T(x,\mu^2) = \int_x^1 \frac{dy}{y} g_1(y,\mu^2) + \widetilde{g}_T(x,\mu^2). \tag{13}$$

For e and $\widetilde{g}_T$, one can write down relations similar to (12).

The Q^2-evolution of the first and second terms in (11) is described separately. The evolution of $\widetilde{h}_L$ is quite complicated. A detailed analysis of (12) leads to the following relation for the n-th moment of $\widetilde{h}_L$[4]:

$$\mathcal{M}_n[\widetilde{h}_L(\mu^2)] = \sum_{k=2}^{[(n+1)/2]} \left(1 - \frac{2k}{n+2}\right) \frac{1}{2M} \langle PS_{\|}|R_{nk}(\mu)|PS_{\|}\rangle, \tag{14}$$

$$R_{nk} = \frac{1}{2}\left[\bar{\psi}\sigma^{\lambda\alpha} n_\lambda i\gamma_5 (in\cdot D)^{k-2} igG_{\nu\alpha} n^\nu (in\cdot D)^{n-k}\psi - (k \to n-k+2)\right]. \tag{15}$$

We note that the number of independent operators $\{R_{nk}\}$ ($k = 2, \cdots, [(n+1)/2]$) increases with n. In the Q^2-evolution, the mixing among $\{R_{nk}\}$ occurs and the renormalization is described by the anomalous dimension matrix $[\gamma_n(g)]_{kl}$ for $\{R_{nk}\}$. If we put the LO anomalous dimension matrix for $\{R_{nk}\}$ as $[\gamma_n(g)]_{kl} = (\alpha_s/2\pi)[X_n]_{kl}$ corresponding to (9), the solution to the renormalization group equation for $\{R_{nk}\}$ takes the following matrix form:

$$\langle PS_{\|}|R_{nk}(Q^2)|PS_{\|}\rangle = \sum_{l=2}^{[(n+1)/2]} \left[L^{X_n/\beta_0}\right]_{kl} \langle PS_{\|}|R_{nl}(\mu^2)|PS_{\|}\rangle, \qquad (16)$$

where $L \equiv \frac{\alpha_s(Q^2)}{\alpha_s(\mu^2)}$. X_n for $\widetilde{h}_L$ was derived in Ref. [38]. The Q^2-evolution for $\widetilde{g}_T$ and e is also described by matrix equation similar to (16), and the solution was obtained in Refs.[39,40] for g_T and in Ref.[41] for e.[42] As is clear from (14) and (16) $\mathcal{M}_n[\widetilde{h}_L(Q^2)]$ and $\mathcal{M}_n[\widetilde{h}_L(\mu^2)]$ are not connected by a simple equation as in the case for the twist-2 distribution (see (7)).[43] Although (16) gives complete prediction for the Q^2 evolution, it is generally difficult to distinguish contribution from many operators in the analysis of experiments.

In order to get a rough idea on the Q^2-evolution of $\widetilde{h}_L$, we plotted the eigenvalues of X_n in Fig. 4 (right). For comparison, we also showed in the same figure the LO anomalous dimension $\gamma_n^{(0)}/2$ for h_1. (Note the differene in convention between (7) and (16).) As is clear from this figure, the Q^2 evolution of $\widetilde{h}_L$ is much faster than that of h_1. (See discussion below.)

It has been shown in Refs.[44,36,41] that at large N_c (the number of colors), a great simplification occurs in the Q^2-evolution of the twist-3 distributions. Recall X_n in (16) is a function of two Casimir operators $C_G = N_c$ and $C_F = \frac{N_c^2-1}{2N_c}$. If one takes $N_c \to \infty$, i.e. $C_F \to N_c/2$, (14) and (16) is reduced to

$$\mathcal{M}_n[\widetilde{h}_L(Q^2)] = L^{\widetilde{\gamma}_n^h/\beta_0}\mathcal{M}_n[\widetilde{h}_L(\mu^2)], \qquad (17)$$

$$\widetilde{\gamma}_n^h = 2N_c\left(\sum_{j=1}^{n}\frac{1}{j} - \frac{1}{4} + \frac{3}{2(n+1)}\right). \qquad (18)$$

This evolution equation is just like those for the twist-2 distributions (see (7)). In Fig. 4 (left), we showed the distribution of the eigenvalues of X_n obtained numerically at $N_c \to \infty$. The solid line is the analytic solution in (18), which shows (18) corresponds to the lowest eigenvalues at $N_c \to \infty$. Since (17) was obtained by a mere replacement $C_F \to N_c/2$ in (16), the correction to the result is of $O(1/N_c^2) \sim 10$ % level, which gives enough accuaracy for practical applications.

This large-N_c simplification is a consequence of the fact that the coefficients of R_{nk} in (14) constitutes the *left* eigenvector of X_n corresponding to the eigenvalue $\widetilde{\gamma}_n^h$ in this limit:

$$\sum_{k=2}^{[(n+1)/2]} \left(1 - \frac{2k}{n+2}\right)[X_n]_{kl} = -\left(1 - \frac{2l}{n+2}\right)\widetilde{\gamma}_n^h, \qquad (19)$$

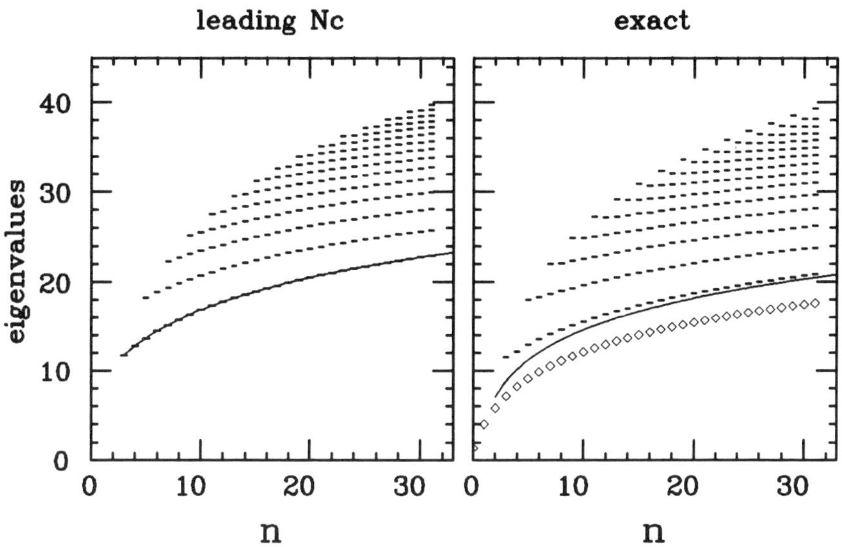

Figure 4: (Right) Complete spectrum of the eigenvalues of the anomalous dimension matrix for $\widetilde{h}_L$ obtained in Ref. [38]. The symbol $\diamond$ denotes the one-loop anomalous dimension for h_1. The solid line is the anomalous dimension (21) at large n. (Left) Spectrum of the eigenvalues of the anomalous dimension matrix for $\widetilde{h}_L$ at large N_c. The solid line denotes the analytic solution given in (18). This figure is taken from Ref. [36].

which implies that all the *right* eigenvectors of X_n except the one corresponding to $\widetilde{\gamma}_n^h$ are orthogonal to the vector consisting of $\left(1-\frac{2k}{n+2}\right)$. This leads to (17).

This large-N_c simplification of the Q^2 evolution was proved for the nonsinglet $\widetilde{g}_T$ in Ref.[44] and for $\widetilde{h}_L$ and e in Ref.[36]. The corresponding anomalous dimensions for $\widetilde{g}_T$ and e are, respectively,

$$\widetilde{\gamma}_n^g = 2N_c\left(\sum_{j=1}^{n}\frac{1}{j}-\frac{1}{4}+\frac{1}{2(n+1)}\right),$$
$$\widetilde{\gamma}_n^e = 2N_c\left(\sum_{j=1}^{n}\frac{1}{j}-\frac{1}{4}-\frac{1}{2(n+1)}\right). \tag{20}$$

Corresponding to three twist-3 distributions in table 1, there are three independent twist-3 fragmentation functions.[9] (Their number is doubled to 6 if one includes final state interactions. See Ref.[9]) It has been shown in Ref.[45] that at large N_c the Q^2 evolution of all these nonsinglet fragmentation functions is also described by a simple evolution equation similar to (17). Therefore the simplification of the twist-3 evolution equation is universal to all twist-3 nonsinglet distribution and fragmentation functions.

To illustrate the actual Q^2 evolution of h_L, we have applied (17) to the bag model calculation of h_L.[46] (Fig. 5) Fig. 5(a) shows the bag calculation of $h_L{}^4$. At the bag scale, purely twist-3 contribution $\widetilde{h}_L$ is comparable to the twist-2 contribution. After the Q^2 evolution to $Q^2 = 10$ GeV^2, h_L is dominated by the twist-2 contribution (Fig. 5(b)). This can be ascribed to two facts: One is the large anomalous dimension (18) compared with the LO anomalous dimension of h_1 ($\diamond$ in the right figure of Fig. 4). The other is the presence of a node for $\widetilde{h}_L(x, Q^2)$, which is taken as model independent due to the constraint $\int_0^1 dx\, \widetilde{h}_L(x, Q^2) = 0$,[47] which is an analogue of the Burkhardt-Cottingham sum rule for $g_2(x, Q^2)$[48]. A similar calculation was done for g_T in Ref.[49]. Using these model calculations, the longitudinal-transverse spin asymmetry, A_{LT}, for the polarized Drell-Yan process was estimated in Ref.[50].

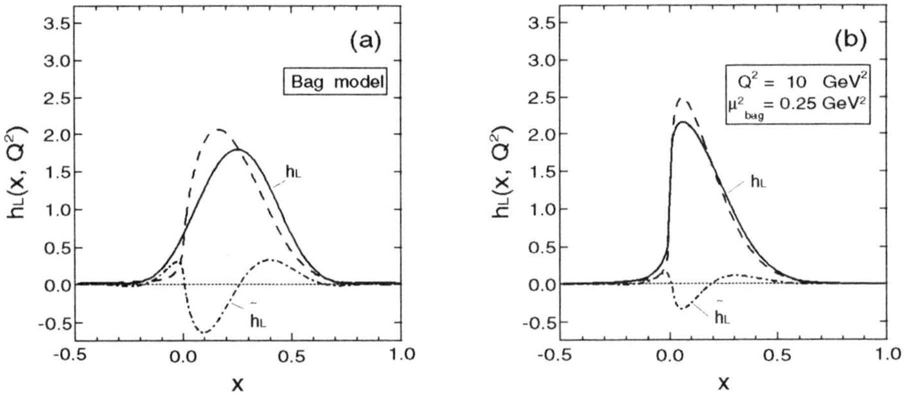

Figure 5: (a) Bag model prediction for h_L. The dashed line represents the twist-2 contribution to h_L. (b) Bag model prediction for h_L evolved to $Q^2 = 10$ GeV^2 assuming the bag scale is $\mu^2 = 0.25$ GeV^2. These figures are taken from Ref. [46]

Another simplification of the twist-3 evolution occurs at $n \to \infty$.[44,36] In this limit, all the twist-3 distributions obey a simple DGLAP equation (17) with a common anomalous dimension which is slightly shifted from (18) and

(20):

$$\gamma_n = 4C_F\left(\sum_{j=1}^{n}\frac{1}{j}-\frac{3}{4}\right)+N_c. \tag{21}$$

This evolution equation satisfies the complete evolution equation to the $O(\ln(n)/n)$ accuracy[44]. In the right figure of Fig. 4, (21) is shown by the solid line. One sees that it is close to the lowest eigenvalues except for small n. Combined with this $n \to \infty$ result, the large-N_c evolution equation in (17) with (18) and (20) for each distribution is valid to $O((1/N_c^2)\ln(n)/n)$ accuracy.

3 Light-cone Distribution Amplitudes of Vector Mesons in QCD

In this section we present a systematic analysis on the light-cone distribution amplitudes (wave functions)[51] of the vector mesons (ρ, ω, ϕ, K^* etc) following our recent work[53]. These amplitudes are relevant for the preasymptotic correction to various exclusive processes producing vector mesons in the final states, such as B meson decay, $B \to \ell\nu V$ (semi-leptonic), $B \to \gamma V$ (radiative), and the electroproduction, $\gamma^* + N \to N' + V$, etc. In particular, we show that the classification and analysis of the light-cone distribution amplitudes for vector mesons can be done in parallel with that of the distribution functions of the nucleon. (Analysis on the light-cone distribution amplitudes for pseudo-scalar mesons is simpler. See e.g. Refs.[51,52].) For definitness, we discuss the ρ^- meson wave functions. Extention to other vector mesons is straightforward.

3.1 *Definition and Classification*

For the ρ^--meson moving in the positive $+\hat{e}_3$ direction, the light-cone wave functions are defined as

$$\phi(u,\mu^2) = P^+\int_{-\infty}^{\infty}\frac{dz^-}{2\pi}e^{iuP\cdot z}\langle 0|\bar{u}(0)\Gamma d(z)|_\mu|\rho^-(P,\lambda)\rangle, \tag{22}$$

where $|\rho^-(P\lambda)\rangle$ stands for the ρ^--meson (mass m_ρ) state with the momentum P and the polarization vector $e_\mu^{(\lambda)}$; $P^2 = m_\rho^2$, $e^{(\lambda)2} = -1$, $P\cdot e^{(\lambda)} = 0$. Γ denotes generic γ matrices and z^- is the only nonzero componet of the space time cordinate z. The variable u in $\phi(u)$ represents a fraction of "+"-momentum P^+ carried by d quark and ϕ has a support on $0 < u < 1$. Here and below the gauge link operator $[0,z] \equiv P\exp\{ig\int_1^0 dt z^\mu A_\mu(tz)\}$ which retores gauge invariance is suppressed for simplicity. The only difference between the wave function (22) and the distribution functions (3) is that the latter is a forward matrix elements while the former is a vacuum-to-meson transition amplitude. In order to classify the wave functions (22), it is convenient to introduce two light-like vectors p and n as was done in section 2.1 in (4). They satisfy the relations $p \cdot n = 1$, $P_\mu = p_\mu + \frac{1}{2}m_\rho^2 n_\mu$ and $e_\mu^{(\lambda)} = (e^{(\lambda)} \cdot n)p_\mu + (e^{(\lambda)} \cdot p)n_\mu + e_{\perp\mu}^{(\lambda)}$. We introduce two coupling constants f_ρ and f_ρ^T by the relation

$$\langle 0|\bar{u}(0)\gamma_\mu d(0)|\rho^-(P\lambda)\rangle = f_\rho m_\rho e_\mu^{(\lambda)}, \tag{23}$$

and

$$\langle 0|\bar{u}(0)\sigma_{\mu\nu} d(0)|\rho^-(P\lambda)\rangle = f_\rho^T \left(e_\mu^{(\lambda)} P_\nu - e_\nu^{(\lambda)} P_\mu\right). \tag{24}$$

With these definitions the classification of (22) can be done based on spin, chirality and twist, as was the case for the distibution functions in the nucleon. The only difference is (i) $e_\mu^{(\lambda)}$ is a vector, while S_μ for the nucleon is an axial vector, and (ii) the wave function (22) should be linear in $e_\mu^{(\lambda)}$, since it is a matrix element between the vacuum and the ρ meson state.

The explicit definitions of the chiral-odd ρ distribution amplitudes are:

$$\begin{aligned}
&\int \frac{d\eta}{2\pi} e^{i\eta u} \langle 0|\bar{u}(0)\sigma_{\mu\nu} d(\eta n)|_\mu|\rho^-(P,\lambda)\rangle \\
&\quad = i f_\rho^T \Big[(e_{\perp\mu}^{(\lambda)} p_\nu - e_{\perp\nu}^{(\lambda)} p_\mu)\phi_\perp(u,\mu^2) + (p_\mu n_\nu - p_\nu n_\mu)(e^{(\lambda)} \cdot n) m_\rho^2 h_\parallel^{(t)}(u,\mu^2) \\
&\qquad + \frac{1}{2}(e_{\perp\mu}^{(\lambda)} n_\nu - e_{\perp\nu}^{(\lambda)} n_\mu) m_\rho^2 h_3(u,\mu^2)\Big],
\end{aligned} \tag{25}$$

and

$$\begin{aligned}
&\int \frac{d\eta}{2\pi} e^{i\eta u} \langle 0|\bar{u}(0) d(\eta n)|_\mu|\rho^-(P,\lambda)\rangle \\
&\quad = \frac{1}{2}\left(f_\rho^T - f_\rho \frac{m_u + m_d}{m_\rho}\right)(e^{(\lambda)} \cdot n) m_\rho^2 \frac{d h_\parallel^{(s)}(u,\mu^2)}{du},
\end{aligned} \tag{26}$$

and the chiral-even distribution amplitudes are defined as

$$\begin{aligned}
&\int \frac{d\eta}{2\pi} e^{i\eta u} \langle 0|\bar{u}(0)\gamma_\mu d(\eta n)|_\mu|\rho^-(P,\lambda)\rangle = f_\rho m_\rho \\
&\times \left[p_\mu (e^{(\lambda)} \cdot n)\phi_\parallel(u,\mu^2) + e_{\perp\mu}^{(\lambda)} g_\perp^{(v)}(u,\mu^2) - \frac{1}{2} n_\mu (e^{(\lambda)} \cdot n) m_\rho^2 g_3(u,\mu^2)\right]
\end{aligned} \tag{27}$$

Twist	2 $O(1)$	3 $O(1/Q)$	4 $O(1/Q^2)$
$e_{\parallel}$	$\phi_{\parallel}$	$h_{\parallel}^{(t)}$, $h_{\parallel}^{(s)}$	g_3
$e_{\perp}$	$\underline{\phi_{\perp}}$	$\underline{g_{\perp}^{(v)}}$, $\underline{g_{\perp}^{(a)}}$	$\underline{h_3}$

Table 2: Spin, twist and chiral classification of the ρ meson distribution amplitudes. Underlined ones are odd.

and

$$\int \frac{d\eta}{2\pi} e^{i\eta u} \langle 0|\bar{u}(0)\gamma_\mu \gamma_5 d(\eta n)|_\mu |\rho^-(P,\lambda)\rangle$$
$$= \frac{i}{4}\left(f_\rho - f_\rho^T \frac{m_u + m_d}{m_\rho}\right) m_\rho \epsilon_\mu{}^{\nu\alpha\beta} e_{\perp\nu}^{(\lambda)} p_\alpha n_\beta \frac{dg_\perp^{(a)}(u,\mu^2)}{du}. \tag{28}$$

All the distribution amplitudes $\phi = \{\phi_\parallel, \phi_\perp, h_\parallel^{(s)}, h_\parallel^{(t)}, g_\perp^{(v)}, g_\perp^{(a)}, h_3, g_3\}$ are normalized as $\int_0^1 du\, \phi(u) = 1$. The appearance of the derivative form for $h_\parallel^{(s)}$ and $g_\perp^{(a)}$ in (26) and (28) is consistent with the relation $\langle 0|\bar{u}(0)\Gamma d(0)|\rho^-(P,\lambda)\rangle = 0$ for $\Gamma = 1, \gamma_\mu\gamma_5$. The various factors in front of $\phi(u)$ in (25)-(28) can be derived by the normalization condition for ϕ and the QCD equation of motion. Dimensional analysis of (25)-(28) determines the twist of each wave function, following the same argumnet in section 2.1. Table 2 summarizes the classification of ϕ's. If one ignore the mass difference between the u and d quarks, the G-parity invariance leads to $\phi(u) = \phi(1-u)$.

Similarly to the case of the twist-3 distribution functions (see (11) and (12)), the twist-3 wave functions $h_\parallel^{(s)}$, $h_\parallel^{(t)}$, $g_\perp^{(v)}$ and $g_\perp^{(a)}$ contain the higher Fock component describing the multi-particle component in the wave function. In order to reveal this, we define three-particle twist-3 quark-antiquark-gluon distribution amplitudes as

$$\int \frac{d\eta}{2\pi} \int \frac{d\zeta}{2\pi} e^{i\eta\alpha_d} e^{i\zeta\alpha_u} \langle 0|\bar{u}(\zeta n)\gamma_\alpha g G_{\mu\nu}(0) d(\eta n)|\rho^-(P,\lambda)\rangle$$
$$= ip_\alpha[p_\mu e_{\perp\nu}^{(\lambda)} - p_\nu e_{\perp\mu}^{(\lambda)}] f_{3\rho}^V \mathcal{V}(\alpha_d, \alpha_u) + \ldots$$
$$\int \frac{d\eta}{2\pi} \int \frac{d\zeta}{2\pi} e^{i\eta\alpha_d} e^{i\zeta\alpha_u} \langle 0|\bar{u}(\zeta n)\gamma_\alpha \gamma_5 g \widetilde{G}_{\mu\nu}(0) d(\eta n)|\rho^-(P,\lambda)\rangle$$
$$= p_\alpha[p_\nu e_{\perp\mu}^{(\lambda)} - p_\mu e_{\perp\nu}^{(\lambda)}] f_{3\rho}^A \mathcal{A}(\alpha_d, \alpha_u) + \ldots$$

$$\int \frac{d\eta}{2\pi} \int \frac{d\zeta}{2\pi} e^{i\eta\alpha_d} e^{i\zeta\alpha_u} \langle 0|\bar{u}(\zeta n)\sigma_{\alpha\beta} g G_{\mu\nu}(0) d(\eta n)|\rho^-(P,\lambda)\rangle$$
$$= \frac{1}{2}(e^{(\lambda)} \cdot n)[p_\alpha p_\mu g^\perp_{\beta\nu} - p_\beta p_\mu g^\perp_{\alpha\nu} - p_\alpha p_\nu g^\perp_{\beta\mu} + p_\beta p_\nu g^\perp_{\alpha\mu}] f^T_{3\rho} m_\rho \mathcal{T}(\alpha_d, \alpha_u) + \dots,$$

where $g^\perp_{\mu\nu} = g_{\mu\nu} - p_\mu n_\nu - p_\nu n_\mu$, the ellipses stand for Lorentz structures of twist higher than three[56], and $\mathcal{V}, \mathcal{A}, \mathcal{T}$ refers in an obvious way to the vector, axial-vector and tensor distributions. α_d, α_u and $1 - \alpha_d - \alpha_u$ are the momentum fractions of d quark, u quark and gluon, respectively, and $\mathcal{V}, \mathcal{A}, \mathcal{T}$ have a support on $\alpha_{u,d} > 0$, $\alpha_u + \alpha_d < 1$. When the quark masses are the same, the G-parity invariance implies that that the function $\mathcal{A}$ is symmetric and the functions $\mathcal{V}$ and $\mathcal{T}$ are antisymmetric under the interchange $\alpha_u \leftrightarrow \alpha_d$. This motivates us to define the coupling constants $f^V_{3\rho}, f^A_{3\rho}, f^T_{3\rho}$ by the normalization conditions for $\{\mathcal{V}, \mathcal{A}, \mathcal{T}\}$:

$$\int_0^1 d\alpha_d \int_0^{1-\alpha_d} d\alpha_u \, (\alpha_d - \alpha_u)\, \mathcal{V}(\alpha_d, \alpha_u) = 1,$$
$$\int_0^1 d\alpha_d \int_0^{1-\alpha_d} d\alpha_u \, \mathcal{A}(\alpha_d, \alpha_u) = 1,$$
$$\int_0^1 d\alpha_d \int_0^{1-\alpha_d} d\alpha_u \, (\alpha_d - \alpha_u)\, \mathcal{T}(\alpha_d, \alpha_u) = 1. \tag{32}$$

Distribution amplitudes defined in this section contain only the light-cone momentum fraction. One might ask what about the intrinsic transverse momentum of quarks which causes power correction to an exclusive process. Are they missing in this formalism? The answer is no. One can show that in calculating the amplitude for an exclusive process the effect of intrinsic transverse momentum of quarks is systematically and Gauge invariantly incorporated by considering the multi-particle (three or more) distribution amplitudes, i.e., it can be reexpressed in the form of higher twist distribution amplitudes.

3.2 QCD Equation of Motion

The two- and three-particle distribution amplitudes introduced in the previous subsection are not independent among one another, but are constrained by the QCD equation of motion. This can be most conveniently done by using the identities among the nonlocal operators defining each amplitude: For example, relevant identities for the odd distribution amplitudes are

$$\frac{\partial}{\partial x_\mu} \{\bar{u}(x)\sigma_{\mu\nu} x^\nu d(-x)\} =$$

$$= i \int_{-1}^{1} dv\, v\, \bar{u}(x) x^{\alpha} \sigma_{\alpha\beta} x^{\mu} g G_{\mu\beta}(vx) d(-x)$$

$$- i x^{\beta} \partial_{\beta} \{\bar{u}(x) d(-x)\} - (m_u - m_d) \bar{u}(x) \not{x}\, d(-x), \tag{33}$$

$$\bar{u}(x) d(-x) - \bar{u}(0) d(0) =$$

$$= \int_0^1 dt \int_{-t}^{t} dv\, \bar{u}(tx) x^{\alpha} \sigma_{\alpha\beta} x^{\mu} g G_{\mu\beta}(vx) d(-tx)$$

$$+ i \int_0^1 dt\, \partial^{\alpha} \left\{\bar{u}(tx) \sigma_{\alpha\beta} x^{\beta} d(-tx)\right\}$$

$$+ i(m_u + m_d) \int_0^1 dt\, \bar{u}(tx) \not{x}\, d(-tx). \tag{34}$$

Here we introduced a shorthand notation for the derivative over the total translation:

$$\partial_{\alpha} \{\bar{u}(tx) \Gamma[tx, -tx] d(-tx)\} \equiv \frac{\partial}{\partial y^{\alpha}} \{\bar{u}(tx + y) \Gamma[tx + y, -tx + y] d(-tx + y)\}\Big|_{y \to 0}, \tag{35}$$

with the generic Dirac matrix structure Γ. Sandwitching these identities by the vacuum and the ρ meson state, we obtain the relation among two- and three-particle distribution amplitudes. We finally obtain for the odd distribution amplitudes as

$$(1 - \delta_+)\, h_{\parallel}^{(s)}(u) = \bar{u} \int_0^u dv\, \frac{1}{\bar{v}}\, \Phi(v) + u \int_u^1 dv\, \frac{1}{v}\, \Phi(v), \tag{36}$$

and

$$h_{\parallel}^{(t)}(u) = \frac{1}{2} \xi \left(\int_0^u dv \frac{1}{\bar{v}}\, \Phi(v) - \int_u^1 dv \frac{1}{v}\, \Phi(v) \right) + \delta_+ \phi_{\parallel}(u)$$

$$+ \zeta_{3\rho}^{T} \frac{d}{du} \int_0^u d\alpha_d \int_0^{\bar{u}} d\alpha_u\, \frac{1}{1 - \alpha_u - \alpha_d}\, \mathcal{T}(\alpha_d, \alpha_u), \tag{37}$$

where

$$\Phi(u) = 2\phi_{\perp}(u) - \delta_+ \left(\phi_{\parallel}(u) - \frac{1}{2} \xi \phi_{\parallel}'(u) \right) + \frac{1}{2} \delta_- \phi_{\parallel}'(u)$$

$$+ \zeta_{3\rho}^{T} \frac{d}{du} \int_0^u d\alpha_d \int_0^{\bar{u}} d\alpha_u\, \frac{1}{1 - \alpha_u - \alpha_d} \left(\alpha_d \frac{d}{d\alpha_d} + \alpha_u \frac{d}{d\alpha_u} - 1 \right) \mathcal{T}(\alpha_d, \alpha_u). \tag{38}$$

with

$$\delta_{\pm} = \frac{f_\rho}{f_\rho^T}\frac{m_u \pm m_d}{m_\rho}, \qquad \zeta_{3\rho}^T = \frac{f_{3\rho}^T}{f_\rho^T m_\rho}. \tag{39}$$

Here and below we use the shorthand notation $\bar{u} = 1 - u$ and $\xi = u - (1-u) = 2u - 1$. According to the various "source" terms on the right-hand side of (36) and (37), one can decompose the solution in an obvious way into three pieces as

$$\begin{aligned} h_\parallel^{(t)}(u) &= h_\parallel^{(t)WW}(u) + h_\parallel^{(t)g}(u) + h_\parallel^{(t)m}(u), \\ h_\parallel^{(s)}(u) &= h_\parallel^{(s)WW}(u) + h_\parallel^{(s)g}(u) + h_\parallel^{(s)m}(u), \end{aligned} \tag{40}$$

where $h_\parallel^{(t)WW}(u)$ and $h_\parallel^{(s)WW}(u)$ denote the "Wandzura-Wilczek" type contributions of twist 2 operators, $h_\parallel^{(t)g}(u)$ and $h_\parallel^{(s)g}(u)$ stand for contributions of three-particle distribution amplitudes and $h_\parallel^{(t)m}(u)$ and $h_\parallel^{(s)m}(u)$ are due to the quark mass corrections. This decomposition can be compared with the relation (11) for the twist-3 distribution. In particular, we get

$$\begin{aligned} h_\parallel^{(t)WW}(u) &= \xi\left(\int_0^u dv \frac{\phi_\perp(v)}{\bar{v}} - \int_u^1 dv \frac{\phi_\perp(v)}{v}\right), \\ h_\parallel^{(s)WW}(u) &= 2\left(\bar{u}\int_0^u dv \frac{\phi_\perp(v)}{\bar{v}} + u\int_u^1 dv \frac{\phi_\perp(v)}{v}\right). \end{aligned} \tag{41}$$

Similarly we obtain for the even distribution amplitudes as

$$\left(1 - \tilde{\delta}_+\right) g_\perp^{(a)}(u) = \bar{u}\int_0^u dv\, \frac{1}{\bar{v}}\, \Psi(v) + u\int_u^1 dv\, \frac{1}{v}\, \Psi(v), \tag{42}$$

and

$$\begin{aligned} g_\perp^{(v)}(u) &= \frac{1}{4}\left[\int_0^u dv\, \frac{1}{\bar{v}}\, \Psi(v) + \int_u^1 dv\, \frac{1}{v}\, \Psi(v)\right] + \tilde{\delta}_+ \phi_\perp(u) \\ &+ \zeta_{3\rho}^A \int_0^u d\alpha_d \int_0^{\bar{u}} d\alpha_u\, \frac{1}{1 - \alpha_d - \alpha_u}\left(\frac{d}{d\alpha_d} + \frac{d}{d\alpha_u}\right)\mathcal{A}(\alpha_d, \alpha_u) \\ &+ \zeta_{3\rho}^V \frac{d}{du}\int_0^u d\alpha_d \int_0^{\bar{u}} d\alpha_u\, \frac{\mathcal{V}(\alpha_d, \alpha_u)}{1 - \alpha_d - \alpha_u}, \end{aligned} \tag{43}$$

where

$$\Psi(u) = 2\phi_{\parallel}(u) + \tilde{\delta}_+ \xi \phi'_{\perp}(u) + \tilde{\delta}_- \phi'_{\perp}(u)$$
$$+2\zeta_{3\rho}^V \frac{d}{du} \int_0^u d\alpha_d \int_0^{\bar{u}} d\alpha_u \frac{1}{1-\alpha_d-\alpha_u} \left(\alpha_d \frac{d}{d\alpha_d} + \alpha_u \frac{d}{d\alpha_u} \right) \mathcal{V}(\alpha_d, \alpha_u)$$
$$+2\zeta_{3\rho}^A \frac{d}{du} \int_0^u d\alpha_d \int_0^{\bar{u}} d\alpha_u \frac{1}{1-\alpha_d-\alpha_u} \left(\alpha_d \frac{d}{d\alpha_d} - \alpha_u \frac{d}{d\alpha_u} \right) \mathcal{A}(\alpha_d, \alpha_u) \quad (44)$$

with

$$\tilde{\delta}_{\pm} \equiv \frac{f_\rho^{T^2}}{f_\rho^2} \delta_{\pm} = \frac{f_\rho^T}{f_\rho} \frac{m_u \pm m_d}{m_\rho}, \qquad \zeta_{3\rho}^{V,A} = \frac{f_{3\rho}^{V,A}}{f_\rho m_\rho}. \quad (45)$$

Eqs. (42) and (43) again allow the decomposition of $g_\perp^{(v)}(u)$ and $g_\perp^{(a)}(u)$ into several terms according to the source terms:

$$g_\perp^{(v)}(u) = g_\perp^{(v)WW}(u) + g_\perp^{(v)g}(u) + g_\perp^{(v)m}(u),$$
$$g_\perp^{(a)}(u) = g_\perp^{(a)WW}(u) + g_\perp^{(a)g}(u) + g_\perp^{(a)m}(u), \quad (46)$$

where $g_\perp^{(v)WW}(u)$ and $g_\perp^{(a)WW}(u)$ denotes the contribution from the twist 2 distribution amplitudes (Wandzura-Wilczek part), $g_\perp^{(v)g}(u)$ and $g_\perp^{(a)g}(u)$ are the contribution from the three-particle distribution amplitudes $\mathcal{V}$ and $\mathcal{A}$.

The relations (36), (37), (42) and (43) are exact in QCD and form a basis for the renormalization and model buildings for the twist-3 wave functions. Wandzura-Wilczek parts in (40) and (46) do not mix with other sources under renormalization. Renormalization and the model building based on the QCD sum rule approach has been carried out in the framework of the conformal expansion for the distribution amplitudes in Refs.[53,54,55].

4 Summary

In the first part of this talk, I discussed the Q^2 evolution of the odd spin-dependent parton distributions $h_1(x, Q^2)$ and $h_L(x, Q^2)$. The Q^2 evolution for the transversity distribution $h_1(x, Q^2)$ was completed in the $\overline{\text{MS}}$ scheme. This means the Q^2 evolution of all the twist-2 distributions has been understood in the level. The resulting Q^2 evolution of $h_1(x, Q^2)$ turned out to cause quite different behavior from the helicity distribution $g_1(x, Q^2)$ in the small x region. The LO Q^2 evolution for the twist-3 distribution $h_L(x, Q^2)$ (and $e(x, Q^2)$) was completed. Although their Q^2 evolution is quite complicated due to the mixing among increasing number of quark-gluon-quark operators, it obeys a simple DGLAP equation similar to the twist-2 distibution in the $N_c \to \infty$ limit, as was the case for the Q^2 evolution of the nonsinglet $g_T(x, Q^2)$ distribution. The same simplification at $N_c \to \infty$ was also proved for the twist-3 fragmentation functions. Therefore this large-N_c simplification was proved to be universal for the twist-3 distribution and fragmentation functions.

In the second part of this talk, I presented a systematic analysis on the light-cone distribution amplitudes for the vector mesons relevant for exclusive processes. Classification can be done in parallel with that of the nucleon's parton distribution functions. Constraint relations among two- and three- particle distribution amplitudes are derived using QCD equation of motion. These relations are exact and must be satisfied for model buildings. Renormalization of the two-particle twist-3 distribution amplitudes can also be performed based on these relations.

Acknowledgements

I would like to thank I.I. Balitsky, P. Ball, V.M. Braun, A. Hayashigaki, Y. Kanazawa, N. Nishiyama and K. Tanaka for the collaboration on the subjects discussed in this talk.

References

1. See, for example, *Spin Structure of the Nucleon* (World Scientific, 1996, eds. T.-A. Shibata et al.); *SPIN96* (World Scientific, 1997, eds. C.W. de Jager et al.).
2. J.C. Collins, D. Soper and G. Sterman, in *Perturbative Quantum Chromodynamics* (World Scientific, Singapore, 1989, ed. A.H. Mueller), and references quoted therein.
3. J.C. Collins and D.E. Soper, Nucl. Phys. **194**, 445 (1982).
4. R.L. Jaffe and X. Ji, Nucl. Phys. **B375**, 527 (1992).
5. J.P. Ralston and D.E. Soper, Nucl. Phys. **B152**, 109 (1979).
6. X. Artru and M. Mekhfi, Z. Phys. **C45** 669 (1990); Nucl. Phys. **A532** (1991) 351c.
7. J.L. Cortes, B. Pire and J.P. Ralston, Z. Phys. **C55**, 409 (1992).
8. R.L. Jaffe and X. Ji, Phys. Rev. Lett **71**, 2547 (1993).
9. X. Ji, Phys. Rev. **D49**, 114 (1994).
10. R.L. Jaffe, Phys. Rev. **D54**, 6581 (1996).
11. R.L. Jaffe, X. Jin and J. Tang, Phys. Rev. Lett. **80**, 1166 (1998).
12. R.L. Jaffe, Comm. Nucl. Part. Phys. **19**, 239 (1990).

13. The Spin Muon Collaboration (D. Adams et al.), Phys. Lett. **B336**, 125 (1994); E143 Collaboration (K. Abe et al.), Phys. Rev. Lett. **76**, 587 (1996).
14. A.V. Manohar, Phys. Rev. Lett. **65**, 2511 (1990); **66**, 289 (1991); X. Ji, Phys. Lett. **B289**, 137 (1992).
15. P.V. Pobylitsa and M.V. Polyakov, Phys. Lett. **B389**, 350 (1996); V. Barone, T. Calarco and A. Drago,Phys. Lett. **B390**, 287 (1997); L. Gamberg, H. Reinhardt and H. Weigel, Phys. Rev. **D58**, 054014 (1998); M. Wakamatsu and T. Kubota, hep-ph/9809443.
16. For a review on h_1, see R.L. Jaffe, hep-ph/9710465, in Proc. of 2nd Topical Workshop on Deep Inelastic Scattering off Polarized Targets: Theory Meets Experiment (DESY-Zeuthen, Sep. 1-5, 1997). DESY 97-200, p167.
17. V.N. Gribov and L.N. Lipatov, Sov. J. Nucl. Phys. **15**, 438 (1972), G. A. Altarelli and G. Parisi, Nucl. Phys. **B126**, 298 (1977), Yu. L. Dokshitzer, Sov. Phys. JETP **46**, 641 (1977).
18. E.G. Floratos, D.A. Ross and C.T. Sachrajda, Nucl. Phys. **B129**, 66 (1977); **B139**, 545 (1978) (E); Nucl. Phys. **B152**, 493 (1979).
19. A. Gonzalez-Arroyo, C. Lopez and F.J. Yndurain, Nucl. Phys. **B153**, 161 (1979).
20. G. Curci, W. Furmanski and R. Petronzio, Nucl. Phys. **B175**, 27 (1980), W. Furmanski and R. Petronzio, Phys. Lett **97B**, 437 (1980).
21. R. Mertig, W.L. van Neerven, Z. Phys. **C70** 637 (1996); W. Vogelsang, Phys. Rev. **D54** 2023 (1996); Nucl. Phys. **B475**, 47 (1996).
22. M. Glück, E. Reya and A. Vogt, Z. Phys. **C53**, 127 (1992); A.D. Martin, W.J. Stirling and R.G. Roberts, Phys. Lett. **B354**, 155 (1995); H.L. Lai, et. al. Phys. Rev. **D51**, 4763 (1995).
23. M. Glück, E. Reya, M. Stratmann and W. Vogelsang, Phys. Rev. **D53**, 4775 (1996); T. Gehrmann and W.J. Stirling, Phys. Rev. **D53**, 6100 (1996); G. Altarelli, R. Ball, S. Forte and G. Ridolfi, Nucl. Phys. **B496**, 337 (1997).
24. W. Vogelsang, Phys. Rev. **D57**, 1886 (1998).
25. A. Hayashigaki, Y. Kanazawa and Y. Koike, Phys. Rev. **D56**, 7350 (1997).
26. A. Hayashigaki, Y. Kanazawa and Y. Koike, hep-ph/9710421, in Proc. of 2nd Topical Workshop on Deep Inelastic Scattering off Polarized Targets: Theory Meets Experiment (DESY-Zeuthen, Sep. 1-5, 1997) DESY 97-200, p. 157.
27. V. Barone, T. Calarco and A. Drago, Phys. Rev. **D56**, 527 (1997).
28. S. Scopetta and V. Vento, Phys. Lett. **B424**, 25 (1996).

29. S. Aoki, M. Doui, T. Hatsuda and Y. Kuramashi, Phys. Rev. **D56**, 433 (1997).
30. R. Kirschner, L. Mankiewicz, A. Schäfer and L. Szymanowski, Z. Phys. **C74**, 501 (1997); R. Kirschner, hep-ph/9710253, in Proc. of 2nd Topical Workshop on Deep Inelastic Scattering off Polarized Targets: Theory Meets Experiment (DESY-Zeuthen, 1997) DESY 97-200, p. 259.
31. J. Soffer, Phys. Rev. Lett. **74**, 1292 (1995).
32. G.R. Goldstein, R.L. Jaffe and X. Ji, Phys. Rev. **D52**, 5006 (1995).
33. C. Bourrely, J. Soffer and O.V. Teryaev, Phys. Lett. **B420**, 375 (1998).
34. A.P. Contogouris, B. Kamal and Z. Merebashvili, Phys. Lett. **B337**, 169 (1994); B. Kamal, Phys. Rev. **D53**, 1142 (1996).
35. O. Martin, A. Schäfer, M. Stratmann, and W. Vogelsang, Phys. Rev. **D57**, 3084 (1998).
36. I.I. Balitsky, V.M. Braun, Y. Koike and K. Tanaka, Phys. Rev. Lett. **77**, 3078 (1996).
37. W. Wandzura and F. Wilczek, Phys. Lett. **B72**, 195 (1977).
38. Y. Koike and K. Tanaka, Phys. Rev. **D51**, 6125 (1995).
39. E.V. Shuryak and A.I. Vainshtein, Nucl. Phys. **B201**, 141 (1982); A.P. Bukhvostov, E.A. Kuraev and L.N. Lipatov, Sov. Phys. JETP **60**, 22 (1984). P.G. Ratcliffe, Nucl. Phys. **B264**, 493 (1986); I.I. Balitsky and V.M. Braun, Nucl. Phys. **B311**, 541 (1988/89); X. Ji and C. Chou, Phys. Rev **D42**, 3637 (1990); D. Müller, Phys. Lett. **B407**, 314 (1997).
40. J. Kodaira, Y. Yasui and T. Uematsu, Phys. Lett. **B344**, 348 (1995); J. Kodaira, Y. Yasui, K. Tanaka and T. Uematsu, Phys. Lett. **B387**, 855 (1996); J. Kodaira, T. Nasuno, H. Tochimura, K.Tanaka, Y. Yasui, Prog. Theor. Phys. **99**, 315 (1998).
41. Y. Koike and N. Nishiyama, Phys. Rev. **D55**, 3068 (1997).
42. In carrying out the renormalization of the twist-3 distributions in the covariant gauge, there arises complication due to the mixing with an equation-of-motion (EOM) operator and a BRS exact operator as well as "alien" operators in the intermediate step. For this point and a method to handle the issue, see Refs. [38,40] and a review J. Kodaira and K. Tanaka, hep-ph/9812449 to appear in Prog. Theor. Phys.
43. For the evolution equation for h_L and e, see A.V. Belitsky and D. Müller, Nucl. Phys. **B503**, 279 (1997).
44. A. Ali, V.M. Braun and G. Hiller, Phys. Lett. **B266**, 117 (1991); see also K. Sasaki, Phys. Rev. **D58**, 094007 (1998).
45. A.V. Belitsky, Phys. Lett. **B312**, 312 (1997); A.V. Belitsky and E.A. Kuraev, Nucl. Phys. **B499**, 301 (1997).
46. Y. Kanazawa and Y. Koike, Phys. Lett. **B403**, 357 (1997).

47. M. Burkardt, in *Proc. 10th International Symposium on High Energy Spin Physics, Hagoya, Japan, Nov. 1992.*(Ed. by Hasegawa et al., Universal Academy Press, Tokyo, 1993) p35.
48. H. Burkhardt and W.N. Cottingham, Ann. Phys. **56**, 453 (1970).
49. M. Stratmann, Z. Phys. **C60**, 763 (1993).
50. Y. Kanazawa, Y. Koike and N. Nishiyama, Phys. Lett. **B430**, 195 (1998).
51. See, for a review, V.L. Chernyak and A.R. Zhitnisky, Phys. Rep. **112**, 173 (1984).
52. V.M. Braun and I.E. Filyanov, Z. Phys. **C48**, 239 (1990) and the references therein.
53. P. Ball, V.M. Braun, Y. Koike and K.Tanaka, Nucl. Phys. **B529**, 323 (1998).
54. Y. Koike, N. Nishiyama and K. Tanaka, Phys. Lett. **B437**, 153 (1998).
55. K. Tanaka, in these proceedings.
56. For the complete structures, see V.M. Braun and P. Ball, hep-ph/9810475.

GAUGE INVARIANT FIELD STRENGTH CORRELATORS : LATTICE STUDIES AND SEMICLASSICAL MODELS

E.-M. ILGENFRITZ
Institute of Theoretical Physics, Kanazawa University,
Kakuma-machi, Kanazawa 920-1192, Japan
e-mail: ilgenfri@hep.s.kanazawa-u.ac.jp

Lattice studies of the field strength correlators are discussed from the following points of view: *(i)* parameters of the instanton liquid model extracted from data obtained by the cooling method for $T = 0$, *(ii)* difficulties of the caloron gas to explain data at $T \leq T_{dec}$, *(iii)* finite T correlators obtained by the RG smoothing method, *(iv)* color correlations between and color coherence inside topological clusters.

This talk summarizes results [1, 2, 3, 4] obtained in collaboration with B. Martemyanov, S. Molodtsov, M. Müller-Preussker, Yu. Simonov and S. Thurner. Field strength correlators, in the Euclidean vacuum and in thermal states, encode information on generic non-perturbative structures of non-Abelian gauge fields in QCD. In the Stochastic Vacuum Model [5], the two-point correlator is used to describe the statistical properties of fields: confinement is directly related to it [6] and, Wick-rotated to Minkowski space, it serves to parametrize high energy hadron-hadron scattering (see *e.g.* Ref. [7]). For heavy quarkonium states, it describes the effect of the gluon condensate on the level splitting (see Ref. [8] and references therein). More fundamentally, the two structure functions of the correlator are inherent to an infrared effective dual formulation of QCD [9]. Here I shall address a few other questions. How do the data constrain instanton models of vacuum structure ? Do they suggest a particular scenario of deconfinement ? Is cooling suitable to measure the correlators ? How can two-point field strength operators be used to explore details of semiclassical structure (in smoothed lattice configurations) that have not been studied before ? The answers are relevant for our perception of vacuum structure and for phenomenological models thereof.

The two-point field strength correlator is a non-local generalization of the gluon condensate, involving the field strength tensor at separate points :

$$D_{\mu\rho,\nu\sigma}(x_1, x_2) = g^2 \, \langle 0|Tr\{G_{\mu\rho}(x_1)S(x_1, x_2)G_{\nu\sigma}(x_2)S^{\dagger}(x_1, x_2)\}|0\rangle \, . \qquad (1)$$

Here $G_{\mu\rho} = T^a G^a_{\mu\rho}$ is the field strength and $S(x_1, x_2)$ is the Schwinger line (parallel transporter) necessary to ensure gauge invariance. The correlator describes the coherence of the gauge field modulo gauge transformations. The

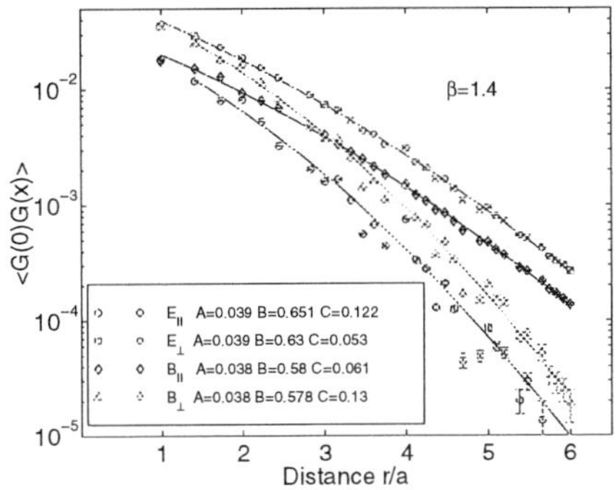

Fig. 1 : Correlation data

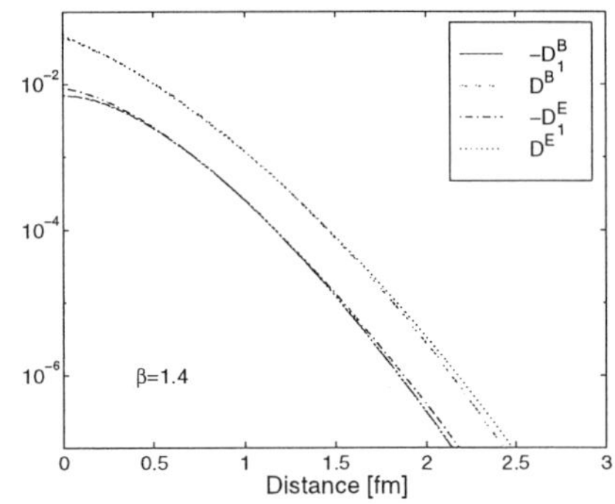

Fig. 2 : Structure functions

local gluon condensate $G_2 = \langle 0|\frac{\alpha_s}{\pi} G^a_{\mu\nu}\ G^a_{\mu\nu}|0\rangle$ is obtained in the limit $|x_1 - x_2| \to 0$ from the regulated correlator. At $T = 0$, Euclidean $O(4)$ invariance admits the decomposition ($x = x_1 - x_2$)

$$D_{\mu\rho,\nu\sigma}(x) = (\delta_{\mu\nu}\delta_{\rho\sigma} - \delta_{\mu\sigma}\delta_{\rho\nu})\left(D(x^2) + D_1(x^2)\right) \qquad (2)$$
$$+ (x_\mu x_\nu \delta_{\rho\sigma} - x_\mu x_\sigma \delta_{\rho\nu} + x_\rho x_\sigma \delta_{\mu\nu} - x_\rho x_\nu \delta_{\mu\sigma}) \frac{\partial D_1(x^2)}{\partial x^2}$$

with two vacuum structure functions $D(x^2)$ and $D_1(x^2)$. D_1 does not contribute to confinement. Perturbatively, both D and D_1 behave like $1/x^4$. Purely (anti)selfdual nonperturbative background fields contribute *only* to D (see Ref. [1] and references therein).

Lattice data for $D_{\mu\rho,\nu\sigma}(x)$ have been obtained in recent years by DiGiacomo et al., for pure Yang Mills theory [10] and full QCD at $T = 0$ [11] and for YM near the deconfinement transition [12]. In practice, the correlators $D_{||}(x^2) = D(x^2) + D_1(x^2) + x^2\ \partial D_1(x^2)/\partial x^2$ and $D_\perp(x^2) = D(x^2) + D_1(x^2)$ have been measured, and cooling has been used to eliminate ultraviolet fluctuations. It has been argued that, due to the diffusive nature of cooling, an appropriate number $n_{cool} \sim O(d^2)$ of cooling iterations must be performed before the correlators can be measured at a distance da (a = lattice spacing). Measuring the correlator $D_{||}((da)^2)$ or $D_\perp((da)^2)$ as the plateau value *vs.* n_{cool} of the corresponding extended operators seems to justify this prescription. Both functions have been fitted according to an ansatz for $D(x^2) = A_0 \exp(-|x|/\lambda_A) + a_0 \exp(-|x|/\lambda_a)/|x|^4$ (and analogously for $D_1(x^2)$), in an attempt to separate the nonperturbative (exponential) part from the perturbative one.

We have discussed [1] these fits from the instanton liquid point of view. Within the one-instanton approximation only D can be described because the instanton is selfdual. The shortest Schwinger line factor has á closed expression in the instanton field. The integration over the instanton position must be performed numerically. The power like shape of the classical instanton field does not give rise to an exponential decay, ranging from $|x| = 0$ to asymptotics, of the nonperturbative part of D. The instanton field strength correlator is flat near $x = 0$ and has a tail $\sim \log|x|/|x|^4$. Various theoretical ideas, however, explain a modified instanton profile at large distance, *e.g.* by feeding back the interaction with the collective instanton medium into the variational instanton field equation [13]. Due to cooling, which requires $x_{min} \sim 3...4\ a$, the correlator at $T = 0$ is presently known only for distances 0.1...1.0 fm. In this region, the best fit to the nonperturbative part of the cooled structure function D by the one-instanton term alone gives a density density $n = 4\ \mathrm{fm}^{-4}$ and size $\rho = 0.3$ fm of instantons in pure YM, to compare with $n = 0.5\ \mathrm{fm}^{-4}$ and $\rho = 0.44$ fm in full QCD (with four flavors). The latter values are only roughly compatible with phenomenological estimates for real QCD. Second order terms in n contribute to D and are leading in D_1. We have analyzed this numerically and found that $D_1 < 0$ is characteristic for the instanton liquid. This is required by the heavy quarkonium fine structure [8], but seemingly in contradiction with the Pisa results (the original fits). We have shown that the quenched data [10] can accomodate negative D_1 as well. With the second order terms taken into account for pure YM theory, the best fit leads to a change towards smaller $n = 3.4\ \mathrm{fm}^{-4}$ and $\rho = 0.27$ fm. The smallness of the correction *a posteriori* justifies the dilute gas approach to pure YM (and even more for full QCD).

The picture becomes more complicated at $T \neq 0$ [14]. The $O(4)$ invariance is lost, and five independent structure functions exist instead of two. All of them separately depend on $|\mathbf{x}|^2$ and $(x^4)^2$. If for example $\mathbf{x} \parallel \mathbf{e}_3$, these are (with $i = 1, 2$) : $D_{||}^E = \langle E_3(\mathbf{x})E_3(0)\rangle$ and $D_{\perp}^E = \langle E_i(\mathbf{x})E_i(0)\rangle$ for the color electric, $D_{||}^B = \langle B_i(\mathbf{x})B_i(0)\rangle$ and $D_{\perp}^B = \langle B_3(\mathbf{x})B_3(0)\rangle$ for the color magnetic field, and an additional electric-magnetic correlator. They are decomposed into the electric and magnetic analogues of D and D_1 : $D_{||}^{(E,B)} = D^{(E,B)} + D_1^{(E,B)} + \mathbf{x}^2\ \partial D_1^{(E,B)}/\partial \mathbf{x}^2$ and $D_{\perp}^{(E,B)} = D^{(E,B)} + D_1^{(E,B)}$. We have attempted a description [2] of the finite T quenched correlator [12] (so far with cooling available only at distance > 0.4 fm !) using the caloron (periodic instanton) solution [15]. Below the deconfinement temperature, where $D^B \approx D^E$ and $D_1^B \approx D_1^E$ has been extracted, a consistent decription of magnetic and electric correlators was not possible in this way. The extraction of a nonperturbative part from the cooled data (by a fit as above) gives $D_{\perp} < D_{||}$ for $\mathbf{E}$ and $\mathbf{B}$, in contrast with the caloron model for $\mathbf{E}$. $D_{||}^B$ has a much shorter correlation length than $D_{\perp}^B$ which is difficult to reconcile with only one ρ. Before drawing

Table 1: Condensates and correlation lengths in pure $SU(2)$ for 4 temperatures $T = t_r\, T_{dec}$. The condensates are $C^{(E,B)} = D^{(E,B)}(0,0)/T_{dec}^4$ and $C_1^{(E,B)} = D_1^{(E,B)}(0,0)/T_{dec}^4$. The integrated correlation lengths $\xi_{int}^{(E,B)}$ of $D^{(E,B)}$ are defined as $\xi_{int} = \int_0^\infty dx D(x)/D(0)$ and given in femtometer.

β	t_r	C^E	C^B	$-C_1^E$	$-C_1^B$	ξ_{int}^E	ξ_{int}^B
1.4	0.71	$1.2\ 10^{-2}$	$1.1\ 10^{-2}$	$2.3\ 10^{-3}$	$1.8\ 10^{-3}$	0.32	0.34
1.5	0.90	$2.4\ 10^{-2}$	$2.3\ 10^{-2}$	$4.5\ 10^{-3}$	$3.5\ 10^{-3}$	0.27	0.29
1.55	1.02	$2.9\ 10^{-2}$	$3.2\ 10^{-2}$	$5.2\ 10^{-3}$	$6.8\ 10^{-3}$	0.27	0.26
1.6	1.15	$3.1\ 10^{-2}$	$4.5\ 10^{-2}$	$2.6\ 10^{-3}$	$1.2\ 10^{-2}$	0.27	0.24

final conclusions, one should note that these data are from a $16^3 \times 4$ lattice such that (isotropic) cooling might be problematic.

Recently a method has been advocated [16, 17] to avoid nonlocal, iterative procedures for suppressing UV fluctuations. Instantons grow (and become more classical) in underrelaxed cooling such that their size must be backward extrapolated. Blocking and inverse blocking (with respect to a perfect action) is used in the alternative method for smoothing of Monte Carlo configurations, preserving long range and semi-local structures *above* the blocking level. So far, this method has been used to analyse instanton and monopole structure. Here I want to discuss first how this method modifies the measured field strength correlators. We have studied this in finite T pure $SU(2)$ theory [4]. Similar to cooling, the renormalization constant becomes $Z \approx 1$ for the gluon fields. Simulations have been done on a $12^3 \times 4$ lattice at four β values near the deconfinement transition. We have measured only equal-time correlators ($x_4 = 0$). Raw data for $\beta = 1.4$ (confinement) are shown in Fig. 1. In the fit we assumed a form $A\exp(-Bx - Cx^2)$ for all four functions and the same limit for $|\mathbf{x}| \to 0$ for $D_{||}$ and $D_\perp$. We have determined the structure functions D^E,D_1^E,D^B and D_1^B (shown in Fig. 2 for $\beta = 1.4$). The $D_1^{(E,B)}$ have been numerically obtained integrating $D_{||}^{(E,B)} - D_\perp^{(E,B)}$ given by the fits and found to be negative. The amplitudes $D^E(0,0)$ and $D^B(0,0)$ grow across the transition (the D_1 are negative and much smaller) while the integrated correlation lengths of D^E and D^B stay constant or shrink at the transition and beyond (see the table for 4 temperatures). The most dramatic change happens to $D_{||}^E$ which becomes extremely short ranged at the transition.

Finally I want to stress that similar observables (with smoothing) are suitable to study new aspects of instanton physics on the lattice [3]. Color correlations play an important role in instanton models, in particular at $T \neq 0$ and with dynamical fermions, but there was no method known how to measure them. Often it is assumed that *different*, locally classical field excitations are *randomly oriented* relative to each other while all components of the fields inside the *same* excitation have a *coherently classical shape.* These assumptions can be critically examined now. Clusters of action and topological density, which are dominantly (anti)selfdual [17] but not yet classical, become visible with smoothing. It is easy to study the distribution of cluster sizes and distances within a cluster pair. Following the considerations above, a next step can be done. At cluster centers z_1, z_2 (where the charge density $|q|$ is maximal) we construct the field strength operators. Connecting them with two Schwinger lines S and S^+ we obtain a gauge invariant operator $\mathcal{D}^{c_1 c_2}_{\mu\rho,\nu\sigma}(z_1, z_2)$ (compare (1)) refering to a cluster pair c_1, c_2 in a given configuration. In order to make the color correlations obvious, we have measured the contracted operator $\mathcal{O}^{c_1 c_2} \propto \mathcal{D}^{c_1 c_2}_{\mu\rho,\mu\rho}(z_1, z_2)$, *i.e. diagonal* correlators summed over all components of **E** and **B**. In order to eliminate the size of the clusters (equivalent to the maximal field strength in an instanton) we have normalized $\mathcal{O}^{c_1 c_2}$ dividing it by $\Pi_{i=1,2}(G^a_{\mu\nu}(z_i) G^a_{\mu\nu}(z_i))^{\frac{1}{2}}$. Summed and normalized in this way it can be considered as cluster overlap in color space. Picking particular field strength components one is able to identify the full relative color orientation matrix between two clusters. This is an (adjoint) matrix $R^{rel}_{c_1 c_2} = R^{-1}_{c_1} R_{c_2}$ where $R^{ab}_{c_i} \propto \mathrm{tr}(T^a U_{c_i} T^b U^+_{c_i})$ refer to single instantons. They rotate the vector potential $A^b_\mu(x, z_i, \rho_i)$ everywhere in the ith instanton. In the absence of color correlations all $SU(N_c)$ matrices U_{c_i} would be distributed according to the Haar measure. In this case the average of the cluster overlaps $\mathcal{O}$ (*i.e.* of the adjoint traces $\mathrm{Tr} R^{rel}_{c_1 c_2}/(N_c^2 - 1)$) over all pairs in an ensemble of configurations would vanish. Fig. 3, however, shows a nonvanishing signal in $\langle \mathcal{O}(x) \rangle$ at small distance $|x|$ which depends on the phase, the type of the topological cluster pair (equal or opposite sign) and on its distance. A quantity similar to $D_{\mu\rho\mu\rho}(z_1, z)$, measured at the center z_1 of a cluster c and another point z, describes the actual degree of coherence within the cluster. A cluster profile function $\mathcal{P}^c$ has been defined dividing by $(G^a_{\mu\nu}(z_1) G^a_{\mu\nu}(z_1))^{\frac{1}{2}}$. This function $\mathcal{P}^c$, averaged over all clusters in the ensemble of configurations, depends on $x = |z - z_1|$ and represents the instanton profile modified by quantum effects. In the confinement phase (for $\beta = 1.4$) we found substantial deviations from the classical instanton shape $\langle \mathcal{P}(x) \rangle \propto \rho^2/(\rho^2 + x^2)^2$ (the solid line in Fig. 4).

To conclude, the field strength correlators are not yet understood for $T \neq 0$ from a semiclassical point of view, but new methods of lattice analysis relying on it are developed to build and check models describing the change of vacuum.

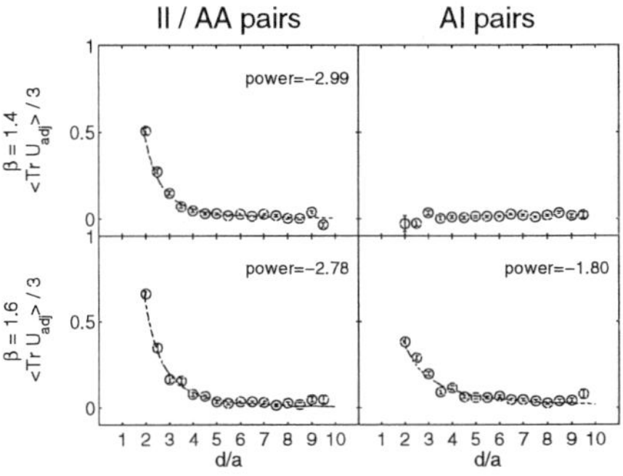

Fig. 3 Color orientations

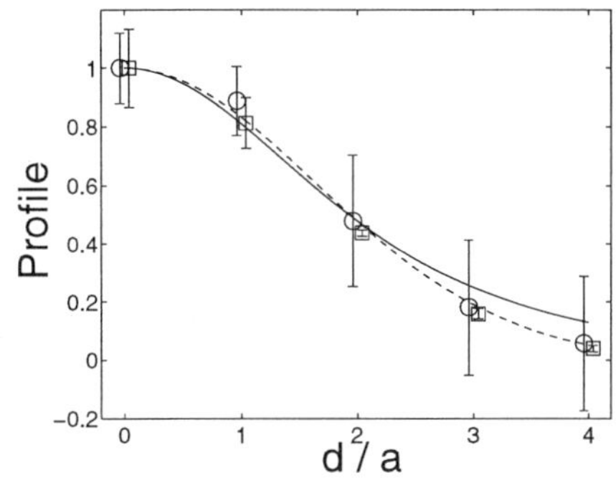

Fig. 4 Instanton profile

References

1. E.-M. Ilgenfritz et al., *Phys. Rev.* D **58**, 114508 (1998).
2. E.-M. Ilgenfritz, B. V. Martemyanov, M. Müller-Preussker, in preparation.
3. E.-M. Ilgenfritz and S. Thurner, hep-lat/9810010.
4. E.-M. Ilgenfritz and S. Thurner, in preparation.
5. H. G. Dosch, *Phys. Lett.* B **190**, 177 (1987); H. G. Dosch and Yu. A. Simonov, *Phys. Lett.* B **205**, 339 (1988); Yu. A. Simonov, *Nucl. Phys.* B **324**, 67 (1989).
6. Yu. A. Simonov, *Usp. Fiz. Nauk* **166**, 337 (1996) [*Phys. Usp.* **39**, 313 (1996)].
7. H. G. Dosch, E. Ferreira, A. Kramer, *Phys. Rev.* D **50**, 1992 (1994).
8. Yu. A. Simonov, S. Titard, F. J. Yndurain, *Phys. Lett.* B **354**, 435 (1995).
9. M. Baker, N. Brambilla, H. G. Dosch, A. Vairo, *Phys. Rev.* D **58**, 034010 (1998); U. Ellwanger, hep-ph/9807380.
10. A. Di Giacomo, E. Meggiolaro, H. Panagopoulos, *Nucl. Phys. Proc. Suppl.* **54A**, 343 (1997).
11. M. D'Elia, A. Di Giacomo, E. Meggiolaro, *Phys. Lett.* B **408**, 315 (1997).
12. A. Di Giacomo, E. Meggiolaro, H. Panagopoulos, *Nucl. Phys.* B **483**, 371 (1997).
13. D. I. Diakonov and V. Yu. Petrov, *Nucl. Phys.* B **245**, 259 (1984).
14. Yu. A. Simonov, *Phys. Atom. Nucl.* **58**, 309 (1995) [hep-ph/9311216].
15. D. J. Gross, R. D. Pisarski, L. G. Yaffe, *Rev. Mod. Phys.* **53**, 43 (1981).
16. M. Feurstein et al., *Nucl. Phys.* B **511**, 421 (1998).
17. E.-M. Ilgenfritz et al., *Phys. Rev.* D **58**, 094502 (1998).

Precritical Chiral Fluctuations in Nuclei[a]

T. HATSUDA
Physics Department, Kyoto University,
Kyoto 606-8502, Japan
E-mail: hatsuda@ruby.scphys.kyoto-u.ac.jp

T. KUNIHIRO
Faculty of Science and Technology, Ryukoku University,
Seta, Otsu-city, 520-2194, Japan
E-mail: kuni@math.ryukoku.ac.jp

Spectral enhancement near $2m_\pi$ threshold in the $I=J=0$ channel in nuclei is shown to be a distinct signal of the partial restoration of chiral symmetry. The relevance of this phenomenon with possible detection of $2\pi^0$ and 2γ in hadron-nucleus and photo-nucleus reactions is discussed.

1 Introduction

The low density theorem in QCD and model calculations suggest that partial restoration of chiral symmetry takes place in nuclear medium.[1] If the quark condensate decreases substantially in nuclear matter, fluctuation of the condensate becomes large in accordance with the general wisdom of statistical physics.[2] This implies that there is a softening of a collective excitation in the scalar-isoscalar channel with the following physical consequences[3]: (i) partial degeneracy of the scalar-isoscalar particle (traditionally called the σ-meson) with the pion, and (ii) decrease of the decay width of σ due to the phase space suppression caused by (i) in the reaction $\sigma \to 2\pi$. Although σ appears only as a broad resonance in the vacuum and is hard to be distinguished from the background[4], it was suggested that it may appear as a sharp resonance at finite temperature (T) because of (i) and (ii) discussed above.[3,5,6]

In this talk, we demonstrate, using a toy model, that the spectral enhancement associated with the partial chiral restoration takes place also at finite baryon density close to $\rho_0 = 0.17\text{fm}^{-3}$. As possible experiments to detect this softening, we will discuss the production of the neutral-dipion ($2\pi^0$) and diphoton (2γ) in reactions with heavy nuclear targets. We also mention the relevance of the softening and the recent measurement of the near-threshold $\pi^+\pi^-$ production in π^+-nucleus reactions.[7]

---

[a]The extended version of this work is given in nucl/th-9810022 (T. Hatsuda, T. Kunihiro and H. Shimizu)

2 Basic idea

Let us first describe the general aspects of the spectral enhancement near the two-pion threshold. Consider the propagator of the σ-meson at rest in the medium : $D_\sigma^{-1}(\omega) = \omega^2 - m_\sigma^2 - \Sigma_\sigma(\omega;\rho)$, where m_σ is the mass of σ in the tree-level, and $\Sigma_\sigma(\omega;\rho)$ is the loop corrections in the vacuum as well as in the medium. The corresponding spectral function reads

$$\rho_\sigma(\omega) = -\frac{1}{\pi}\frac{\mathrm{Im}\Sigma_\sigma}{(\omega^2 - m_\sigma^2 - \mathrm{Re}\Sigma_\sigma)^2 + (\mathrm{Im}\Sigma_\sigma)^2}. \tag{1}$$

Near the two-pion threshold, the phase space factor gives $\mathrm{Im}\Sigma_\sigma \propto \theta(\omega - 2m_\pi)\ (1 - 4m_\pi^2/\omega^2)^{1/2}$ in the one-loop order. On the other hand, partial restoration of chiral symmetry indicates that m_σ^* ("effective mass" of σ defined by $\mathrm{Re}D_\sigma^{-1}(\omega = m_\sigma^*) = 0$) approaches to m_π. Therefore, there exists a density ρ_c at which $\mathrm{Re}D_\sigma^{-1}(\omega = 2m_\pi)$ vanishes even before the complete σ-π degeneracy takes place. At $\rho = \rho_c$, the spectral function is solely dictated by the imaginary part of the self-energy;

$$\rho_\sigma(\omega \simeq 2m_\pi) = -[\pi\ \mathrm{Im}\Sigma_\sigma]^{-1} \propto \theta(\omega - 2m_\pi)\ (1 - \frac{4m_\pi^2}{\omega^2})^{-1/2}. \tag{2}$$

This implies that, even if there is no sharp resonance in the scalar channel in the vacuum, there arises a mild (integrable) singularity just above the threshold in the medium. This is a general phenomenon associated with the partial restoration of chiral symmetry.

3 Linear sigma model

Let us now evaluate $\rho_\sigma(\omega)$ in a *toy* model, namely the SU(2) linear σ-model:

$$\mathcal{L} = \frac{1}{4}\mathrm{tr}[\partial M \partial M^\dagger - \mu^2 M M^\dagger - \frac{2\lambda}{4!}(MM^\dagger)^2 - h(M + M^\dagger)], \tag{3}$$

where tr is for the flavor index and $M = \sigma + i\vec{\tau}\cdot\vec{\pi}$. Although the model is not a precise low energy representation of QCD, [8] it is known to describe the pion dynamics qualitatively well up to 1GeV. [9]

The coupling constants μ^2, λ and h, which are determined in the vacuum to reproduce $f_\pi = 93$ MeV, $m_\pi = 140$ MeV as well as the s-wave π-π scattering phase shift in the one-loop order, are recapitulated for two characteristic cases in Table 1 in which m_σ^{peak} is defined as a peak position of $\rho_\sigma(\omega)$.

Let us now consider the σ-meson embedded in nuclear matter. The interaction of M with the nucleon N in SU(2) chiral symmetry is modeled as

$$\mathcal{L}_I(N, M) = -g\chi\bar{N}U_5N - m_0\bar{N}U_5N, \tag{4}$$

Table 1: Parameters for $m_\sigma^{peak} = 550$, 750 MeV taken from ref.6.

	m_σ^{peak} (MeV)	$\sqrt{-\mu^2}$ (MeV)	$\lambda/4\pi$	$h^{1/3}$ (MeV)
(I)	550	284	5.81	123
(II)	750	375	9.71	124

where we have used a polar representation $\sigma + i\vec{\tau}\cdot\vec{\pi}\gamma_5 \equiv \chi U_5$ for convenience. The first term in (4) with g is a standard chiral invariant coupling in the linear σ model. The second term with a new parameter m_0 is also chiral invariant and non-singular, but is not usually taken into account in the literatures.

After the dynamical breaking of chiral symmetry in the vacuum ($\langle\sigma\rangle_{\rm vac} \equiv \sigma_0 \neq 0$), the coupling of $\tilde{\sigma} = \sigma - \sigma_0$ and π with N is dictated by $g_{\rm s} \equiv g$ and $g_{\rm p} \equiv g_{\rm s} + m_0/\sigma_0$), respectively. Because of m_0, the standard constraint $g_s = g_p$ can be relaxed without conflicting with chiral symmetry. $g_{\rm p}$ is constrained as $g_{\rm p} = m_N/\sigma_0 \simeq m_N/f_\pi = g_{\pi N} = 13.5$, while $g_{\rm s}$ is independent from $g_{\rm p}$ and can be treated as a free parameter. This freedom partially avoid the well-known problem that $g_s = g_p$ combined with eq.(3) does not reproduce the known nuclear matter properties in the mean-field level. [10]

4 One-loop estimate

Let us parametrize the chiral condensate in the medium as

$$\langle\sigma\rangle \equiv \sigma_0 \; \Phi(\rho). \tag{5}$$

In the linear density approximation, $\Phi(\rho) = 1 - C\rho/\rho_0$ with $C = (g_{\rm s}/\sigma_0 m_\sigma^2)\rho_0$. The plausible value of $\Phi(\rho = \rho_0)$ is $0.7 \sim 0.9$. [1]

The one-loop corrections to the self-energy for σ can be read off from the diagrams in Fig.1: $\Sigma_\sigma(\omega;\rho) = \Sigma_{\rm vac}^A + \Sigma_{\rm vac}^B + \Sigma_{MF}(\rho) + \Sigma_{ph}(\rho)$. $\Sigma_{\rm vac}^A$ ($\Sigma_{\rm vac}^B$) corresponds to Fig.1A and Fig.1B , respectively. The explicit form of $\Sigma_{\rm vac}^{A+B}$ is given in Appendix A of ref.6. $\Sigma_{MF}(\rho)$ corresponds to the mean field correction in the nuclear matter (Fig.1C). $\Sigma_{ph}(\rho)$ is a correction from the nuclear particle-hole excitation. We take only the density dependent part in these diagrams and neglect the problematic vacuum-loops of the nucleon. [10]

The leading term in the mean-field part is easily estimated as

$$\Sigma_{MF}(\rho) = \lambda\sigma_0 \, (\langle\sigma\rangle - \sigma_0) = -\lambda\sigma_0^2 \, (1 - \Phi(\rho)), \tag{6}$$

Leading term in the particle-hole part (Fig.1D) starts from $O(\rho^{5/3})$ and is not more than a few % of $\Sigma_{MF}(\rho)$ at $\rho = \rho_0$. This is in contrast to the case of the pion, where both $\Sigma_{MF}(\rho)$ and $\Sigma_{ph}(\rho)$ are proportional to ρ and cancel

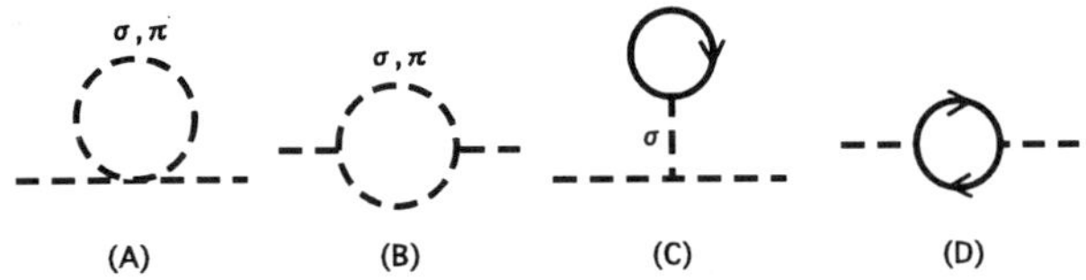

Figure 1: One-loop self energy. The dashed (solid) line denotes σ or π (N).

with each other due to chiral symmetry. Because of this cancellation, we can neglect the two-loop contribution related to the medium modification of the low-momentum pion near the two-pion threshold.

Up to the order we are considering, $\mathrm{Im}\Sigma_\sigma$ solely comes from $\mathrm{Im}\Sigma^B_{\mathrm{vac}}$, since there is no Landau damping and scalar-vector mixing for the σ-meson at rest in nuclear matter. For $2m_\pi \leq \omega \leq 2m_\sigma$,

$$\mathrm{Im}\Sigma_\sigma(\omega;\rho) = \mathrm{Im}\Sigma^B_{\mathrm{vac}} = -\frac{\lambda^2}{96\pi}\sigma_0^2\sqrt{1-\frac{4m_{0\pi}^2}{\omega^2}}. \tag{7}$$

As we have already discussed, the threshold peak is expected to be prominent when $\mathrm{Re}D_\sigma^{-1} = \omega^2 - m_\sigma^2 - \mathrm{Re}\Sigma_\sigma = 0$. In terms of Φ, this condition is rewritten as $\Phi(\rho_c) = 0.74$ (for case (I) in table 1), and 0.76 (for case (II) in Table 1). The numbers in the right hand side are rather insensitive to the parameters in Table 1 as far as the physical quantities in the vacuum are fixed. In the linear density formula $\Phi(\rho) \simeq 1 - 0.2\rho/\rho_0$, we obtain $\rho_c \simeq 1.25\rho_0$. The spectral functions together with $\mathrm{Re}D_\sigma^{-1}(\omega)$ for two cases (I) and (II) are shown in Fig.2. The characteristic enhancements just above the $2m_\pi$ threshold are seen for $\rho \simeq \rho_c$.

We emphasize here that there are two reason behind the enhancement: partial restoration of chiral symmetry ($m_\sigma^* \to m_\pi$), and the cusp structure of $\mathrm{Re}D^{-1}(\omega = 2m_\pi$. Both features can be seen in the lower panels of Fig.2: Although the cusp is not prominent at zero density, it eventually hits the real axis at $\rho = \rho_c$ because $\mathrm{Re}D^{-1}(\omega)$ increases associated with $m_\sigma^* \to 2m_\pi$. This is a general phenomena for systems where the internal symmetry is partially restored in the medium. Another important observation is that, even at densities well below the point where m_σ^* and m_π are degenerate, one can expect a large enhancement of $\rho_\sigma(\omega \simeq 2m_\pi)$.

5 Experiments

To confirm the threshold enhancement associated with the partial chiral restoration, measuring $2\pi^0$ and 2γ with hadron/photon beams off the heavy nuclear

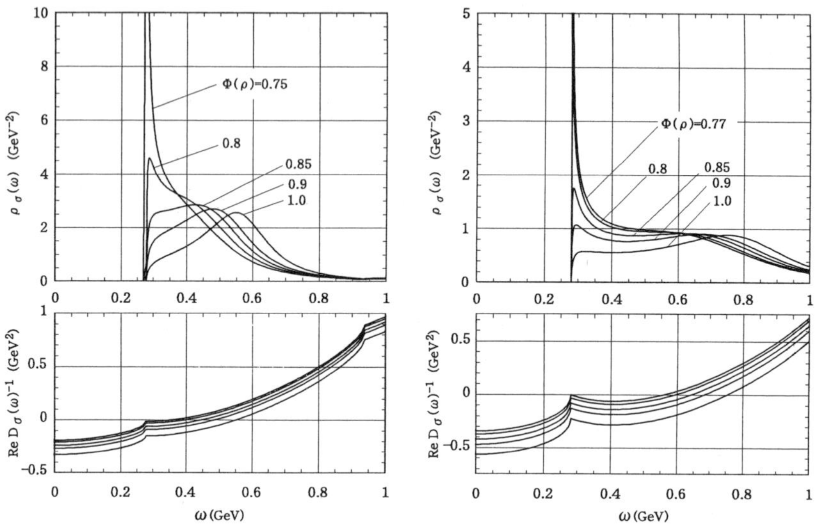

Figure 2: Left: Spectral function for σ and the real part of D_σ^{-1} for several values of $\Phi = \langle\sigma\rangle/\sigma_0$ with $m_\sigma^{peak} = 550$ MeV (case (I) in Table 1). In the lower panel, Φ decreases from bottom to top. Right: Same for $m_\sigma^{peak} = 750$ MeV (case (II) in Table 1.)

targets should be most appropriate. By measuring $\sigma \to 2\pi^0 \to 4\gamma$, [11] one can avoid the possible $I = J = 1$ background from the ρ meson inherent in the $\pi^+\pi^-$ measurement. Measuring the electromagnetic decay $\sigma \to 2\gamma$ is also important because of the small final state interactions. [3] There is also a possibility that one can detect dilepton through the scalar-vector mixing in matter: $\sigma \to \gamma^* \to e^+e^-$. [12]

To enhance the production cross section of the critical fluctuation in the σ-channel, (d, ^{3}He) reactions may be useful as the case of η and ω mesic nuclei.[13] To cover the spectral function in the range $2m_\pi < \omega < 750$ MeV, the incident kinetic energies of d in the laboratory system is estimated as $1.1\text{GeV} < E < 10$ GeV,

Recently CHAOS collaboration[7] measured the $\pi^+\pi^\pm$ invariant mass distribution $M^A_{\pi^+\pi^\pm}$ in the reaction $A(\pi^+, \pi^+\pi^\pm)X$ with the mass number A ranging from 2 to 208: They observed that the yield for $M^A_{\pi^+\pi^-}$ near the $2m_\pi$ threshold is close to zero for $A = 2$, but increases dramatically with increasing A. They identified that the $\pi^+\pi^-$ pairs in this range of $M^A_{\pi^+\pi^-}$ is in the $I = J = 0$ state. Attempts so far in hadronic models without considering the partial chiral restoration failed to reproduce this enhancement. [14,15] On the other hand, A dependence of the the invariant mass distribution of the CHAOS data near $2m_\pi$ threshold has close resemblance to our model calculation in Fig.2, which suggests that this experiment may already provide a hint about how the (partial) restoration of chiral symmetry manifests itself at finite density.

6 Summary

In summary, we have shown that the spectral function in the $I = J = 0$ channel has a large enhancement near the $2m_\pi$ threshold even at nuclear matter density due to the partial chiral restoration. Detection of the dipion and diphoton spectral distribution in the reactions of hadron/photon with heavy nucleus is suitable to confirm the idea of partial chiral restoration in nuclei.

References

1. T. Hatsuda and T. Kunihiro, Phys. Rep. **247**, 221 (1994). G. E. Brown and M. Rho, Phys. Rep. **269**, 333 (1996).
2. P. C. Hohenberg and B. I. Halperin, Rev. Mod. Phys. **49**, 435 (1977). S. Sachdov, `cond-mat/9705266`.
3. T. Hatsuda and T. Kunihiro, Phys. Lett. **145**(1984) 7; Phys. Rev. Lett. **55**, 158 (1985); Phys. Lett. **B185**, 304 (1987).
4. Particle Data Group, Eur. Phys. J. **C** 363, 390 (1998).
5. H. A. Weldon, Phys. Lett. **B274**, 133 (1992). C. Song and V. Koch, Phys. Lett. **B404**, 1 (1997).
6. S. Chiku and T. Hatsuda, Phys. Rev. **D58**, 076001 (1998).
7. CHAOS Collaboration, Phys. Rev. Lett. **77**, 603 (1996).
8. J. Gasser and H. Leutwyler, Ann. Phys. **158**, 142 (1984).
9. L. H. Chan and R. W. Haymaker, Phys. Rev. **D7**, 402 (1973); ibid. **D10**, 4170 (1974).
10. B. D. Serot and J. D. Walecka, Int. J. Mod. Phys. **E6**, 515 (1997).
11. H. Shimizu, in Proceedings of *XV RCNP Osaka International Symposium, Nuclear Physics Frontiers with Electro-weak Probes (FRONTIER 96)*, (Osaka, March 7-8, 1996), p.161.
12. T. Kunihiro, Prog. Theor. Phys. Supplement **120**, 75 (1995).
13. R.S. Hayano, S. Hirenzaki and A. Gillitzer, `nucl-th/9806012`.
14. R. Rapp, J. W. Durso and J. Wambach, Nucl. Phys. **A596**, 436 (1996).
15. H. C. Chiang, E. Oset, and M. J. Vicente-Vacas, Nucl. Phys. **A644**, 77 (1998).

Lattice gauge theory studies of the gluon propagator

D. B. Leinweber, J. I. Skullerud,[a] and A. G. Williams
CSSM and *Department of Physics and Mathematical Physics,*
The University of Adelaide, SA 5005, Australia
http://www.physics.adelaide.edu.au/cssm/

The gluon propagator in Landau gauge is calculated in quenched QCD on a large ($32^3 \times 64$) lattice at $\beta = 6.0$. In order to assess finite volume and finite lattice spacing artefacts, we also calculate the propagator on a smaller volume for two different values of the lattice spacing. Cuts on the data are imposed to minimise these errors. Scaling of the gluon propagator is demonstrated between $\beta = 6.0$ and $\beta = 6.2$. The tensor structure is measured and found to be in good agreement with the Landau gauge form, except at very small momentum values, where some small finite volume errors persist. New structure seen in the infrared region survives conservative cuts to the lattice data, and serves to exclude a number of models that have appeared in the literature. A number of functional forms for the momentum dependence of the propagator are investigated. The form $D(q^2) = D_{\rm IR} + D_{\rm UV}$, where $D_{\rm IR}(q^2) \propto (q^4 + M^4)^{-1}$ and $D_{\rm UV}$ is an infrared regulated one-loop asymptotic form, is found to provide an adequate description of the data over the entire momentum region studied.

1 Introduction

The infrared behaviour of the gluon propagator is important for an understanding of confinement. Previous conjectures range from a strong divergence [1] to a propagator that vanishes in the infrared [2,3].

Lattice QCD should in principle be able to resolve this issue by first-principles, model-independent calculations. However, lattice studies have been inconclusive up to now,[4,5] since they have not been able to access sufficiently low momenta. The lower limit of the available momenta on the lattice is given by $q_{\rm min} = 2\pi/L$, where L is the length of the lattice. Here we will report results using a lattice with a length of 3.3 fm in the spatial directions and 6.7 fm in the time direction. This gives us access to momenta as small as 400 MeV.

2 Lattice formalism

The gluon field A_μ can be extracted from the link variables $U_\mu(x)$ using

$$U_\mu(x) = e^{ig_0 a A_\mu(x+\hat{\mu}/2)} + \mathcal{O}(a^3)\,. \qquad (1)$$

[a]UKQCD Collaboration

Inverting and Fourier transforming this, we obtain

$$A_\mu(\hat{q}) \equiv \sum_x e^{-i\hat{q}\cdot(x+\hat{\mu}/2)} A_\mu(x+\hat{\mu}/2)$$
$$= \frac{e^{-i\hat{q}_\mu a/2}}{2ig_0 a}\left[\left(U_\mu(\hat{q}) - U_\mu^\dagger(-\hat{q})\right) - \frac{1}{3}\mathrm{Tr}\,\left(U_\mu(\hat{q}) - U_\mu^\dagger(-\hat{q})\right)\right], \qquad (2)$$

where $U_\mu(\hat{q}) \equiv \sum_x e^{-i\hat{q}x} U_\mu(x)$ and $A_\mu(\hat{q}) \equiv t^a A_\mu^a(\hat{q})$. The available momentum values $\hat{q}$ are given by

$$\hat{q}_\mu = 2\pi n_\mu/(aN_\mu), \qquad n_\mu = 0, \ldots, N_\mu - 1 \qquad (3)$$

where N_μ is the number of points in the μ direction. The gluon propagator $D_{\mu\nu}^{ab}(\hat{q})$ is defined as

$$D_{\mu\nu}^{ab}(\hat{q}) = \langle A_\mu^a(\hat{q}) A_\nu^b(-\hat{q})\rangle / V\,. \qquad (4)$$

In the continuum Landau gauge, the propagator has the structure

$$D_{\mu\nu}^{ab}(q) = \delta^{ab}(\delta_{\mu\nu} - \frac{q_\mu q_\nu}{q^2})D(q^2)\,. \qquad (5)$$

At tree level, $D(q^2)$ will have the form

$$D^{(0)}(q^2) = 1/q^2\,. \qquad (6)$$

On the lattice, this becomes

$$D^{(0)}(\hat{q}) = 1/\sum_\mu \left(\frac{2}{a}\sin\frac{\hat{q}_\mu a}{2}\right)\,. \qquad (7)$$

Since QCD is asymptotically free, we expect that up to logarithmic corrections, $q^2 D(q^2) \to 1$ in the ultraviolet. Hence we define the new momentum variable q by

$$q_\mu \equiv \frac{2}{a}\sin\frac{\hat{q}_\mu a}{2}, \qquad (8)$$

and work with this throughout.

The Landau gauge can be implemented on the lattice by minimising the functional

$$F_U^L[g] = 1 - \sum_{\mu,x} \Re\mathrm{Tr}\, U_\mu^g(x)\,, \qquad (9)$$

Name	β	a^{-1} (GeV)	Volume	$N_{\rm conf}$	$a\hat{q}_{\rm max}$	$\theta_{\rm max}$
Small	6.0	1.885	$16^3 \times 48$	125	$2\pi/4$	10^{-12}
Large	6.0	1.885	$32^3 \times 64$	75	$2\pi/4$	10^{-12}
Fine	6.2	2.63	$24^3 \times 48$	223	$2\pi/4$	10^{-12}

Table 1: Simulation parameters. The lattice spacing is taken from the string tension,[7] with the 'physical' value taken to be $\sqrt{\sigma} = 427$ MeV.

where $U^g_\mu(x) = g(x)U_\mu(x)g^\dagger(x+\hat{\mu})$, with respect to gauge transformations g. It can be shown that a stationary point of $F^L_U[g]$ satisfies the lattice version of the Landau gauge condition,

$$\theta = \frac{1}{VN_c}\sum_x \theta(x) = \frac{1}{VN_c}\sum_x \mathrm{Tr}\left(\Delta(x)\Delta^\dagger(x)\right) = 0 \tag{10}$$

where $\Delta(x) = \sum_\mu \left(A_\mu(x+\hat{\mu}) - A_\mu(x)\right)$ is the lattice derivative of the gluon field.

The (bare) lattice gluon propagator is related to the renormalised continuum propagator $D_R(q;\mu)$ via

$$D^L(qa) = Z_3(\mu,a)D_R(q;\mu)\,. \tag{11}$$

The renormalisation constant $Z_3(\mu,a)$ can be found by imposing a momentum subtraction renormalisation condition $D_R(q)|_{q^2=\mu^2} = \frac{1}{\mu^2}$.

The asymptotic behaviour of the renormalised gluon propagator in the continuum is given to one-loop level by

$$D_R(q) \equiv D_{\rm bare}(q)/Z_3 = \frac{1}{q^2}\left(\frac{1}{2}\ln(q^2/\Lambda^2)\right)^{-d_D} \tag{12}$$

with $d_D = 13/44$ for quenched QCD in the Landau gauge.

3 Simulation Parameters, Finite Size Effects and Anisotropies

We have analysed three lattices, with different values for the volume and lattice spacing. The details are given in table 1. In the following, we are particularly interested in the deviation of the gluon propagator from the tree level form. We will therefore factor out the tree level behaviour and plot $q^2D(q^2)$ rather than $D(q^2)$ itself.

Fig. 1 shows the gluon propagator as a function of qa for the small and large lattices, with momenta in different directions plotted separately. For

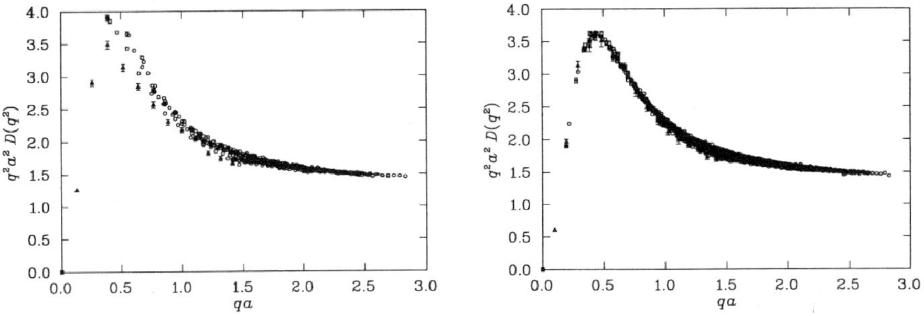

Figure 1: Componentwise data, for the small lattice (left) and the large lattice (right). The filled triangles denote momenta directed along the time axis, while the filled squares denote momenta directed along one of the spatial axes.

low momentum values on the small lattice, there are large discrepancies due to finite size effects between points representing momenta along the time axis and those representing momenta along the spatial axes. These discrepancies are absent from the data from the large lattice, indicating that finite size effects here are under control.

However, at higher momenta, there are anisotropies which remain for the large lattice data, and which are of approximately the same magnitude for the two lattices. In order to eliminate these anisotropies, which arise from finite lattice spacing errors, we select momenta lying within a cylinder of radius $\Delta\hat{q}a = 2 \times 2\pi/32$ along the 4-dimensional diagonals.[8]

4 Tensor structure

By studying the tensor structure of the gluon propagator, we may be able to determine how well the Landau gauge condition is satisfied, and also discover violations of continuum rotational invariance. The continuum tensor structure (5) follows from the condition $q_\mu A_\mu = 0$. This translates directly to the lattice provided we use the symmetric definition (1) of the gluon field.

The tensor structure may be measured directly by taking the ratios of different components of $D_{\mu\nu}(q)$ for the same value of q. The results for the small lattice are summarised in table 2, and compared to what one would expect from (5), and to what one would obtain by replacing q with $\hat{q}$ in (5). The results are similar for the two other lattices. It is clear from table 2 that our numerical data are consistent with the expectation from (5), for these momentum values.

However, when we go to lower momentum values, we find violations of the form (5), arising from finite volume errors. Violations of the order 10–15% persist for the lowest momentum points even on the large lattice, as can be seen from table 3. This gives us a more rigorous test of finite volume errors than what we could obtain by inspection in section 3.

Table 2: Tensor structure for the small lattice. $\hat{q}$ is in units of $2\pi/L_s$, where L_s is the spatial length of the lattice. The theoretical predictions are the values for the ratios one obtains from (5), and from (5) with $q \to \hat{q}$. The numbers in brackets are the statistical uncertainties in the last digit(s). Where no error is quoted, the statistical uncertainty is less than 10^{-6}.

		Theoretical prediction		This simulation
$\hat{q}$	Components	Using $\hat{q}$	Using q	
[2,1,0,0]	(1,1)/(1,2)	-0.5	-0.509796	-0.509796
	(1,1)/(2,2)	0.25	0.259892	0.259892
	(1,1)/(3,3)	0.2	0.206281	0.204(8)
	(1,2)/(2,2)	-0.5	-0.509796	-0.509796
[4,1,0,0]	(1,1)/(1,2)	-0.25	-0.275899	-0.275899
	(1,1)/(3,3)	0.05882	0.070736	0.076(3
	(1,2)/(2,2)	-0.25	-0.275899	-0.275899
[4,2,1,0]	(1,1)/(1,2)	-0.625	-0.681848	-0.678(9)
	(1,1)/(2,3)	-2.5	-2.47137	-2.3(4)
	(1,2)/(2,2)	-0.4706	-0.502914	-0.500(6)

Table 3: Tensor structure for low momentum values, on the large lattice. All the ratios in this table should be 1 according to (5). Finite volume effects show up in differences between the tensor components in the time (4) direction and those in the spatial directions.

$[\hat{q}_x, \hat{q}_y, \hat{q}_z, \hat{q}_t]$	Components	Measured ratio
[1,0,0,0]	(2,2)/(4,4)	1.13(6)
	(3,3)/(4,4)	1.17(6)
[0,1,0,0]	(1,1)/(4,4)	1.13(7)
	(3,3)/(4,4)	1.17(8)
[2,0,0,0]	(2,2)/(4,4)	1.10(7)
	(3,3)/(4,4)	1.14(6)
[0,2,0,0]	(1,1)/(4,4)	1.09(6)
	(3,3)/(4,4)	1.02(7)

5 Scaling behaviour

Since the renormalised propagator $D_R(q;\mu)$ is independent of the lattice spacing, we can use (11) to derive a simple, q-independent expression for the ratio of the unrenormalised lattice gluon propagators at the same value of q:

$$\frac{D_c(qa_c)}{D_f(qa_f)} = \frac{Z_3(\mu,a_c)D_R(q;\mu)/a_c^2}{Z_3(\mu,a_f)D_R(q;\mu)/a_f^2} = \frac{Z_c}{Z_f}\frac{a_f^2}{a_c^2} \tag{13}$$

where the subscript f denotes the finer lattice ($\beta = 6.2$ in this study) and the subscript c denotes the coarser lattice ($\beta = 6.0$). We can use this relation to study directly the scaling properties of the lattice gluon propagator by matching the data for the two values of β. This matching can be performed by adjusting the values for the ratios $R_Z = Z_f/Z_c$ and $R_a = a_f/a_c$ until the two sets of data lie on the same curve.[9]

In the leftmost plot of fig. 2 we present the result of the matching using $\hat{q}$ as the momentum variable. The minimum value for χ^2/dof of about 1.7 is obtained for $R_a \sim 0.815$. This value for R_a is considerably higher than the value of 0.716 ± 0.040 obtained from an analysis of the static quark potential.[7] From this discrepancy, as well as the relatively high value for χ^2/dof, we may conclude that the gluon propagator, taken as a function of $\hat{q}$, does not exhibit scaling behaviour for the values of β considered here.

The rightmost plot of fig. 2 shows the result of the matching using q of (8) as the momentum variable. We can see immediately that this gives much more satisfactory values both for χ^2/dof and for R_a. The minimum value for χ^2/dof of 0.6 is obtained for $R_a = 0.745$. Taking a confidence interval where $\chi^2/\text{dof} < \chi^2_{min} + 1$ gives us an estimate of $R_a = 0.745^{+32}_{-37}$, which is fully compatible with the value[7] of 0.716 ± 0.040. The corresponding estimate for the ratio of the renormalisation constants is $R_Z = 1.038^{+26}_{-21}$.

6 Model Fits

We have demonstrated scaling in our lattice data over the entire range of q^2 considered, and will now proceed with model fits. We introduce an infrared-regulated version $L(q^2,M)$ of the one-loop logarithmic correction given by Eq. (12) in order to ensure that these models have the correct leading ultraviolet behavior. This is given by

$$L(q^2,M) \equiv \left(\frac{1}{2}\ln\frac{q^2+M^2}{M^2}\right)^{-d_D} . \tag{14}$$

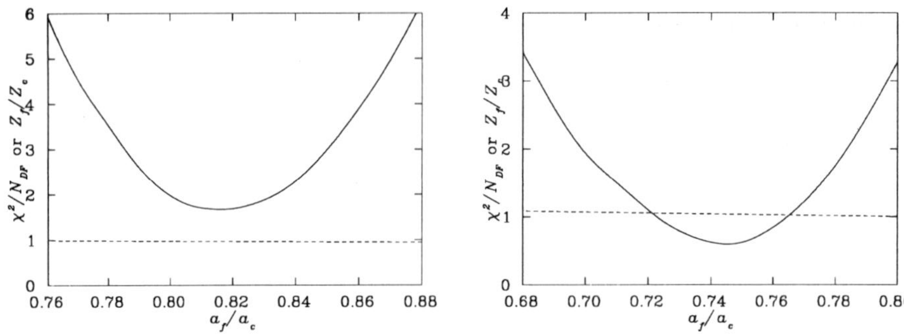

Figure 2: χ^2 per degree of freedom as a function of the ratio of lattice spacings for matching the small and fine lattice data according to (13), using $\hat{q}$ (left) and q (right) as the momentum variable. The dashed line indicates the ratio R_Z of the renormalisation constants.

For simplicity of presentation of the models, all model formulae are to be understood as functions of dimensionless quantities (scaled by the appropriate powers of the lattice spacing a). The following functional forms have been considered:

$$D(q^2) = \frac{Zq^2}{q^4+M^4} L(q^2, M) \qquad \text{(Gribov [2])} \qquad (15)$$

$$D(q^2) = Z\frac{q^2}{q^4+2A^2q^2+M^4} L(q^2, M) \qquad \text{(Stingl [3])} \qquad (16)$$

$$D(q^2) = \frac{Z}{(q^2)^{1+\alpha}+M^2} \qquad \text{(Marenzoni } et\ al\text{ [5])} \qquad (17)$$

$$D(q^2) = Z\left[(q^2 + M^2(q^2))\ln\frac{q^2+4M^2(q^2)}{\Lambda^2}\right]^{-1} \qquad \text{(Cornwall [6])} \qquad (18)$$

$$\text{where} \quad M(q^2) = M\left\{\ln\frac{q^2+4M^2}{\Lambda^2}\Big/\ln\frac{4M^2}{\Lambda^2}\right\}^{-6/11}$$

$$D(q^2) = Z\left(\frac{A}{(q^2+M_{\rm IR}^2)^{1+\alpha}} + \frac{1}{q^2+M_{\rm UV}^2}L(q^2, M_{\rm UV})\right) \qquad \text{(Model A)} \qquad (19)$$

$$D(q^2) = Z\left(\frac{A}{(q^2)^{1+\alpha}+(M_{\rm IR}^2)^{1+\alpha}} + \frac{1}{q^2+M_{\rm UV}^2}L(q^2, M_{\rm UV})\right) \qquad \text{(Model B)} \qquad (20)$$

$$D(q^2) = Z\left(Ae^{-(q^2/M_{\rm IR}^2)^\alpha} + \frac{1}{q^2+M_{\rm UV}^2}L(q^2, M_{\rm UV})\right) \qquad \text{(Model C)} \qquad (21)$$

We have also considered Models A′ and B′, which are constructed from Models A and B respectively by setting $M_{\rm UV} = M_{\rm IR} = M$, as well as models with specific values for the exponent α. Equations (15) and (16) are modified in order to exhibit the asymptotic behaviour of (12). Models A and B are constructed as generalisations of (17) with the correct dimension and asymptotic behaviour.

All models were fitted to the large lattice data using the cylindrical cut. The lowest momentum value was excluded, as the volume dependence of this point could not be assessed. In order to balance the sensitivity of the fit between the high- and low-momentum region, nearby data points within $\Delta(qa) < 0.05$ were averaged.

χ^2 per degree of freedom and parameter values for fits to all these models are shown in table 4. It is clear that Model B accounts for the data better than any of the other models. The best fit to this model is illustrated in fig. 3. However, the parameter values for this model are very unstable, as indicated by the large errors in the estimates for M_{UV}, A and Z in table 4.

This instability can be largely removed by setting $M_{\mathrm{UV}} = M_{\mathrm{IR}}$, as indicated by the small errors for all the parameters in Models A′ and B′ in table 4. This advantage is traded off against a considerably worse χ^2 in the infrared region. However, as shown in fig. 4, χ^2/N_{df} drops rapidly once the first 4–6 points are discarded. These are the same points that the analysis of the tensor structure in section 4 showed to be sensitive to finite volume errors at the level of 10–15%. It is also interesting to observe that at the lowest momenta, the data points decrease slightly as the volume increases. It can be noted [9] that the parameter values do not change when these points are included.

Since the parameter α in Models B and B′ come out to be very close to 1, we have also considered fixing this parameter to 1, yielding an inverse-quartic type form. The resulting model has essentially the same behaviour as Model B′, as we can see from fig. 4. When the first 6 points are discarded, we get a very good fit, which is illustrated in fig. 3.

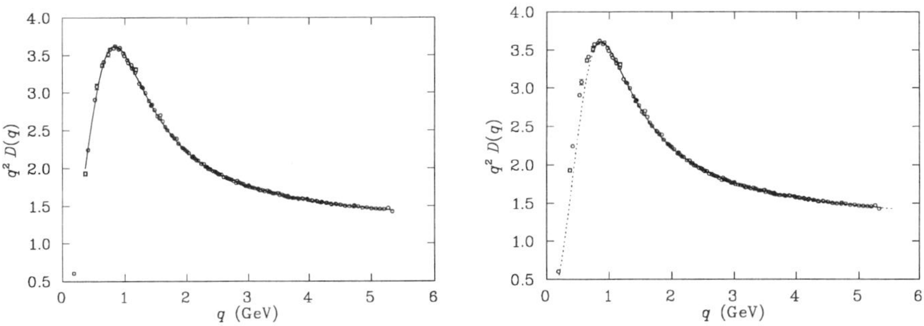

Figure 3: The gluon propagator multiplied by q^2, with nearby points averaged. In the left-hand plot, the line illustrates our best fit to Model B, including all points shown except the one at lowest momentum value. The line in the right hand plot is obtained by fitting all points excluding the first 7 to Model B′ with $\alpha = 1$. This fit undershoots the data in the infrared, as indicated by the dotted line. The scale is taken from the string tension [7].

Model	χ^2/dof	Z	A	$M_{\rm IR}$	$M_{\rm UV}$ or Λ	α
Gribov	1972	2.19^{+31}_{-15}		0.23^{+14}_{-1}		
Stingl	1998	2.2	0	0.23		
Marenzoni	163	2.41^{+0}_{-12}		0.14^{+4}_{-14}		0.29^{+6}_{-2}
Cornwall	50.3	6.5^{+7}_{-9}		0.24^{+3}_{-16}	0.27^{+7}_{-7}	
Model A	2.96	1.54^{+10}_{-20}	1.24^{+21}_{-21}	0.46^{+2}_{-14}	0.96^{+47}_{-17}	1.31^{+16}_{-43}
Model A′	3.73	1.71^{+9}_{-0}	0.84^{+0}_{-29}	0.48^{+2}_{-17}		1.52^{+12}_{-37}
Model B	1.57	1.78^{+45}_{-20}	0.49^{+17}_{-6}	0.43^{+5}_{-1}	0.20^{+37}_{-19}	0.95^{+7}_{-5}
Model B′	4.00	1.62^{+3}_{-4}	0.58^{+5}_{-1}	0.40^{+9}_{-2}		0.92^{+17}_{-1}
B′; $\alpha = 1$	6.68	1.66^{+1}_{-3}	0.54^{+4}_{-1}	0.42^{+2}_{-2}		
Model C	47	2.09^{+30}_{-12}	29^{+166}_{-2}	0.22^{+0}_{-16}	0.14^{+11}_{-10}	0.49^{+0}_{-16}

Table 4: Parameter values for fits to models (15)–(21). The values quoted are for fits to the entire set of data. The errors denote the uncertainties in the last digit(s) of the parameter values which result from varying the fitting range. The fitting ranges considered when evaluating the uncertainties are those with a minimum of 40 points included and with the minimum value for qa no larger than 1.03 (point number 40).

7 Discussion and Outlook

We have evaluated the gluon propagator on an asymmetric lattice with a large physical volume. By studying the anisotropies in the data, and comparing the data with those from a smaller lattice, we have been able to conclude that finite size effects are under control on the large lattice. We have demonstrated scaling of the gluon propagator between $\beta = 6.0$ and $\beta = 6.2$.

A clear turnover in the behaviour of $q^2 D(q^2)$ has been observed at $q \sim 1$GeV, indicating that the gluon propagator diverges less rapidly than $1/q^2$ in the infrared, and may be infrared finite or vanishing. The data are consistent with a functional form $D(q^2) = D_{\rm IR} + D_{\rm UV}$, where

$$D_{\rm IR} \propto \frac{1}{q^4 + M^4} \tag{22}$$

$M = (790 \pm 40)$ MeV, and $D_{\rm UV}$ is the asymptotic form given by (12). A more detailed analysis [9] of the asymptotic behaviour reveals that the one-loop formula (12) remains insufficient at $q^2 = 50\text{GeV}^2$.

Issues for future study include the effect of Gribov copies and of dynamical fermions. We also hope to use improved actions to perform realistic simulations at larger lattice spacings. This would enable us to evaluate the gluon propagator on larger physical volumes, giving access to lower momentum values.

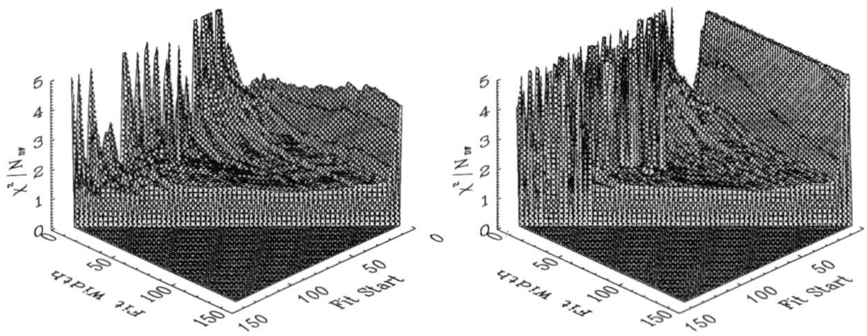

Figure 4: χ^2 per degree of freedom for fits of Model B′ (left) and Model B′ with $\alpha = 1$(right). The "Fit start" axis indicates the starting point for the fit, while the "Fit width" axis indicates the number of points included in the fit. The most inclusive fits are in the near right-hand corner, with the smallest value for the starting point and the largest number of points included. We see that both models are stable over a wide variety of fitting ranges, but that χ^2/N_{df} increases sharply as the 4–6 most infrared points are included in the fit.

Acknowledgments

The numerical work was mainly performed on a Cray T3D at EPCC, University of Edinburgh, using UKQCD Collaboration time under PPARC Grant GR/K41663. Financial support from the Australian Research Council is gratefully acknowledged. We thank Claudio Parrinello for stimulating discussions.

1. N. Brown and M.R. Pennington, *Phys. Rev.* **D 39**, 2723 (1989)
2. V.N. Gribov, *Nucl. Phys.* **B 139**, 19 (1978); D. Zwanziger, *Nucl. Phys.* **B 378**, 525 (1992)
3. M. Stingl, *Phys. Rev.* **D 34**, 3863 (1986); *Phys. Rev.* **D 36**, 651 (1987)
4. C. Bernard, C. Parrinello, A. Soni, *Phys. Rev.* **D49**, 1585 (1994)
5. P. Marenzoni, G. Martinelli, N. Stella, *Nucl. Phys.* **B 455**, 339 (1995); P. Marenzoni *et al*, *Phys. Lett.* **B 318**, 511 (1993)
6. J. Cornwall, *Phys. Rev.* **D 26**, 1453 (1982)
7. G.S. Bali and K. Schilling, *Phys. Rev.* **D 47**, 661 (1993)
8. D.B. Leinweber, J.I. Skullerud, A.G. Williams, C. Parrinello, *Phys. Rev.* **D 58**, 031501 (1998)
9. D.B. Leinweber, J.I. Skullerud, A.G. Williams, C. Parrinello, ADP-98-72/T339 (hep-lat/9811027)

Solar Nuclear Fusion Reaction by Chiral Perturbation Theory

Dong-Pil Min
Physics Department, Seoul National University
Seoul, 151-742, Korea[#1]
and
Department de Física Theòrica, Universitat de València
46100 Burjassot (València), Spain
E-mail: dpmin@phya.snu.ac.kr

The proton burning solar process $p + p \to d + e^+ + \nu_e$ is taken to show how the effective quantum field theory works in nuclear physics. The result represents a model-independent calculation consistent with low-energy effective theory of QCD, which supports within the theoretical uncertainties the previous calculation. The present method demonstrates how the traditional nuclear physics calculation using potential is related with the modern effective field theory method. In addition, it also illustrates how the effective theory behaves when an absence of a *relevant* degree of freedom becomes not remediable, i.e., at the limit of the traditional method.

1 Introduction

The proton fusion reaction $p + p \to d + e^+ + \nu_e$ which plays an important role for stellar evolution and for the solar-neutrino problem, has quite a long history of investigation[1,4,3,2,5]. However, due to the general mistrust and unsettling feeling on the hadronic interaction involved in the calculation because of lack of systematicity and transparent linkage with QCD, this reaction often has been an object of doubt by those who wish to find further clue to the solar neutrino problems[6]. Current progress on the effective theories of matter now provides a systematic approach to mark the period on this trend.

The success of the Chiral Perturbation Theory in the low energy phenomena is now widely accepted and the many demonstrations by the theoretical calculation based on this QCD effective langrangian are available. In this talk, I will try to carry the essentials of those calculations and discuss the power of the effective theory of QCD in the low energy phenomena such as solar proton burning process.

The chiral perturbation theory is based on the chiral symmetry as well as other fundamental symmetries of QCD.[7,8] The degrees of freedom with which the effective field theory is constructed are hadrons, not the quark and gluon. But both descriptions can bear the same class of solution[9]. The merits of the

---

#1 Permanent Address

chiral perturbation theory are that we can obtain the solution to the precision which we want by means of the expansion method, if the convergence is guaranteed. And this convergence is dictated by the counting rule of the theory. The ratio of the momentum of the physical process under investigation to the value of the cutoff scale or the chiral scale is the basic quantity of the expansion. What is the typical momentum scale of $p+p \to d+e^+ + \nu_e$ in the Sun?

The core temperature of the Sun is believed to be about $T_C = 15.5\times 10^6\ K$, for which the corresponding nucleon momentum is $p \sim \sqrt{2m_p T_C} \sim 1.6$ MeV. At these low energies, the reaction is dominated by the ${}^1S_0 \to d$ transition, involving only the transition operators that carry orbital angular momentum $L=(0,2)$, spin $S=1$ and total angular momentum $J=1$. On the other hand, the momentum carried by the leptons, $\vec{k} = \vec{p}_e + \vec{p}_\nu$, is also very small, $0 \le |\vec{k}| \le 0.78$ MeV. We can therefore safely ignore leptonic terms of $\mathcal{O}(|\vec{k}|^2)$ and work only to $\mathcal{O}(|\vec{k}|)$. It is immediately an obvious expectation that the expansion would converge very fast because the expansion will be based on the ratio of these low momentum of order of 1 MeV to the momentum cutoff or chiral scale, which is bigger than few hundred MeV. We will show that this is actually respected.

Can we understand why some calculations give surprising results? It is also very intriguing that the chiral perturbation theory teaches us the reason why we can make the wrong estimation. To explain it, we rely on the other examples in which the gauge symmetry plays an important role. In the chiral perturbation theory, those symmetries of QCD are respected order by order, such that one should consider all terms of given order to be able to discuss the numerical outcome. In other words, if the full terms of given order are not calculated out, the number can be all arbitrary, not to compare with the data.[10,11] Sometimes the one-loop order diagrams are of the same order in chiral counting with other vertex renormalization and counter terms, one should take all terms into account.

Basic strategy of our calculation is as follows: We rely on the "Weinberg theorem" (see, e.g., Ref. ([9]).), and separate the relevant transition amplitude into the "reducible" and "irreducible" terms. Bound states and resonances involve reducible graphs that are kinematically enhanced due eventually to the almost on-shell nucleons in the graph. To incorporate this feature, we need to sum up a given set of reducible terms to all orders of the chain, a procedure equivalent to solving the Schrödinger equation.[12] The vertices that appear as kernels in the chain are calculated to a specific order in chiral expansion. This process of solving Schrödinger equation can be replaced by using the most well known accurate wavefunctions like the Argonne v_{18} wavefunction.[15] This procedure can be called as "hybrid" calculation [13,10,14]. This process includes the error by one order higher than the chiral order considered by the irreducible graphs, which means that it is consistent to the chiral perturbation. However, it is also possible to respect the chiral order from the beginning and calculate the wavefunction to the chiral order which we want to keep, either by using the Power Divergence Subtraction method or by using the cutoff method.[16,17] For this, reader is invited to refer to the lecture of Rho in this proceeding. Our approach [17] is rather transparent in physical interpretation, and will be explained in short. We will give the comparison between the results with hybrid calculation and the cutoff method.

2 Effective Field Theory: Heavy Baryon Chiral Perturbation Theory

The irreducible terms for the electroweak hadronic current follow the counting rule in which the ratio of the typical momentum scale probed in the process, Q, to the chiral scale or the cutoff scale, Λ_χ is the expansion parameter. We will not denote Λ_χ explicitly in the following unless the ambiguity arises. A Feynman diagram for an A-nucleon processes goes as Q^ν multiplied by a slowly varying function of Q. Now for a Feynman graph with L loops, N_E external fields and C separately connected pieces, the exponent ν is given by

$$\nu = 4 - A - 2C + 2L - N_E + \sum_i \nu_i\ , \quad \text{with}\ \nu_i = d_i + e_i + \frac{n_i}{2} - 2, \tag{1}$$

where n_i is the number of nucleon lines, d_i the number of derivatives or powers of m_π and e_i the number of external fields at the i-th vertex. The effective Lagrangian can be ordered according to ν_i:

$$\mathcal{L}_{eff} = \mathcal{L}_0 + \mathcal{L}_1 + \mathcal{L}_2 + \cdots, \tag{2}$$

where $\mathcal{L}_n$ has vertices with $\nu_i = n$.

2.1 Current: Irreducible parts

The leading order(LO) and the next-to-leading order(NLO) order come from the one-body Gamow-Teller term with $\nu = \nu_0 \equiv -3$ (i.e., $L = 0$, $C = A = 2$, $N_E = 1$ and $\nu_i = 0$), and from the weak magnetism (WM) term with $\nu = \nu_0 + 1$ involving $\nu_i = 1$ vertex. They read

$$\vec{J}^-_{\rm 1B} = \vec{J}^-_{\rm GT} + \vec{J}^-_{\rm WM} + \cdots, \tag{3}$$

with

$$\vec{J}_{\rm GT}^{-} = -g_A \sum_l \tau_l^- \vec{\sigma}_l, \tag{4}$$

$$\vec{J}_{\rm WM}^{-} = \frac{i\mu_V}{2m_N}\vec{k} \times \sum_l \tau_l^- \vec{\sigma}_l. \tag{5}$$

with $g_A \simeq 1.2601(25)$, $\mu_V \equiv \mu_p - \mu_n \simeq 4.7059$ the nucleon isovector magnetic moment and m_N the nucleon mass. To proceed to higher orders, it is convenient to do chiral counting relative to the leading order. Expressed in this way, the leading order one-body current eq.(4) is taken to be $\sim \mathcal{O}(1)$, and the next-to-leading order term eq.(5) is $\sim \mathcal{O}(Q)$. The two-body exchange current can enter at the next-to-next-to-leading order (NNLO), that is at $\mathcal{O}(Q^2)$. Because of the selection rule, however, there is no term of this order contributing to the process. Hence the two-body exchange current at tree order enters at $\mathcal{O}(Q^3)$. Their overall importance relative to the one-body term should be considered diminished here. Calculation of the two-body Gamow-Teller operator can be found in Refs.([18,5]). We will denote this part by the J_{2B}^-, so that the total *weak* current becomes $\vec{J}^- = \vec{J}_{1B}^- + \vec{J}_{2B}^-$.

It is convenient to define the reduced matrix element of the weak current as

$$\mathcal{M} \equiv \mathcal{M}_{\rm 1B} + \mathcal{M}_{\rm 2B} \tag{6}$$

such that

$$\begin{aligned}\frac{\langle\psi_d|\vec{J}^-|\psi_{pp}\rangle}{\sqrt{8\pi}g_A A_S N_{pp}} &= \left(\hat{s}_d - i\frac{\mu_V}{2g_A m_N}\vec{k}\times\hat{s}_d\right)\mathcal{M}_{\rm 1B} + \hat{s}_d\mathcal{M}_{\rm 2B} \\ &= \left(\hat{s}_d - i\frac{\mu_V}{2g_A m_N}\vec{k}\times\hat{s}_d\right)\mathcal{M} + \mathcal{O}\left(Q^4, Q^2|\vec{k}|, |\vec{k}|^2\right), \end{aligned}\tag{7}$$

where $\hat{s}_d$ is the spin polarization vector of the deuteron and the normalization factors A_S and N_{pp} of the deuteron and the pp scattering state, ψ_d and ψ_{pp}, respectively, which will be defined below. Although the reshuffling of terms in the last line introduces a term of $\mathcal{O}(Q^3|\vec{k}|)$ (*i.e.*, the product of the WM term and two-body term), the error made thereby should be negligible.

2.2 *Wavefunction: Reducible part*

Now let us calculate wavefunctions. To treat the potential, we shall integrate out all fields other than the matter fields for nucleon and pion and treat to next-to-leading order (NLO). The potential has then the one pion exchange and the contact interactions used for the nucleon-nucleon channel plus the Coulomb interaction. Our *unregulated* potential has the form

$$V_{N+C}(\vec{q}) = -\tau_1 \cdot \tau_2 \frac{g_A^2}{4f_\pi^2} \frac{\sigma_1 \cdot \mathbf{q}\, \sigma_2 \cdot \mathbf{q}}{\mathbf{q}^2 + m_\pi^2} + \frac{4\pi}{M}\left[C_0 + (C_2\delta^{ij} + D_2\sigma^{ij})q^i q^j\right] + Z_1 Z_2 \frac{\alpha}{\mathbf{q}^2} \quad (8)$$

with

$$\sigma^{ij} = \frac{3}{\sqrt{8}}\left(\frac{\sigma_1^i\sigma_2^j + \sigma_1^j\sigma_2^i}{2} - \frac{\delta^{ij}}{3}\sigma_1 \cdot \sigma_2\right), \quad (9)$$

where $\mathbf{q}$ is the momentum transferred, $g_A \simeq 1.26$ the axial-vector coupling constant, $f_\pi \simeq 93$ MeV the pion decay constant, $M \simeq 940$ MeV the nucleon mass, and $Z_1 Z_2 = 1$ for the pp channel and zero otherwise. The parameters C's and D_2 are defined for each channel. The D_2 term, however, is effective only for the spin-triplet channel. Thus, there are two parameters (C_0 and C_2) for the pp $^1S_0$ channels, and three (C_0, C_2 and D_2) for the $^3S_1$ channel.

The first (nonlocal) term is the pion exchange involving the Goldstone boson and hence completely known. The (local) terms in the square brackets represent the effect inherited from the degrees of freedom that have been integrated out, for instance the heavy mesons whose masses are much higher than the cutoff. It should be noted that when the pion is also integrated out, its effect will appear in the local terms, modifying the constants. This means that even if we drop the first term of (8), we still have part, if not most, of pionic effects in the theory.

We use the Gaussian regulator as

$$V(\vec{r}) = \int \frac{d^3\vec{q}}{(2\pi)^3} S_\Lambda(\vec{q}^2)\, e^{i\vec{q}\cdot\vec{r}}\, V(\vec{q}) \quad (10)$$

with

$$S_\Lambda(\vec{q}^2) = \exp\left(-\frac{\vec{q}^2}{2\Lambda^2}\right) \quad (11)$$

where Λ is the cut-off. The cut-off Λ defines a regime in which the effective theory is applicable. It is not a free parameter. By the rule of (cut-off) effective field theory, physical quantities should not depend sensitively on the exact form of the cut-off nor on the exact value of Λ. We will discuss this shortly.

Inserting eqs.(8, 11) into eq.(10), we obtain

$$V(\mathbf{r}) = \alpha_\pi \frac{\tau_1 \cdot \tau_2}{3}\left[\sigma_1 \cdot \sigma_2\, v_\Lambda(r, m_\pi) + S_{12}(\hat{r})\, t_\Lambda(r, m_\pi)\right]$$

$$+\frac{4\pi}{M}\left[C_0' - C_2\nabla_{\mathbf{r}}^2 - \frac{S_{12}(\hat{r})}{\sqrt{8}}D_2 r\frac{\partial}{\partial r}\frac{1}{r}\frac{\partial}{\partial r}\right]\delta_\Lambda^3(\mathbf{r})$$
$$+ Z_1 Z_2 \alpha\, v_\Lambda(r,0) \tag{12}$$

with $S_{12}(\hat{r}) \equiv 3\sigma_1\cdot\hat{r}\,\sigma_2\cdot\hat{r} - \sigma_1\cdot\sigma_2$,

$$\delta_\Lambda^3(\mathbf{r}) = \int\frac{d^3\mathbf{q}}{(2\pi)^3}\,\mathrm{e}^{i\mathbf{q}\cdot\mathbf{r}}\,S_\Lambda(\mathbf{q}^2), \tag{13}$$

$$v_\Lambda(r,m) = \frac{4\pi}{S_\Lambda(-m^2)}\int\frac{d^3\mathbf{q}}{(2\pi)^3}\,\mathrm{e}^{i\mathbf{q}\cdot\mathbf{r}}\,\frac{S_\Lambda(\mathbf{q})}{\mathbf{q}^2+m^2}, \tag{14}$$

$$t_\Lambda(r,m) = \frac{r}{m^2}\frac{\partial}{\partial r}\frac{1}{r}\frac{\partial}{\partial r}v_\Lambda(r,m) \tag{15}$$

and

$$\alpha_\pi \equiv S_\Lambda(-m_\pi^2)\frac{g_A^2 m_\pi^2}{16\pi f_\pi^2}. \tag{16}$$

The factor $S_\Lambda^{-1}(-m_\pi^2)$ in the definition of $v_\Lambda(r,m)$ is introduced so that $v_\Lambda(r,m)$ and $t_\Lambda(r,m)$ become Yukawa functions in the large-r limit:

$$\lim_{\Lambda r\gg 1} v_\Lambda(r,m) = \frac{\mathrm{e}^{-mr}}{r},$$
$$\lim_{\Lambda r\gg 1} t_\Lambda(r,m) = \left(1+\frac{3}{mr}+\frac{3}{(mr)^2}\right)\frac{\mathrm{e}^{-mr}}{r}. \tag{17}$$

In eq.(12), we have absorbed the $\delta_\Lambda^3(\mathbf{r})$ term appearing in the one-pion-exchange potential into the nucleon contact interactions by redefining C_0 as

$$C_0' \equiv C_0 - \frac{M\alpha_\pi}{3m_\pi^2 S_\Lambda(-m_\pi^2)}\tau_1\cdot\tau_2\,\sigma_1\cdot\sigma_2. \tag{18}$$

Hereafter we will omit the prime on C_0. As for the one-pion-exchange potential, we set $\alpha_\pi = 0.075$ following[15,19]. Let us write the potential for each channel of interest.

- **The np and pp $^1S_0$ channel**: Here tensor interactions are ineffective and so the potential reduces, with eq.(22), to

$$V(\mathbf{r};{}^1S_0) = -\alpha_\pi\, v_\Lambda(r,m_\pi) + \frac{4\pi}{M}\left[C_0^{1S0} - \Lambda^2(\Lambda^2 r^2 - 3)\,C_2^{1S0}\right]\delta_\Lambda^3(\mathbf{r})$$
$$+ Z_1 Z_2\alpha\, v_\Lambda(r,0). \tag{19}$$

The coefficients $C_{0,2}^{1S0}$ differ in the pp channel from those in the np channel due to the charge-independence breaking interactions.

• **The deuteron $^3S_1 - {}^3D_1$ coupled channel**: The potential is of the form

$$V(\mathbf{r}; {}^3S_1 - {}^3D_1) = V_C^d(r) + S_{12}(\hat{r})\, V_T^d(r) \tag{20}$$

where, due to eq.(22),

$$V_C^d(r) = -\alpha_\pi\, v_\Lambda(r, m_\pi) + \frac{4\pi}{M}\left[C_0^{3S1} - \Lambda^2(\Lambda^2 r^2 - 3)\, C_2^{3S1}\right] \delta_\Lambda^3(\mathbf{r}),$$
$$V_T^d(r) = -\alpha_\pi\, t_\Lambda(r, m_\pi) - \frac{4\pi}{M}\frac{D_2^{3S1}}{\sqrt{8}}\, \Lambda^4 r^2\, \delta_\Lambda^3(\mathbf{r}). \tag{21}$$

With the regulator (11), it is a simple matter to work out the explicit forms of the quantities we need:

$$\nabla_{\mathbf{r}}^2 \delta_\Lambda^3(\mathbf{r}) = \Lambda^2(\Lambda^2 r^2 - 3)\, \delta_\Lambda^3(\mathbf{r}), \quad r\frac{\partial}{\partial r}\frac{1}{r}\frac{\partial}{\partial r}\delta_\Lambda^3(\mathbf{r}) = \Lambda^4 r^2\, \delta_\Lambda^3(\mathbf{r}) \tag{22}$$

and

$$\delta_\Lambda^3(\mathbf{r}) = \frac{\Lambda^3 \exp\left[-\frac{\Lambda^2 r^2}{2}\right]}{(2\pi)^{\frac{3}{2}}}, \tag{23}$$
$$v_\Lambda(r, m) = \frac{1}{2r}\left[\mathrm{e}^{-mr}\,\mathrm{erfc}\left(\frac{-\Lambda r + \frac{m}{\Lambda}}{\sqrt{2}}\right) - (r \to -r)\right], \tag{24}$$
$$v_\Lambda(r, 0) = \frac{1}{r}\,\mathrm{erf}\left(\frac{\Lambda r}{\sqrt{2}}\right) \tag{25}$$

where $\mathrm{erf}(x)$ and $\mathrm{erfc}(x) \equiv 1 - \mathrm{erf}(x)$ are the error functions,

$$\mathrm{erf}(x) \equiv \frac{2}{\sqrt{\pi}}\int_0^x dt\, \mathrm{e}^{-t^2}. \tag{26}$$

Finally the radial functions, $u_0(r)$ for $^1S_0$ and $u_d(r)$ and $w_d(r)$ for $^3S_1 - {}^3D_1$, are obtained by solving the Schrödinger equations

$$u_0''(r) + M\left[E - V(r; {}^1S_0)\right] u_0(r) = 0, \tag{27}$$

and

$$u_d''(r) + M\left[E - V_C^d(r)\right] u_d(r) = \sqrt{8} M V_T^d(r) w_d(r),$$
$$w_d''(r) - \frac{6}{r^2} w_d(r) + M\left[E - V_C^d(r) + 2V_T^d(r)\right] w_d(r) = \sqrt{8} M V_T^d(r) u_d(r), \tag{28}$$

with the boundary condition

$$u_0(0) = u_d(0) = w_d(0) = 0. \tag{29}$$

It follows from eqs.(3) and (7) that the single-particle (reduced) matrix element simplifies to

$$\mathcal{M}_{1\mathrm{B}} = \int_0^\infty dr\, u_d(r) u_{pp}(r). \tag{30}$$

2.3 Renormalization

For a finite cutoff Λ, only finite quantities enter. However renormalization is required to render the theory consistent (see Ref.([20]) for details). In particular, certain cutoff-dependent terms have to be absorbed into the constants before determining them from empirical data. We proceed as follows. For a given cutoff and for each channel, we determine the constants C's and D's in (8) by relating them to the low-energy scattering parameters appearing in the effective-range formula.

• **The** pp $^1S_0$ **channel**: In this channel, the infinite-ranged Coulomb potential must be taken into account. We define the phase shift δ^C and consequently the low-energy scattering parameters a^C and r^C. The expressions obtained in Ref.([5]) for the low-energy scattering parameters are

$$a^C = \frac{1}{m_p \alpha} \left. \frac{u'_{\mathrm{C+N}}(r)\phi_2(r) - u_{\mathrm{C+N}}(r)\phi'_2(r)}{u'_{\mathrm{C+N}}(r)\phi_1(r) - u_{\mathrm{C+N}}(r)\phi'_1(r)} \right|_{r \geq R}, \tag{31}$$

$$r^C = 2\int_0^\infty dr\, \left[\phi_C^2(r) - u_{\mathrm{C+N}}^2(r)\right] \tag{32}$$

with

$$\phi_C(r) = \phi_1(r) - \frac{\phi_2(r)}{m_p \alpha\, a^C}, \tag{33}$$

where $m_p \simeq M$ is the proton mass, $u_{\mathrm{C+N}}(r)$ is the radial wavefunction that approaches the function $\phi_C(r)$ asymptotically, and $\phi_1(r)$ and $\phi_2(r)$ are the radial wavefunctions calculated with only the Coulomb interaction at zero energy. In fixing C_0^{pp} and C_2^{pp}, we shall use the following experimental values (extracted from the Nijmegen multi-energy analysis [19]) of a^C and r^C:

$$a^C_{\mathrm{exp}} = -7.8196(26)\ \mathrm{fm}, \quad r^C_{\mathrm{exp}} = 2.790(14)\ \mathrm{fm}. \tag{34}$$

• **The deuteron $^3S_1-^3D_1$ channel**: In this channel, we have the choice of renormalizing the theory either at zero energy or at the deuteron pole position $E = -B_d$, where B_d is the deuteron binding energy. Here we shall adopt the latter. The coefficient D_2^d will be determined by the experimentally well-known deuteron D/S ratio η_d. Near the deuteron pole position, the effective-range formula has the form

$$p \cot \delta_d = -\gamma + \frac{1}{2}\rho_d(p^2+\gamma^2) + \mathcal{O}\left[(p^2+\gamma^2)^2\right] \tag{35}$$

where δ_d is the S-wave eigenphase and $\gamma \equiv \sqrt{MB_d}$ is the deuteron wave number. Thus the quantities B_d, ρ_d and η_d are inputs to determine the three constants C_0^d, C_2^d and D_2^d. Since by construction our potential is energy-independent, the wavefunction normalization factor A_s is not an independent quantity. It is related to the input parameters by

$$A_s^2 = \frac{2\gamma}{(1+\eta_d^2)(1-\gamma\rho_d)}. \tag{36}$$

In our numerical calculation, we shall use the following experimental values for B_d, A_s and η_d:

$$B_d^{\rm exp} = 2.224575(9)\ {\rm MeV}, \quad A_s^{\rm exp} = 0.8846(8)\ {\rm fm}^{\frac{1}{2}}, \quad \eta_d^{\rm exp} = 0.0256(4). \tag{37}$$

From eq.(36) we get $\rho_d = 1.7635(46)$ fm.

3 Test of wavefunctions and hybrid method

The renormalization process permits us to determine in fact the constants of C_0, C_2 and D_2 depending on the channels at the zero momentum limits. With those constants, we can calculate the phase shift of *pp* scattering at relatively high momentum and some properties of deuteron without any free parameter. We compute the deuteron static properties such as the charge radius r_d, the quadrupole moment Q_d and the magnetic moment μ_d, in Tab.1. The precise value of the cutoff is of course not determined by the theory but the consistency of the scheme requires that the results should be insensitive to its precise value provided that it is picked judiciously: e.g., around the pion mass if the pion is integrated out and around the two-pion mass if the pion is present.

We can remark that the quality of wavefunctions are extremely good. This result is from the first principle of effective field theory. This can be compared with the *hybrid* method[5], in which the wavefunctions are taken from the well-known phenomenological models. The present results justify our former

Table 1: The next-to-leading order (NLO) results *with* the pion field. It is the same as the Table 2 of Rho's lecture, supplemented with values of coefficients of potential.

Λ (MeV)	200	250	300	350	400	500
C_0^{pp} (fm)	−1.30	−1.25	−1.18	−0.88	0.23	16.95
C_2^{pp}(fm^3)	0.11	0.27	0.35	0.42	0.53	1.21
C_0^d (fm)	−1.32	−1.03	−0.77	−0.54	−0.25	1.22
C_2^d(fm^3)	−0.24	0.06	0.14	0.18	0.20	0.28
D_2^d(fm^3)	0.90	0.46	0.24	0.12	0.04	−0.05
r_d (fm)	1.963	1.965	1.966	1.967	1.968	1.968
Q_d (fm^2)	0.261	0.268	0.272	0.273	0.274	0.274
P_D (%)	3.16	4.11	4.77	5.16	5.35	5.39
μ_d	0.862	0.856	0.853	0.850	0.849	0.849
$\mathcal{M}_{\rm GT}$ (fm)	4.884	4.874	4.867	4.862	4.859	4.854

approaches. Before discussing the higher order corrections, we are in position to discuss the limit of the effective theory. It is quite interesting to consider the potential with pion field integrated out. Its effect will be absorbed into C_0, C_2 and D_2. Therefore the resulting potential will be the same as eq.(8) except the pion exchange term. We could repeat the renormalization process and calculate the wavefunctions without pion degrees of freedom. This will be, in fact, the type of potential that the traditional nuclear physics has been employing except that we here have the chiral constraints incorporated. The results is given in Tab.2. Quality of agreements is very satisfactory but some properties of scattering phase shift at energies approaching to the cutoff value show the limit of the model[17,7].

4 Result and Discussion

For comparison with Ref.([2]), we focus on the low-energy cross-section factor $S_{pp}(E)$[21],

$$S_{pp}(E) \equiv \sigma(E) E e^{2\pi\eta} \tag{38}$$

where $\sigma(E)$ is the cross section and $\eta = \frac{m_p \alpha}{2p}$. Explicitly, it is of the form

$$S_{pp}(E) = (1+\delta_{2\rm B})^2 \frac{6}{\pi}\, m_p \alpha\, G_V^2 g_A^2 \frac{\Lambda^2}{\gamma^3}\, m_e^5 f(E_0+E), \tag{39}$$

where $f(E_0 + E)$ with $E_0 \equiv 2m_p - m_d$ is the phase volume defined in Ref.([21]) with however the WM term contribution as well as the deuteron recoil taken

Table 2: The next-to-leading order (NLO) results *without* the pion field. This is the same as the Table 1 of Rho's lecture, supplemented with values of coefficients of potential.

Λ (MeV)	140	150	175	200	225	250
C_0^{pp} (fm)	−1.97	−1.97	−1.99	−2.00	−1.96	−1.83
C_2^{pp}(fm^3)	−0.19	0.07	0.46	0.67	0.79	0.90
C_0^d (fm)	−1.88	−1.78	−1.39	−0.74	0.35	2.30
C_2^d(fm^3)	−1.96	−1.34	−0.50	−0.15	0.01	0.08
D_2^d(fm^3)	3.65	3.34	2.82	2.54	2.45	2.54
r_d (fm)	1.973	1.972	1.974	1.978	1.983	1.987
Q_d (fm^2)	0.259	0.268	0.287	0.302	0.312	0.319
P_D (%)	2.32	2.83	4.34	6.14	8.09	9.90
μ_d	0.867	0.864	0.855	0.845	0.834	0.823
$\mathcal{M}_{\rm GT}$ (fm)	4.887	4.881	4.864	4.846	4.827	4.810

into account. By plugging in the numbers, we arrive at

$$S_{pp}^{\chi PT}(0) = 4.04 \times 10^{-25}\,\text{MeV-barn} \,. \tag{40}$$

This is to be compared with that of KB

$$S_{pp}^{\rm KB}(0) = 3.89 \times 10^{-25}\,\text{MeV-barn}. \tag{41}$$

This result justifies a cut-off effective field theory of low-energy QCD as in the case of the np capture[17], and supports the canonical value of Bahcall et al.[2], and does not support the "relativistic field theory model" result of Ivanov et al.[6].

We have seen that the effective field theory can be linked with the traditional method of nuclear physics, producing the better level of precisions due to the systematic treatment. It also makes easy to understand the underlying physical picture in terms of the physical variables employed and integrated-out.

Acknowledgments

This work is based on the works with K. Kubodera, M. Rho, and T.-S. Park. This work was supported in part by sabbatical grant of Seoul National University, Korea and by the Programa de profesores invitados, Universitat de Valencia, Spain. Author is very grateful to the discussions and hospitality of Professor V. Vento during his stay in Valencia.

1. H. Bethe, & C. L. Critchfield, Phys. Rev.**54** (1938) 248
2. M. Kamionkowski and J.N. Bahcall, Astrophys. J. **420** (1994) 884.
3. J.N. Bahcall, in *Unsolved problems in astrophysics*, eds. by J.N. Bahcall and J.P. Ostriker (Princeton University Press, Princeton, 1997).
4. For a recent review, see E.G. Adelberger et al, Rev. Mod. Phys. **70** (1998) 1265.
5. T.-S. Park, K. Kubodera, D.-P. Min and M. Rho, Astrophys. J. **507** (1998) 443.
6. A. N. Ivanov, N.I. Troitskaya, M. Faber, & H. Oberhummer, Nucl. Phys. **A617** (1997) 414; Erratum-ibid.**A618**(1997) 509; Erratum-ibid.**A625** (1997) 896A; Oberhummer, H., Ivanov, A. N., Troitskaya, N. I., & Faber, M. 1997, preprint [astro-ph/9705119]
7. M. Rho, this proceeding.
8. S. Weinberg, Phys. Lett. **B 251** (1990) 288; Nucl. Phys. **B 363** (1991) 3; Phys. Lett. **B 295** (1992) 114.
9. See, e.g., S. Weinberg, *The Quantum Theory of Fields II* (Cambridge Press, 1996); Nature **386** (1997) 234.
10. T-S. Park, D.-P. Min and M . Rho, Phys. Repts. **233** (1993) 341.
11. Shung-ichi Ando, this proceeding.
12. C. Ordonez, L. Ray and U. van Kolck, Phys. Rev. Lett. **72** (1994) 1982; Phys. Rev. **C 53** (1996) 2086.
13. T.-S. Park, D.-P. Min and M. Rho, Phys. Rev. Lett. **74** (1995) 4153; Nucl. Phys. **A 596** (1996) 515.
14. T.-S. Park, K. Kubodera, D.-P. Min and M. Rho, Phys. Rev. **C 58** (1998) R637.
15. R.B. Wiringa, V.G.J. Stoks and R. Schiavilla, Phys. Rev. **C 51** (1995) 38 and references therein.
16. D.B. Kaplan, M.J. Savage and M.B. Wise, Phys. Lett. **B 424** (1998) 390; Nucl. Phys. **B 534** (1998) 329; nucl-th/9804032.
17. T.-S. Park, K. Kubodera, D.-P. Min and M. Rho, "The power of effective field theories for nuclei," nucl-th/9807054
18. T.-S. Park, H. Jung and D.-P. Min, Phys. Lett. **B 409** (1997) 26.
19. V.G.J. Stoks, R.A.M. Klomp, M.C.M. Rentmeester and J.J. de Swart, Phys. Rev. **C 48** (1993) 792.
20. J.F. Donoghue and B.R. Holstein, hep-ph/9803312; J.F. Donoghue, B.R. Holstein and B. Borasoy, hep-ph/9804281.
21. J. N. Bahcall, Nucl. Phys. **75** (1966) 10.

SUPERCONDUCTIVITY IN QUARK MATTER — PHASE DIAGRAM OF QUARK MATTER —

M. IWASAKI
Department of Material Science, Kochi University,
Akebono-Cho 2-5-1, Kochi 780-8520 Japan
E-mail: miwasaki@cc.kochi-u.ac.jp

Color-superconductivity in quark matter is investigated by mean field (BCS) theory. Using the Nambu formalism we derive gap equation at finite temperature. It is shown that the superconductivity disappear at high-density due to the asymptotic freedom and also at high-temperature due to the thermal motion; their critical values are $\rho_c \sim 20\text{fm}^{-3}$ and $T_c \sim 30\text{GeV}$

1 Introduction

It has been recognized that superconductivity is rather popular phenomenon in many Fermi systems at low temperature. Quarks in hadrons are also Fermions so that the many-quark system (quark matter) may be superconducting state in the ground state. In order to show the existence of an attractive force between two quarks, let us consider the one-gluon exchange force,

$$V_{\text{OGE}} = \frac{\alpha_s}{4}(\vec{\lambda_1} \cdot \vec{\lambda_2})(\gamma_1 \cdot \ gamma_2)\frac{1}{r}. \tag{1}$$

The s-wave state seems most favorable for binding of two quarks because of the Coulomb-type force. The expectation value of the color matrix $\langle \vec{\lambda_1} \cdot \vec{\lambda_2} \rangle$ is 4/3 in the color-symmetric state and (-8/3) in the color-antisymmetric state so that we should take the latter case. Therefore Fermi statistic leads to the spin-parallel (symmetric) state. The expectation value of the spin matrix $\langle \gamma_1 \cdot \gamma_2 \rangle$ is

$$\langle \gamma_1 \cdot \gamma_2 \rangle = 1 - \langle \vec{\alpha}_1 \rangle_{\mathbf{k}} \cdot \langle \vec{\alpha}_2 \rangle_{-\mathbf{k}} \tag{2}$$

Noting that the Dirac matrix $\vec{\alpha}$ means a velocity operator of the quark, the last term in the right-hand side becomes negative. Thus we get $\langle V_{\text{OGE}} \rangle < 0$ and it has been shown that there appears an attractive force in the s-wave, color-triplet and spin-parallel state.

Such color-superconductivity of quark matter was proposed by Bailin and Love [1] long time ago. Although their treatment was Ginzburg-Landau (phenomenological) theory, Iwado and the present author [2] recently constructed a BCS (microscopic) theory for the quark matter and discussed its stability.

It is the purpose of this paper to discuss the color-superconductivity of quark matter by using a mean field (BCS) approximation developped by Nambu.[4]

Then we extend the theory to finite temperature and study a phase diagram of the quark matter. To this end we take the QCD-like theory together with the Higashijima-Miransky approximation [3] and derive the gap equation. Thc numerical calculations are done at various density and temperature so that we will obtain a phase diagram of the quark matter. It should be noted that a lack of confinement in our model restricts ourselves to the quark matter with larger density than $\rho \sim 1\mathrm{fm}^{-3}$.

2 BCS theory by Nambu formalism

We start with the QCD Lagrangian with single flavor:

$$\mathcal{L} = \psi^+(i\partial_t - H_0 + \mu)\psi + \mathcal{L}_{int}, \tag{3}$$

where H_0 is the Dirac Hamiltonian of quarks and μ chemical potential. We will develop a generalized mean field (BCS) theory according to Nambu. [4] Let us introduce a mean field Lagrangian,

$$\mathcal{L}_{mf} = \psi^+(i\partial_t - H_0 + \mu)\psi + \Delta\psi^+\psi^+ + \Delta^+\psi\psi, \tag{4}$$

where Δ is a pairing field which is determined later self-consistently. As was shown in the previous section, two quarks have an attractive force in the s-wave, color-triplet and spin-parallel state. This leads to the following mean field:

$$\Delta_{\alpha,\beta} = \delta_{s,-s'}\hat{\epsilon}_{c,c'}\Delta(k) \equiv \delta_{s,-s'}\hat{\Delta}(k), \tag{5}$$

where the single particle state with momentum $\mathbf{k}$ ($-\mathbf{k}$), helicity s (s') and color c (c') is denoted by α (β) and the skew-symmetric tensor $\hat{\epsilon}_{c,c'}$ is defined by $\hat{\epsilon}_{c,c'} = -\hat{\epsilon}_{c',c}$ and $\hat{\epsilon}_{1,2} = \hat{\epsilon}_{2,3} = \hat{\epsilon}_{3,1} = 1$. This expression represents our Cooper pair is in the s-wave, color-triplet and spin-parallel state.

Now we use the Nambu formalism, [4] which seems to be most suitable for our purpose. Let us introduce a 6-component spinor,

$$\Psi(k) \equiv \begin{pmatrix} c_{k,s} \\ c^+_{-k,-s} \end{pmatrix}. \tag{6}$$

where the color suffixes are omitted. By using this new spinor, the mean field Lagrangian Eq. (4) is rewritten as

$$\int d^4x \mathcal{L}_{mf} = \int \frac{d^4k}{(2\pi)^4} \sum_s \Psi^+(k) \begin{pmatrix} k_0 - \varepsilon_k & \hat{\Delta}_k \\ \hat{\Delta}^+_k & k_0 + \varepsilon_k \end{pmatrix} \Psi(k), \tag{7}$$

where $\varepsilon_k = \sqrt{k^2+m^2} - \mu = e_k - \mu$ is the single particle energy. The propagator for the new spinor is defined by

$$G(k) = \begin{pmatrix} k_0 - \varepsilon_k & \hat{\Delta}_k \\ \hat{\Delta}_k^+ & k_0 + \varepsilon_k \end{pmatrix}^{-1} = \begin{pmatrix} k_0 + \varepsilon_k & -\hat{\Delta}_k \\ \hat{\Delta}_k^+ & k_0 - \varepsilon_k \end{pmatrix} (k_0^2 - \varepsilon_k^2 - \hat{\Delta}_k^+ \hat{\Delta}_k)^{-1}. \quad (8)$$

The non-diagonal element, which is called an anomalous propagator,

$$F(k) \equiv G(k)_{1,2} = -\hat{\Delta}_k (k_0^2 - \varepsilon_k^2 - \hat{\Delta}_k^+ \hat{\Delta}_k)^{-1}, \quad (9)$$

is transformed into

$$F(k) = -\hat{\Delta}_k (k_0^2 - \varepsilon_k^2 - 3\Delta_k^2)^{-1} = \frac{-\hat{\Delta}_k}{k_0^2 - E_k^2}, \quad (10)$$

with the help of the identity, $\hat{\epsilon}^3 = -3\hat{\epsilon}$. It should be noted that the $E_k \equiv \sqrt{\varepsilon_k^2 + 3\Delta_k^2}$ means the quasi-particle energy of the mean field Hamiltonian.

3 Gap Equation

We turn now to the problem of fixing the mean field or the energy gap Δ_k. To this end, we take the residual interaction defined by

$$\mathcal{L}_{int}^{'} = \mathcal{L} - \mathcal{L}_{mf}. \quad (11)$$

The self-consistency in the Nambu formalism requires that the quasi-particle energy E_p is unaffected by the residual interaction which means that the self-energy $\Sigma(p)$ of the quasi-particle should vanish,

$$\Sigma(p) = 0. \quad (12)$$

We can use the Feynman rules to calculate the self-energy. Instead to calculate it exactly, we are satisfied to calculate it to first order in $\mathcal{L}_{int}^{'}$ but the coupling constant is replaced by the running one (Higashijima-Miransky approximation).

$$\alpha_s(q^2) = \frac{4\pi}{9\log{(q^2+p_c^2)/\Lambda^2}} \quad (13)$$

where q^2 should be larger one of k^2 and p^2 and Λ is the scale parameter of QCD. The another parameter p_c has been introduced to make the coupling constant finite in the infrared region. This fact leads to a lack of confinement of this model. Then we obtain the self-consistent (gap) equation, that is,

$$\hat{\Delta}(p) = \int \frac{d^4k}{(2\pi)^4} \vec{\Gamma}^\mu F(k) \vec{\Gamma}^\nu D_{\mu\nu}(p-k) \quad (14)$$

where the quark-gluon interaction is represented by

$$\mathcal{L}_{int} = \bar{\psi}\vec{\Gamma}^{\mu}\psi\vec{A}_{\mu}, \tag{15}$$

where $D_{\mu\nu}$ is a gluon propagator.

Here two assumptions are made: (1) static approximation, $\Delta(q^2) \to \Delta(\mathbf{q}^2)$ and (2) the quark-gluon interaction does not couple with the negative energy states. Then the gap equation (14) is rewritten as

$$\Delta(p) = \frac{4\pi}{3}\int \frac{d^3k}{(2\pi)^3}\alpha_s(q)\frac{(e_k e_p + \mathbf{k}\cdot\mathbf{p})\Delta(k)}{e_k e_p|\mathbf{k}-\mathbf{p}|^2 E_k}. \tag{16}$$

This equation is an integral equation to determine the energy gap or the mean field.

Next we extend the gap equation to finite temperature T. This can be carried out as usual manner [5] in our formalism: the Green's function $G(\mathbf{k}, k_0)$ is replaced by the temperature Green's function $G(\mathbf{k}, i\omega_n)$ where ω_n denotes a discrete variable $(2n+1)\pi T$. The gap equation at finite temperature is represented by

$$\Delta(p) = \int \frac{d^3k}{(2\pi)^3} iT\sum_n \vec{\Gamma}F(\mathbf{k}, i\omega_n)\vec{\Gamma}D_{\mu\nu}(p-k). \tag{17}$$

Following the standard procedure, we transform the summation in the integrand into the integration of complex variable ω by using the following formula,

$$\sum_n F(i\omega_n) = \frac{i}{2\pi T}\oint_C \frac{F(\omega)}{e^{\omega/T}+1}d\omega. \tag{18}$$

Then using Cauchy's theorem, we obtain the final result:

$$\Delta(p) = \frac{4\pi}{3}\int \frac{d^3k}{(2\pi)^3}\alpha_s(q)\frac{(e_k e_p + \mathbf{k}\cdot\mathbf{p})\Delta(k)}{e_k e_p|\mathbf{k}-\mathbf{p}|^2 E_k}(1-2f_k), \tag{19}$$

where $f_k = (e^{\varepsilon_k/T}+1)^{-1}$ is the usual Fermi distribution function for the quasi-particles. This equation is the same as Eq. (16) except for the appearance of the last factor, which means that the thermal motion is against the superconductivity.

4 Numerical Results

Our task is to calculate the gap equation at various densities and temperatures and investigate the superconducting region in the (ρ, T) plane. Before the computation, one modification should be done; we take into account an effective mass in the gluon propagator because gluons are surrounded by the medium. The color-electric and magnetic masses are given by [6]

$$M_e^2 = g^2(\frac{N_c}{3} + \frac{N_f}{6})T^2 + \frac{\mu^2}{2\pi^2} \qquad M_m^2 = 0 \tag{20}$$

respectively. The use is made of $N_c = N_f = 3$ and μ is taken as the Fermi energy. As for the other parameters, the following values are used:

$$m = 10\text{MeV} \quad \Lambda = 400\text{MeV} \quad p_c^2 = 1.5\Lambda^2 \tag{21}$$

These are the same values as those taken by Higashijima.

The numerical results are following:

1) The gap energy $\Delta(k)$ has a peak around the Fermi momentum and the order of magnitude is about 10 MeV at $\rho = 1\text{fm}^{-3}$ and $T = 0$.

2) When the density increases ($T = 0$), the energy gap at the Fermi momentum decreases very slowly and vanishes at about $\rho \sim 20\text{fm}^{-3}$. This behavior is due to the running coupling constant; the coupling constant at the Fermi momentum decreases with increasing density by the asymptotic freedom. (Note that $k_F \propto \rho^(1/3)$)

3) When the temperature increases ($\rho = 1\text{fm}^{-3}$), the energy gap at the Fermi momentum decreases and vanishes at about $T \sim 30\text{MeV}$ due to the thermal motion. The random motion of heat causes this destruction of Cooper pairs.

Thus we have found the superconducting region in the $(T - \rho)$ plane; the region is bounded by the conditions of $\rho < 20\text{fm}^{-3}$ and $T < 30\text{MeV}$. Outside the boundary, the quark matter is in the normal phase (Fermi gas).

5 Discussions and Summary

In this section we consider the signals of the color-superconductivity assuming its existence.[7] It is most probable that quark matter would exist in the core of a compact star. For example the core of a neutron star may be composed of u-quark matter and d-quark one. If they are in super phase, one may expect that Josephson current flows between them. The tunneling current is induced by the weak interaction,

$$d \longleftrightarrow u + e^- + \bar{\nu_e}. \tag{22}$$

This super current must give much effects on transport phenomena, such as the cooling, explosion and radiation of stars.

There is another interesting characteristic of our BCS theory; the present Cooper pair is spin-parallel. This implies that a Cooper pair is more stable under a magnetic field than a spin-singlet pair. Since a strong magnetic field might exist inside neutron stars, this property may play an important role in neutron star physics in the future.

In conclusion, we have discussed the color-superconductivity of quark matter. The BCS theory has been extended to finite temperature and we have obtained the gap equation with the running coupling constant. From the numerical calculation it is found that the superconductivity is realized in the finite region of the $(\rho - T)$ plane and disappear at high-density and/or high-temperature due to the asymptotic freedom and thermal motion respectively.

References

1. Bailin and Love, *Nucl. Phys.* B **205**, 325 (1982).
2. M.Iwasaki and T.Iwado, *Phys. Lett.* B **350**, 163 (1995).
3. K.Higashijima, *Phys. Rev.* D **29**, 1228 (1984).
4. Y,Nambu, *Phys. Rev.* **117**, 648 (1960).
5. J.R.Schriefer, *Theory of Superconductivity*, (Addison Wesley, Singapore, 1988).
6. J.I.Kapusta, *Finite-Temperature Field Theory*, (Cambridge University Press, Cambridge, 1989).
7. M.Alford, K.Rajagopal and F.Wilczek, *Phys. Lett.* B **422**, 247 (1998).

A NEW RENORMALIZATION GROUP FOR HAMILTONIAN FIELD THEORY

ROBERT J. PERRY AND SÉRGIO SZPIGEL
Physics Department, The Ohio State University, Columbus, OH 43210, USA
E-mail: perry.6@osu.edu, szpigel@pacific.mps.ohio-state.edu

The Schrödinger equation with a two-dimensional δ-function potential is a simple example of an asymptotically free theory that undergoes dimensional transmutation. Renormalization requires the introduction of a mass scale, which can be lowered perturbatively until an infrared cutoff produced by non-perturbative effects such as bound state formation is encountered. We outline the effective field theory and similarity renormalization group techniques for producing renormalized cutoff hamiltonians, and illustrate the control of logarithmic and inverse-power-law errors both techniques provide.

1 Motivation and Outline

This article follows a more complete discussion in a longer article being prepared for publication.[1] Elsewhere in these Proceedings the possibility of deriving a valid constituent approximation in light-front QCD is discussed.[2] A simple procedure that produces a renormalized light-front QCD hamiltonian starts with the computation of a cutoff hamiltonian using a similarity renormalization group (SRG) [3,4,5,6] and coupling coherence.[7,8,9] In this article we illustrate this step using the two-dimensional δ-function in a one-body Schrödinger equation.[10] We also discuss a related approach, effective field theory (EFT).[11,12]

Locality is the source of perturbative divergences in field theory, and it produces similar effects in simple one-body quantum mechanics problems. Locality leads us to consider potentials consisting of δ-functions and their derivatives. These potentials produce divergences that require the introduction of a mass scale. Changing this mass scale should be equivalent to changing a dimensionless coupling constant (*i.e.*, dimensional transmutation) for a scale-invariant local hamiltonian. The two-dimensional δ-function problem is solved in the second section, providing the *data* for our error analysis of EFT and SRG calculations.

Problems requiring renormalization result from an arbitrarily large number of degrees of freedom being coupled. This is clear in momentum space, where the off-diagonal matrix elements of a δ-function are constants. Divergences result from an infinite number of scales being directly coupled, and renormalization requires a cutoff λ to regulate the divergences. Physical results such as binding energies should be independent of λ, so there should exist a line of cutoff hamiltonians H_λ that all produce the same physical results. A renormalization group (RG) is built from transformations, $\mathcal{T}(\lambda, \lambda')$, that change the cutoff. $\mathcal{T}$ connects hamiltonians at different scales and in the SRG $\mathcal{T}$ is typically a unitary transformation.

In EFT $\mathcal{T}$ exists in principle, but H_λ is determined at each scale by constraining the couplings in front of a finite number of local operators to fit data. A simple scale transformation leads to the classification of these operators as relevant (powers of the cutoff appear), marginal (the cutoff appears only in logarithms that are absorbed in the running marginal coupling), and irrelevant (inverse powers of the cutoff appear). The same operators arise in the SRG, where their strengths are approximated by expansions in powers of a running marginal coupling. For the simple two-dimensional δ-function there are no relevant operators and only one marginal operator. There are always an infinite number of irrelevant operators, and these become increasingly important as the cutoff approaches a non-perturbative energy scale in the problem.

The easiest way to clarify these points is by example. We show a complete solution of the two-dimensional δ-function problem, introducing the K-matrix to compute scattering phase shifts. There is one bound state, and its binding energy sets the scale for non-perturbative physics. The EFT hamiltonian is produced by truncating the series of irrelevant operators, and then fitting a finite number of remaining couplings to an equal number of low-energy scattering constraints. This leads to power-law suppression of errors in the predicted binding energy, with the expansion clearly breaking down as the cutoff approaches the binding energy. The SRG hamiltonian is produced by solving a differential equation for H_λ subject to the constraint that all irrelevant couplings are analytic functions of the single marginal coupling.

2 The two-dimensional δ-function potential

Consider the Schrödinger equation in two dimensions with an attractive Dirac δ-function potential:[10]

$$-\nabla_{\mathbf{r}}^2 \Psi(\mathbf{r}) - \alpha_0 \, \delta^{(2)}(\mathbf{r}) \, \Psi(\mathbf{r}) = E \, \Psi(\mathbf{r}) \, . \tag{1}$$

The coupling α_0 is dimensionless, so the hamiltonian is scale invariant (i.e., there is no intrinsic length or energy scale). It is relatively straightforward to solve this equation in position space by introducing a convenient distribution function to regulate the δ-function, but we will work in momentum space to stay close to the EFT and SRG calculations.

The Schrödinger equation in momentum space is,

$$p^2\ \Phi(\mathbf{p}) - \frac{\alpha_0}{(2\pi)^2} \int d^2q\ \Phi(\mathbf{q}) = E\ \Phi(\mathbf{p})\ . \tag{2}$$

As a consequence of scale invariance, if there is a bound state solution to Eq.(2) then it will admit solutions for any $E < 0$. This corresponds to a continuum of bound states with energies extending down to $-\infty$, so the system is not bounded from below. By rearranging the terms in the Schrödinger equation we obtain

$$\Phi(\mathbf{p}) = \frac{\alpha_0}{2\pi} \frac{\Psi(0)}{(p^2 + E_0)}\ , \tag{3}$$

where $\Psi(0)$ is the position space wave-function at the origin and $E_0 = -E > 0$ is the binding energy.

To obtain the eigenvalue condition for the binding energy, we can integrate both sides of Eq. (3):

$$1 = \frac{\alpha_0}{2\pi} \int_0^\infty dp\, p\, \frac{1}{(p^2 + E_0)}\ . \tag{4}$$

The integral on the r.h.s. diverges logarithmically, so the problem is ill-defined.

Renormalization requires that we regulate the divergence and absorb cut-off dependence in the coupling. First, we regulate the integral with a momentum cutoff, obtaining

$$1 = \frac{\alpha_0}{2\pi} \int_0^\Lambda dp\, p\, \frac{1}{(p^2 + E_0)} = \frac{\alpha_0}{4\pi} \ln\left(1 + \frac{\Lambda^2}{E_0}\right)\ , \tag{5}$$

so that

$$E_0 = \frac{\Lambda^2}{e^{-\frac{4\pi}{\alpha_0}} - 1}\ . \tag{6}$$

Clearly, if the coupling α_0 is fixed then $E_0 \to \infty$ as $\Lambda \to \infty$. In order to eliminate the divergence and produce a finite, well-defined bound state we can renormalize the theory by demanding that the coupling runs with the cutoff Λ in such a way that the binding energy remains fixed as the cutoff is removed:

$$\alpha_0 \to \alpha_\Lambda = \frac{4\pi}{\ln\left(1 + \frac{\Lambda^2}{E_0}\right)}\ . \tag{7}$$

The dimensionless renormalized running coupling α_Λ that characterizes the strength of the interaction is therefore replaced by a new (dimensionful) parameter E_0, the binding energy of the system. This is a simple example of dimensional transmutation: even though the original "bare" hamiltonian is scale invariant, renormalization leads to a scale that characterizes the physical observables. Note that E_0 can be chosen arbitrarily, fixing the energy scale of the underlying (renormalized) theory. It is also interesting to note that the renormalized running coupling α_Λ vanishes as $\Lambda \to \infty$ and so the theory is asymptotically free.

This renormalized hamiltonian can be used to compute other observables. The usual prescription for the calculations is to obtain the solutions with the cutoff in place and then let it approach ∞, using the running coupling in Eq. 7. If an exact calculation can be implemented, the final results should be independent of the regularization and renormalization schemes. As an example, we calculate the scattering wave function,

$$\Phi_k(\mathbf{p}) = \delta^{(2)}(\mathbf{p} - \mathbf{k}) + \frac{\alpha_\Lambda}{2\pi} \frac{\Psi(0)}{(p^2 - k^2 + i\,\epsilon)}\,, \tag{8}$$

where $k = \sqrt{E}$. Integrating both sides over $\mathbf{p}$ with a cutoff Λ in place, we obtain

$$\Psi(0) = \frac{1}{2\pi}\left[1 - \frac{\alpha_\Lambda}{4\pi}\ln\left(1 + \frac{\Lambda^2}{-k^2 + i\epsilon}\right)\right]^{-1}; \tag{9}$$

thus,

$$\alpha_\Lambda\ \Psi(0) = \frac{1}{2\pi}\left[\frac{1}{4\pi}\ln\left(1 + \frac{\Lambda^2}{E_0}\right) - \frac{1}{4\pi}\ln\left(1 + \frac{\Lambda^2}{-k^2 + i\epsilon}\right)\right]^{-1}. \tag{10}$$

In the limit $\Lambda \to \infty$ we obtain:

$$\alpha_\Lambda\ \Psi(0) = \frac{2}{\ln\left(\frac{k^2}{E_0}\right) - i\,\pi}\,. \tag{11}$$

The resulting scattering wavefunction is then given by

$$\Phi_k(\mathbf{p}) = \delta^{(2)}(\mathbf{p} - \mathbf{k}) + \frac{1}{2\pi}\frac{2}{(p^2 - k^2)}\left[\ln\left(\frac{k^2}{E_0}\right) - i\,\pi\right]^{-1}. \tag{12}$$

Only S-wave scattering occurs, because the centrifugal barrier completely screens the δ-function potential for higher angular momentum.

For our numerical calculations we used the K-matrix, which has mixed boundary conditions that lead to a principal value. The K-matrix Lippmann-Schwinger equation with the renormalized potential is given by:

$$K(\mathbf{p},\mathbf{p}';k) = V(\mathbf{p},\mathbf{p}') + \mathcal{P}\int d^2q\, \frac{V(\mathbf{p},\mathbf{q})}{k^2-q^2}\, K(\mathbf{q},\mathbf{p}';k)\,. \tag{13}$$

Since only S-wave scattering takes place we can integrate over the angular variable, obtaining

$$K^{(l=0)}(p,p';k) = V^{(l=0)}(p,p') + \mathcal{P}\int_0^\Lambda dq\, q\, \frac{V^{(l=0)}(p,q)}{k^2-q^2}\, K^{(l=0)}(q,p';k)\,, \tag{14}$$

where

$$V^{(l=0)}(p,p') = -\frac{\alpha_\Lambda}{2\pi}\,. \tag{15}$$

The Lippmann-Schwinger equation for the "on-shell" K-matrix is given by:

$$K^{(l=0)}(k) = -\frac{\alpha_\Lambda}{2\pi} - \frac{\alpha_\Lambda}{2\pi}\, K^{(l=0)}(k)\, \mathcal{P}\int_0^\Lambda dq\, q\, \frac{1}{k^2-q^2}. \tag{16}$$

Solving this equation and taking the limit $\Lambda \to \infty$, we obtain the exact on-shell K-matrix:

$$K_0(k) = -\frac{2}{\ln\left(\frac{k^2}{E_0}\right)}\,. \tag{17}$$

Using

$$k\cot\delta_0(k) = -\frac{2\,k}{\pi}\frac{1}{K_0(k)}\,, \tag{18}$$

we can obtain the exact phase-shifts:

$$\cot\delta_0 = \frac{1}{\pi}\ln\left(\frac{k^2}{E_0}\right)\,. \tag{19}$$

3 Effective field theory

3.1 EFT formalism

EFT provides a systematic procedure for improving the approximate description of composite effective degrees of freedom and their dynamics by increasing the number of local effective interactions. The effects of high-energy degrees of freedom are incorporated in effective interactions whose strength is adjusted to fit appropriate low-energy results. The keys to maintaining predictive power in EFT are approximate locality and an expansion in powers of a small ratio of low-to-high energy scales or equivalently, short-to-long distance scales.

We follow Lepage's treatment of EFT.[11] For simplicity we use a separable hamiltonian with a cutoff λ that smoothly regulates interactions involving high energy states:

$$V_\lambda(p,p') = \left[C_0(\lambda) + C_2(\lambda)\,\frac{(p^2+p'^2)}{2\lambda^2} + C_4(\lambda)\,\frac{(p^4+p'^4)}{4\lambda^4} + C_4'(\lambda)\,\frac{p^2\,p'^2}{2\lambda^4} + \ldots\right] \times e^{-\frac{p^2}{2\lambda^2}}\,e^{-\frac{p'^2}{2\lambda^2}}\,. \tag{20}$$

The parameters C_i in the expansion can be determined by fitting data for low-energy processes. Here we use as *data* the values for the inverse on-shell K-matrix, calculated with the exact theory. We follow the method described by Steele and Furnstahl [13] with the parameters fixed so that EFT gives the same momentum expansion for the inverse on-shell K-matrix as the exact theory to a given order.

The Lippmann-Schwinger equation for the K-matrix with is

$$K_\lambda(p,p';k) = V_\lambda(p,p') + \mathcal{P}\int d^2q\;\frac{V_\lambda(p,q)}{k^2-q^2}\,K_\lambda(q,p';k)\,, \tag{21}$$

where $\mathcal{P}$ denotes the principal value. The equation refers to the S-wave potential (angular variable integrated out) and $d^2q = q\,dq$. To obtain the K-matrix we solve the Lippmann-Schwinger equation nonperturbatively.[14,15,16] The effective potential given in Eq.(20) (truncated at order p^4) can be written in the form

$$V_\lambda(p,p') = e^{-\frac{p^2}{2\lambda^2}}\,e^{-\frac{p'^2}{2\lambda^2}} \sum_{i,j=0}^{2} p^{2i}\,\Lambda_{ij}\,p'^{2j}\,, \tag{22}$$

where Λ_{ij} are the matrix elements of

$$\mathbf{\Lambda} = \begin{pmatrix} C_0(\lambda) & C_2(\lambda)/2\lambda^2 & C_4(\lambda)/4\lambda^4 \\ C_2(\lambda)/2\lambda^2 & C_4'(\lambda)/2\lambda^4 & 0 \\ C_4(\lambda)/4\lambda^4 & 0 & 0 \end{pmatrix}. \tag{23}$$

The solution of the Lippmann-Schwinger equation can then be written in the form

$$K_\lambda(p', p; k) = e^{-\frac{p^2}{2\lambda^2}} \, e^{-\frac{p'^2}{2\lambda^2}} \sum_{i,j=0}^{1} p^{2i} \, \tau_{ij}(k) \, p'^{2j} \, . \tag{24}$$

The unknown matrix $\tau(k)$ satisfies the equation

$$\tau(\mathbf{k}) = \mathbf{\Lambda} + \mathbf{\Lambda} \, \mathcal{I}(k) \, \tau(\mathbf{k}) \, , \tag{25}$$

where

$$\mathcal{I} = \begin{pmatrix} I_0(k) & I_1(k) & I_2(k) \\ I_1(k) & I_2(k) & I_3(k) \\ I_2(k) & I_3(k) & I_4(k) \end{pmatrix} , \tag{26}$$

with

$$I_n(k) = \mathcal{P} \int dq \, q^{2n+D-1} \, \frac{1}{k^2 - q^2} \, e^{-\frac{q^2}{\lambda^2}} \, . \tag{27}$$

Solving Eq. (25) for τ analytically and substituting the solution in Eq. (24) we obtain the effective K-matrix.

It is straightforward to extend the calculation to higher orders. We fit the difference between the effective and exact inverse on-shell K-matrix ($p^2 = p'^2 = k^2$) to an interpolating polynomial in k^2/λ^2 to the highest possible order,

$$\Delta \left[\frac{1}{K} \right] = A_0 + A_2 \, \frac{k^2}{\lambda^2} + A_4 \, \frac{k^4}{\lambda^4} + \cdots \, . \tag{28}$$

The coefficients A_i are minimized with respect to the variation in the parameters C_i of the effective potential. The number of coefficients that can be minimized is given by the number of parameters appearing in the effective potential.

By adjusting only the parameter $C_0(\lambda)$ we should eliminate the leading error. As each term is added to the effective potential, followed by the adjustment of the respective parameter, we expect the errors in the on-shell K-matrix to be systematically reduced by *powers of* k^2/λ^2. Thus, in a log-log plot for $\Delta[1/K]$ we expect to obtain straight lines with slope given by the dominant power of k/λ in the error.

The expansion given by Eq.(28) becomes invalid at best when the momenta involved are of the same order as the cutoff, where one expects the short distance effects to be directly resolved. This point corresponds to the radius of convergence of the effective theory. Using the effective potential (truncated at a given order) with the parameters fixed by fitting the inverse on-shell K-matrix we diagonalize the effective hamiltonian numerically, obtaining a bound-state energy that can be compared to the exact result.

3.2 EFT results for the two-dimension δ-function

The EFT power counting scheme is simple. For the leading-order prediction, we use the effective potential with one parameter. Evaluating the difference $\Delta\,[1/K^{(0)}] = 1/K^{(0)}_\lambda - 1/K_0$ and expanding it in powers of k/λ we obtain:

$$\Delta\left[\frac{1}{K^{(0)}}\right] = \left[\frac{1}{C_0} - \frac{\gamma}{2} + \frac{1}{2}\ln\left(\frac{\lambda^2}{E_0}\right)\right] + \left[\frac{1}{C_0} - \frac{1}{2}\right]\frac{k^2}{\lambda^2} + \mathcal{O}\left(\frac{k^4}{\lambda^4}\right) . \tag{29}$$

By choosing

$$C_0(\lambda) = \frac{-2}{\ln\left(\frac{\lambda^2}{E_0}\right) - \gamma} , \tag{30}$$

the leading logarithmic error is eliminated and the remaining errors are dominated by the term

$$\mathcal{O}\left(\frac{k^2}{\lambda^2}\right) = \frac{1}{2}\left[\gamma - \ln\left(\frac{\lambda^2}{E_0}\right) - 1\right]\frac{k^2}{\lambda^2} . \tag{31}$$

To evaluate the errors in the binding energy we use the effective potential with one, two and four parameters adjusted to fit the inverse K-matrix. We solve the Schrödinger equation numerically for different values of the cutoff λ. In Fig. 1 we show in a log-log plot the absolute values for the relative errors in the bound-state energy as a function of E_0/λ^2. As expected, we see straight lines with slope given by the dominant power of E_0/λ^2 in the error. Note that the dominant logarithmic errors are completely absorbed by adjusting the parameter C_0 and when more parameters are adjusted we obtain power-law improvement.

4 The similarity renormalization group

4.1 SRG formalism

The similarity renormalization group [3,4] is made of similarity transformations that run a smooth cutoff on the magnitude of off-diagonal matrix elements. As the cutoff is lowered, the hamiltonian is forced toward the diagonal. It is straightforward to show that an infinitesimal transformation produces a

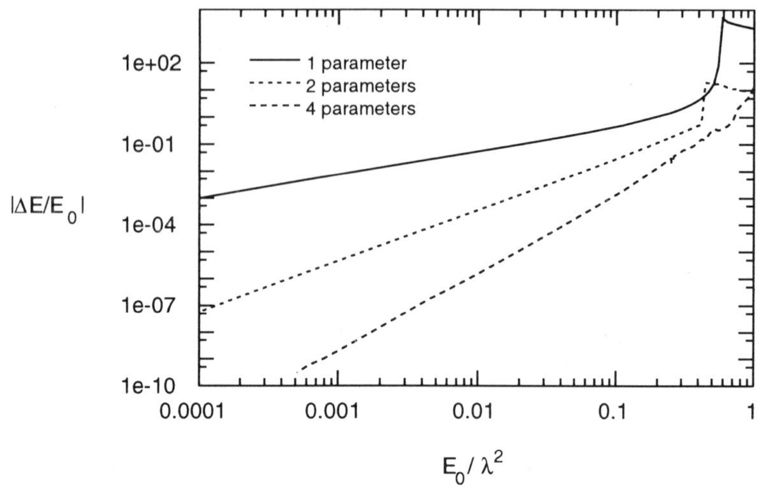

Figure 1. The EFT error in the binding energy for the two-dimensional delta-function using one, two and four parameters. The exact theory is fixed by choosing $E_0 = 1$.

differential equation governing the evolution of the hamiltonian. Głazek and Wilson considered very general transformations in their original work,[3] but we use the simple transformation developed by Wegner.[4] The infinitesimal transformation is

$$\frac{dH_s}{ds} = [H_s, [H_s, h] \; . \tag{32}$$

Here $H_s = h + V_s$, where h is the free hamiltonian, evolves with a flow parameter s that ranges from 0 to ∞. The flow-parameter has dimensions $1/(\text{energy})^2$ and can be expressed in terms of the similarity cutoff λ by the relation $s = 1/\lambda^2$. Wegner advocates using the full diagonal part of the hamiltonian instead of using h in the transformation.

The reduced interaction, $\overline{V}_{sij}$ (the interaction with a gaussian similarity cutoff factored out) is defined by

$$V_{sij} = e^{-s\Delta_{ij}^2} \, \overline{V}_{sij} \; , \tag{33}$$

where $\Delta_{ij} = p_i^2 - p_j^2$ for our simple transformation. The gaussian factor forces the interaction toward the diagonal to leading order, and the equation for the evolution of $\overline{V}$ starts at second order.

The flow equation for the reduced interaction is

$$\frac{d\overline{V}_{sij}}{ds} = \sum_k (\Delta_{ik} + \Delta_{jk})\ \overline{V}_{sik}\ \overline{V}_{skj}\ e^{-2s\Delta_{ik}\Delta_{jk}}\ , \tag{34}$$

where we use $\Delta_{ij}^2 - \Delta_{ik}^2 - \Delta_{jk}^2 = -2\Delta_{ik}\Delta_{jk}$.

To solve this equation we impose a boundary condition, $H_s|_{s\to s_0} \equiv H_{s_0}$, where s_0 can be expressed in terms of the large similarity cutoff as $s_0 = 1/\Lambda^2$. Next we make a perturbative expansion,

$$\overline{V}_s = \overline{V}_s^{(1)} + \overline{V}_s^{(2)} + \cdots\ , \tag{35}$$

where the superscript implies the order in the running canonical coupling. All counterterms are determined using coupling coherence, which assumes $\overline{V}$ is a perturbative function of α.

At first order we have

$$\frac{d\overline{V}_{s_{ij}}^{(1)}}{ds} = 0\ , \tag{36}$$

which implies

$$\overline{V}_{sij}^{(1)} = \overline{V}_{s_0 ij}\ , \tag{37}$$

where s is the final scale. Because of the dimensions of the flow parameter we have $s > s_0$, corresponding to a smaller cutoff. The "no cutoff limit" corresponds to $s_0 \longrightarrow 0$.

At second order we have

$$\frac{d\overline{V}_{sij}^{(2)}}{ds} = \sum_k (\Delta_{ik} + \Delta_{jk})\ \overline{V}_{s_0 ik}\ \overline{V}_{s_0 kj}\ e^{-2s\Delta_{ik}\Delta_{jk}}\ . \tag{38}$$

Integrating, we obtain

$$\overline{V}_{sij}^{(2)} = \frac{1}{2}\sum_k \overline{V}_{s_0 ik}\ \overline{V}_{s_0 kj}\ \left(\frac{1}{\Delta_{ik}} + \frac{1}{\Delta_{jk}}\right) \times$$
$$\times \left[e^{-2s_0\Delta_{ik}\Delta_{jk}} - e^{-2s\Delta_{ik}\Delta_{jk}}\right]\ . \tag{39}$$

The renormalized hamiltonian for the non-relativistic delta-function potential is given by

$$H_\lambda(\mathbf{p},\mathbf{p}') = p^2\delta^{(2)}(\mathbf{p}-\mathbf{p}') + e^{-\frac{(p^2-p'^2)^2}{\lambda^4}}\ \left[\bar{V}_\lambda^{(1)}(\mathbf{p},\mathbf{p}') + \bar{V}_\lambda^{(2)}(\mathbf{p},\mathbf{p}') + ...\right]\ , \tag{40}$$

where

$$\bar{V}_\lambda^{(1)}(\mathbf{p},\mathbf{p}') = -\frac{\alpha_{\lambda,i}}{(2\pi)^2}\,, \tag{41}$$

$$\bar{V}_\lambda^{(2)}(\mathbf{p},\mathbf{p}') = \alpha_{\lambda,i}^2\; F_s^{(2)}(\mathbf{p},\mathbf{p}')\,, \tag{42}$$

$$\bar{V}_\lambda^{(n)}(\mathbf{p},\mathbf{p}') = \alpha_{\lambda,i}^n\; F_s^{(n)}(\mathbf{p},\mathbf{p}')\,. \tag{43}$$

Here λ is a momentum cutoff related to the flow parameter by $s = 1/\lambda^4$ and i denotes the order of the calculation for the running coupling.

The renormalized hamiltonian can be used to compute eigenvalues and eigenstates. Since the hamiltonian is derived perturbatively we expect cutoff dependent errors in the observables. Formally, we can regroup the terms in the renormalized hamiltonian and write it as a momentum expansion similar to the one for the EFT in Eq. (20). The difference is that the expansion parameters are perturbative functions of the running coupling α_λ.

There are two interdependent sources of errors in the perturbative similarity renormalization group. First, α_λ is approximated and this leads to inverse-logarithmic errors. Second, the coefficients of all irrelevant operators are approximated perturbatively, and this leads to products of inverse-logarithmic and inverse-power-law errors.

In our examples we focus on the bound state errors. We fix the coupling at one scale and use the flow-equation to obtain the coupling as a function of the cutoff λ to a given order. We then perform a sequence of bound-state calculations with better approximations for the hamiltonian. Once the sources of errors are identified, it becomes relatively simple to analyze order-by-order how such errors scale with λ. In principle, to completely eliminate the errors proportional to some power m in the momentum expansion we should use the similarity hamiltonian with the exact running coupling (renormalized to all orders) and include the contributions up to $\mathcal{O}(p^m/\lambda^m)$ coming from all effective interactions (all orders in α_λ). We should emphasize that in a realistic calculation we would fit the coupling α_λ to an observable. This nonperturbative renormalization eliminates the dominant source of errors we display in SRG calculations in this article. We choose to renormalize the coupling perturbatively here because the only observable we compute is the single bound state energy of a delta-function potential, and fitting this energy would prevent us from displaying errors.

4.2 SRG results for two-dimensional δ-function

Głazek and Wilson have studied a discretized analog of the two-dimensional δ function, which allowed them to perform both perturbative and nonperturbative transformation.[5] Their perturbative results are similar to ours. In the two-dimensional case the canonical hamiltonian in momentum space with a delta-function potential can be written as

$$H(\mathbf{p},\mathbf{p}') = h(\mathbf{p},\mathbf{p}') + V(\mathbf{p},\mathbf{p}') \; , \tag{44}$$

where $h(\mathbf{p},\mathbf{p}') = p^2\delta^{(2)}(\mathbf{p}-\mathbf{p}')$ corresponds to the free hamiltonian and $V(\mathbf{p},\mathbf{p}') = -\alpha_0/(2\pi)^2$ corresponds to the Fourier transform of the delta-function potential.

Integrating out the angular variable, the flow equation obtained with Wegner's transformation in terms of matrix elements in the basis of free states is given by

$$\frac{dV_s(p,p')}{ds} = -(p^2-p'^2)^2\, V_s(p,p') - \int_0^\infty dk\, k\, (2k^2-p^2-p'^2)\, V_s(p,k)\, V_s(k,p') \; . \tag{45}$$

In principle, we can set the boundary condition at $s=0$ (no cutoff), i.e,

$$H_{s=0}(p,p') = H(p,p') = p^2\delta^{(1)}(p-p') - \frac{\alpha_0}{2\pi} \; . \tag{46}$$

However, the hamiltonian with no cutoff produces logarithmic divergences. The boundary condition must be imposed at some other point, leading to dimensional transmutation. The reduced interaction $\bar{V}_s(p,p')$ is defined such that

$$V_s(p,p') = e^{-s(p^2-p'^2)^2}\, \bar{V}_s(p,p') \; . \tag{47}$$

Assuming that h is cutoff independent we obtain the flow equation for the reduced interaction,

$$\begin{aligned}\frac{d\bar{V}_s}{ds} = -e^{-2s\, p^2p'^2} \int_0^\infty & dk\, k\, (2k^2-p^2-p'^2)\, e^{-2s[k^4-k^2(p^2+p'^2)]} \\ & \times \bar{V}_s(p,k)\, \bar{V}_s(k,p') \; .\end{aligned} \tag{48}$$

This equation is solved using a perturbative expansion, starting with

$$\bar{V}_s^{(1)}(p,p') = -\frac{\alpha_s}{2\pi} \; . \tag{49}$$

We assume a coupling-coherent solution in the form of an expansion in powers of $\alpha_s/2\pi$, satisfying the constraint that the operators $F_s^{(n)}(p,p')$ vanish when $p=p'=0$,

$$\bar{V}_s(p,p') = -\frac{\alpha_s}{2\pi} + \sum_{n=2}^{\infty} \left(\frac{\alpha_s}{2\pi}\right)^n F_s^{(n)}(p,p') \; . \tag{50}$$

Note that the expansion parameter is $\alpha_s/2\pi$.

Using the solution Eq.(50) in Eq.(48) we obtain

$$\frac{d\bar{V}_s}{ds} = -\frac{1}{(2\pi)^2}\frac{d\alpha_s}{ds} + \sum_{n=2}^{\infty}\frac{1}{(2\pi)^n}\left[n\,\alpha_s^{n-1}\frac{d\alpha_s}{ds}F_s^{(n)}(p,p') + \alpha_s^n\frac{dF_s^{(n)}(p,p')}{ds}\right]$$

$$= \int_0^{\infty} dk\, k\,(2k^2 - p^2 - p'^2)\, e^{-2s[p^2p'^2 + k^4 - k^2(p^2+p'^2)]}$$

$$\times\left[-\frac{\alpha_s}{2\pi} + \sum_{n=2}^{\infty}\left(\frac{\alpha_s}{2\pi}\right)^n F_s^{(n)}(p,k)\right]\left[-\frac{\alpha_s}{2\pi} + \sum_{m=2}^{\infty}\left(\frac{\alpha_s}{2\pi}\right)^m F_s^{(m)}(k,p')\right] \quad (51)$$

This equation is solved iteratively order-by-order in $\alpha_s/2\pi$. Again, if $\alpha_s/2\pi$ is small the operator $\bar{V}_s^{(1)}(p,p')$ can be identified as the dominant term in the expansion of $\bar{V}_s(p,p')$ in powers of p and p'. This operator corresponds to a marginal operator (since the coupling is dimensionless and there is no implicit mass scale). The higher-order terms correspond to irrelevant operators.

For simplicity we use the interaction with a large cutoff on all momenta, Λ, to define the *exact* theory. We define

$$\alpha_{s_0=0} = \alpha_\Lambda = \frac{4\pi}{\ln\left(1 + \frac{\Lambda^2}{E_0}\right)}, \quad (52)$$

and set all irrelevant operators to zero at $s_0 = 0$. Note that the coupling α_{λ_0} is fixed at $\lambda_0 = \infty$ by fitting the exact binding energy. With this definition the similarity hamiltonian with no cutoff becomes well-defined and we can set all of the similarity transformation boundary conditions at $s_0 = 0$. All integrals below are cut off ($p \ll \Lambda$)

At second-order we have

$$-\frac{1}{2\pi}\frac{d\alpha_{s,\Lambda}}{ds} + \frac{1}{(2\pi)^2}\alpha_{s,\Lambda}^2\frac{dF_{s,\Lambda}^{(2)}(p,p')}{ds} = -\alpha_{s,\Lambda}^2\, I_{s,\Lambda}^{(2)}(p,p') \quad (53)$$

where

$$I_{s,\Lambda}^{(2)}(p,p') = \frac{1}{(2\pi)^2}\int_0^{\Lambda} dk\, k\,(2k^2 - p^2 - p'^2)\, e^{-2s[p^2p'^2 + k^4 - k^2(p^2+p'^2)]}$$

$$= \frac{1}{(2\pi)^2}\frac{e^{-2s\,p^2p'^2}}{4s}. \quad (54)$$

The resulting second-order running coupling and irrelevant operator are given respectively by

$$\alpha_{s,\Lambda,2} = \frac{\alpha_\Lambda}{1 - \frac{\alpha_\Lambda}{8\pi}\left[\gamma + \ln(2s\Lambda^4) - \mathrm{Ei}(-2s\Lambda^4)\right]} \quad (55)$$

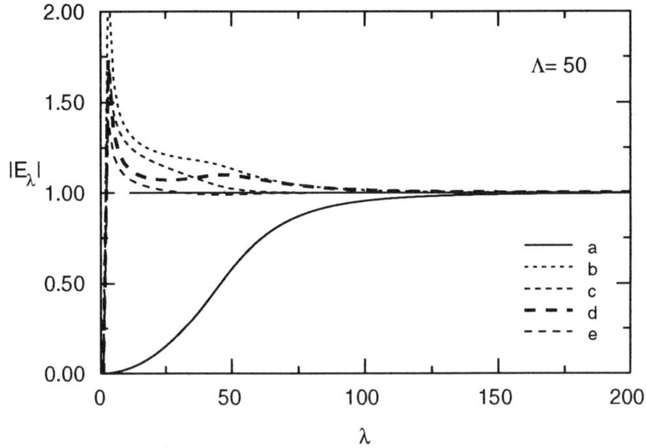

Figure 2. The SRG bound-state energy using various approximations for the similarity hamiltonian. The exact theory is fixed by choosing $\Lambda = 50$ and $E_0 = 1$.

and

$$\begin{aligned} F^{(2)}_{s,\Lambda}(p,p') &= \frac{1}{4}\left[\gamma + \ln(2s\, p^2\, p'^2) - \mathrm{Ei}(-2s\, p^2\, p'^2)\right] \\ &+ \frac{1}{4}\left[\gamma + \ln(2s\Lambda^4) - \mathrm{Ei}(-2s\Lambda^4)\right] \\ &- \frac{1}{4}\left[\gamma + \ln\left(s\left[(p^2-\Lambda^2)^2 + (p'^2-\Lambda^2)^2 - (p^2-p'^2)^2\right]\right)\right. \\ &\qquad \left. - \mathrm{Ei}\left(-s\left[(p^2-\Lambda^2)^2 + (p'^2-\Lambda^2)^2 - (p^2-p'^2)^2\right]\right)\right] . \end{aligned} \tag{56}$$

In Fig. 2 we show the binding-energy as a function of the cutoff and in Fig. 3 we show the errors in the binding-energy obtained using the following approximations for the potential with $\Lambda = 50$:

(a) marginal operator with coupling (α_{λ_0}),

$$V_{\lambda,\Lambda}(p,p') = -\frac{\alpha_\Lambda}{2\pi}\, e^{-\frac{(p^2-p'^2)^2}{\lambda^4}} \; ; \tag{57}$$

(b) marginal operator with running coupling renormalized to second-order ($\alpha_{\lambda,\Lambda,2}$),

$$V_{\lambda,\Lambda}(p,p') = -\frac{\alpha_{\lambda,\Lambda,2}}{2\pi}\, e^{-\frac{(p^2-p'^2)^2}{\lambda^4}} \; ; \tag{58}$$

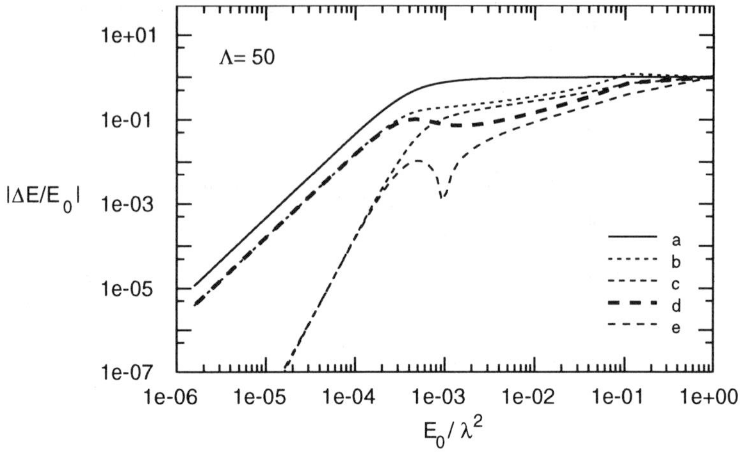

Figure 3. The SRG errors in the binding energy using various approximations for the similarity hamiltonian. The exact theory is fixed by choosing $\Lambda = 50$ and $E_0 = 1$.

(c) marginal operator plus second-order irrelevant operator with running coupling renormalized to second-order $(\alpha_{\lambda,\Lambda,2}, F^{(2)}_\lambda)$,

$$V_{\lambda,\Lambda}(p,p') = \left[-\frac{\alpha_{\lambda,\Lambda,2}}{2\pi} + \left(\frac{\alpha_{\lambda,\Lambda,2}}{2\pi}\right)^2 F^{(2)}_{\lambda,\Lambda}(p,p') \right] e^{-\frac{(p^2-p'^2)^2}{\lambda^4}} \; ; \tag{59}$$

(d) marginal operator with running coupling renormalized to third-order $(\alpha_{\lambda,\Lambda,3})$,

$$V_{\lambda,\Lambda}(p,p') = -\frac{\alpha_{\lambda,\Lambda,3}}{2\pi} \, e^{-\frac{(p^2-p'^2)^2}{\lambda^4}} \; ; \tag{60}$$

(e) marginal operator plus second-order irrelevant operator with running coupling renormalized to third-order $\alpha_{\lambda,\Lambda,3}, F^{(2)}_\lambda)$,

$$V_{\lambda,\Lambda}(p,p') = \left[-\frac{\alpha_{\lambda,\Lambda,3}}{2\pi} + \left(\frac{\alpha_{\lambda,\Lambda,3}}{2\pi}\right)^2 F^{(2)}_{\lambda,\Lambda}(p,p') \right] e^{-\frac{(p^2-p'^2)^2}{\lambda^4}} \; . \tag{61}$$

In Fig. 2 we clearly see all of the binding energies approaching the exact result, which we choose to be 1. If the coupling does not run (a) the energy falls to zero as the cutoff is lowered. When the coupling is allowed to run to second order (b) the binding energy deviates from the exact result more slowly as the cutoff is lowered, and subsequent improved approximations further remove dependence on λ.

Fig. 3 is a log-log plot of the relative error in the binding energy as a function of E_0/λ^2. When $\lambda \gg \Lambda$ the errors scale like inverse powers of E_0/λ^2. They are forced to approach the correct result as $\lambda \to \infty$, and the exact running coupling deviates from the approximate running coupling as a power of Λ/λ in this region, even when the approximate coupling is not allowed to run. The coefficients of these powers are large, and the analysis changes as $\lambda \to \Lambda$. The errors become logarithmic for (a), but they scale as an inverse-logarithm squared in (b) where the coupling runs to second-order. This removes the dominant inverse-logarithmic errors, and we can clearly see further improvement when the coupling runs to third-order in (d). The addition of irrelevant operators in (c) and (e) are important, but the third-order error in the running coupling becomes as large as the error from the leading irrelevant operators, which shows up when (d) crosses (c). When the second-order irrelevant operators are added with the coupling running to third order, we obtain approximate power-law improvement over (b).

5 Conclusion

Renormalization not only removes unphysical divergences, it also allows us to work with finite cutoffs while controlling the errors they introduce. Finite cutoffs are often required for nonperturbative calculations, and the precise control of the errors they introduce is essential.

In light-front QCD we can use the SRG to compute the hamiltonian order-by-order in the canonical QCD coupling, α_λ. We are free to choose λ so that multi-parton high-energy states effectively decouple from few-parton low-energy states, as long as this happens before α_λ becomes too large. We adjust the coupling non-perturbatively by fitting low-energy data. This should remove all inverse-logarithmic errors, leaving only powers of inverse-logarithms and inverse-powers of the cutoff. We can systematically lower the errors by adding irrelevant operators, all of which are fixed by coupling coherence.

Acknowledgments

One of us (R.J.P.) is very grateful to the hosts of the 1998 YITP-Workshop on QCD and Hadron Physics. The opportunity to see Kyoto and to interact with so many excellent young Japanese physicists was fantastic. We would also like to acknowledge many useful discussions with Brent Allen, Dick Furnstahl, Stan Głazek, Roger Kylin, Rick Mohr, Jim Steele, and Ken Wilson. This work was supported by National Science Foundation grant PHY-9800964, and S.S. is a CNPq-Brazil fellow (proc. 204790/88-3).

References

1. Sérgio Szpigel and Robert J. Perry, in preparation.
2. Robert J. Perry, these Proceedings (1999).
3. S. D. Głazek and K.G. Wilson, Phys. Rev. D **48**, 5863 (1993); S. D. Głazek and K.G. Wilson, Phys. Rev. D **49**, 4214 (1994).
4. F. Wegner, Ann. Physik **3**, 77 (1994).
5. S. D. Głazek and K.G. Wilson, Phys. Rev. D **57**, 3558 (1998).
6. T. S. Walhout, hep-th/9806097 (1998).
7. R. Oehme and W. Zimmermann, Commun. Math. Phys. **97**, 569 (1985); R. Oehme, K. Sibold and W. Zimmermann, Phys. Lett. B **147**, 115 (1984).
8. R. J. Perry and K. G. Wilson, Nucl. Phys. B **403**, 587 (1993); R. J. Perry, Ann. Phys. **232**, 116 (1994).
9. B. H. Allen and R. J. Perry, Phys. Rev. D **58**, 125017 (1998), hep-th/9804136; R. D. Kylin, B. H. Allen, and R. J. Perry, hep-th/9812080.
10. R. Jackiw, *in M. A. B. Beg Memorial Volume*, A. Ali and P. Hoodbhoy, eds. (World Scientific, Singapore, 1991).
11. G. P. Lepage, nucl-th/9706029.
12. S. Weinberg, Physica A **96**, 327 (1979); H. Georgi, Ann. Rev. Nucl. Part. Sci. **43**, 209 (1993); A. V. Manohar, hep-ph/9606222; M. B. Wise, hep-ph/9707522; D. B. Kaplan, nucl-th/9506035; U. van Kolck, nucl-th/9808007.
13. J. V. Steele and R. J. Furnstahl, Nucl. Phys. **A637**,46 (1998), nucl-th/9802069; J. V. Steele, nucl-th/9807022.
14. D. R. Phillips, S. R. Beane and T. D. Cohen, Ann. Phys. **263**, 255 (1998), hep-th/9706070.
15. M. C. Birse, nucl-th/9804028; nucl-th/9708435; hep-ph/9807302, hep-ph/9808398.
16. J. Gegelia, nucl-th/9805008.

Theta in the LF massive Schwinger model

Koji Harada
Department of Physics, Kyushu University
Fukuoka 812-8581
E-mail: koji@higgs.phys.kyushu-u.ac.jp

The massive Schwinger model quantized on the light front in the theta vacuum is investigated without explicitly introducing the zero-mode degrees of freedom. We carefully take into account a set of interactions which usually ignored in a naive light-front field theory and show that they renormalize the strength of the external field. The dependence of the meson mass on the theta parameter as well as the fermion mass is investigated. The results are consistent with the well-known ones. Some preliminary numerical results are also presented.

1 Introduction

Light-front (LF) field theory has a very simple vacuum. Because of this property, there has been much interest in LF field theory as a useful framework in studying non-perturbative physics such as bound states[1]. On the other hand, however, we KNOW that the vacuum (of QCD for example) is complicated. A natural question is: How can we understand complicated vacuum structures such as spontaneous symmetry breaking in the LF framework?

It is widely accepted that vacuum structures will be understood as "zero-mode" physics, i.e., the modes with zero longitudinal momentum $p^+ \equiv (p^0 + p^3)/\sqrt{2} = 0$ are responsible to vacuum structures. We need to extract some information on vacuum from the zero-mode sector. There are two approaches to this end:

1. We can take into account the dynamics of the zero modes *explicitly* by, for example, putting the system in a LF box (Discretized Light-Cone Quantization (DLCQ)[2]).

   Unfortunately, however, it is extremely difficult to actually carry out the analysis in most cases.

2. Alternatively, we can take it into account *effectively* by introducing some new interactions which represent the net effects of the zero modes.

   Painfully, it is also difficult to correctly estimate the effects in many cases.

At present, we do not know any powerful way of extracting the information from the zero-mode sector. We need to study the problem in many ways to get better understanding.

In this talk, we concentrate on the simplest vacuum structure: the theta vacuum in the massive Schwinger model (QED_{1+1} with massive fermion). The first approach has been taken[3]. It has been shown that the theta vacuum can be understood as dynamics of gauge field zero modes. The vacuum can be described by a simple one-dimensional Schrödinger equation in a periodic potential. The theta is nothing but the Bloch momentum. Unfortunately, however, we were unable to calculate the theta dependence of the meson mass because of certain subtlety of taking the "continuum" limit. Here, we take the second approach. We consider the massive Schwinger model in a constant background electric field and obtain the effective interactions which represent the physics of the zero modes. It turns out that the net effect is to *renormalize* the parameter.

This work is based on the collaboration with Matthias Burkardt[4].

2 Theta vacuum as effective background field

2.1 known results and motivation

We consider $(1+1)$-dimensional QED,

$$\mathcal{L} = -\frac{1}{4}F_{\mu\nu}F^{\mu\nu} + \bar{\psi}[\gamma^{\mu}(i\partial_{\mu} - eA_{\mu}) - m]\psi, \tag{1}$$

in the $A^{+}=0$ gauge. By integrating LF Gauss law,

$$-\partial_{-}^{2}A^{-} = ej^{+}, \quad j^{+} \equiv \sqrt{2} : \psi_{R}^{\dagger}\psi_{R} :, \tag{2}$$

we obtain

$$A^{-}(x^{-}) = -\frac{e}{2}\int_{-\infty}^{\infty} dy^{-}|x^{-} - y^{-}|j^{+}(y^{-}) - \frac{e\theta}{2\pi}x^{-}, \tag{3}$$

up to an irrelevant constant. θ is a vacuum angle. From the analysis by Coleman[5] in the equal-time formulation, we expect:

1. In the strong coupling region $m \ll e/\sqrt{\pi} = 1$, there are a few number of bound states (mesons). The (lightest) meson mass M depends on θ as

$$M^{2} = 1 + 2e^{\gamma}m\cos\theta + \mathcal{O}(m^{2}). \tag{4}$$

2. In the weak coupling region $m \ll 1$, there are "half-asymptotic" states at $|\theta| = \pi$.

3. There is no trace of such half-asymptotic states in the strong coupling limit. There must be a phase transition. It corresponds to spontaneous breakdown of charge conjugation symmetry.

What we want to do is to understand these things in the LF framework.

Note that the theta term leads to an additional interaction in the Hamiltonian,

$$\delta P^- = -\frac{e^2\theta}{2\pi}\int dx^- x^- j^+(x^-). \tag{5}$$

By substituting the representation in terms of creation/annihilation operators, we get

$$\delta P^- = -\frac{ie^2\theta}{2\pi}\int_0^\infty \frac{dk^+dp^+}{2\pi\sqrt{k^+p^+}}\Big\{\delta'(p^+-k^+)[b^\dagger(k^+)b(p^+)-d^\dagger(k^+)d(p^+)]$$
$$-\delta'(k^++p^+)b^\dagger(k^+)d^\dagger(p^+)+\delta'(k^++p^+)d(k^+)b(p^+)\Big\}. \tag{6}$$

Note that the support of $\delta'(k^+ + p^+)$ is at the edge of the integration region. The last two terms are therefore ignored in many (naive) calculations. In the following we carefully examine if these terms may have non-vanishing contributions.

2.2 Matrix elements of δP^-

In this section, we consider the meson state in the two-body (valence) approximation and its couplings to the external field. The meson state with momentum $\mathcal{P}$ in the valence approximation is given by

$$|\mathcal{P}\rangle = \int_0^{\mathcal{P}} \frac{dk_1dk_2}{2\pi\sqrt{k_1k_2}}\delta(k_1+k_2-\mathcal{P})\psi(k_1,k_2)b^\dagger(k_1)d^\dagger(k_2)|0\rangle. \tag{7}$$

Tree coupling

The meson state may couple to the external field directly. It is easy to see that such coupling can be correctly described by the last term of the following equation

$$M^2\Phi(x) = \frac{m^2-1}{x(1-x)}\Phi(x) + \int_0^1 dy\left(1-\frac{1}{(x-y)^2}\right)\Phi(y) + \underbrace{i\theta\frac{\partial}{\partial x}\Phi(x)}_{new\ term}, \tag{8}$$

where $\Phi(x) = \psi(k, \mathcal{P}-k)$, $x = k/\mathcal{P}$ and $e/\sqrt{\pi} = 1$.

Vacuum polarization

The meson state may also couple to the external field through vacuum polarization. Fig. 1 shows an example of a particular class of such diagrams which we call "chain diagrams." There may be other kind of diagrams which contains vacuum polarization. For example there should be diagrams in which the external field couplles to the middle of the vacuum polarization (non-valence diagrams). We would like to distinguish the chain diagrams from the others due to the following observation:

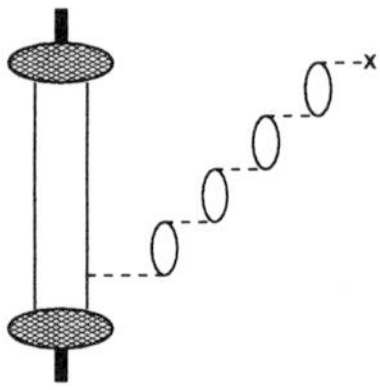

Fig. 1 Chain diagram.

- If we put $\delta P^- = 0$, the whole chain diagrams disappear, while the non-valence diagrams survive except for the external legs.
- Chain diagrams do not appear in the usual perturbative calculations; It comes out from the interplay between δP^- and the interactions which are already included in the Hamiltonian.
- Non-valence diagrams would be included if we extend eq.(8) to include higher Fock (e.g., non-valence) states. Note, however, that at least at $\theta = 0$ the contributions from the higher Fock states are know to be small in the strong coupling region.)

Do these chain diagrams have non-zero contributions? Because of the $1/(q^+)^2$ factor of the (instantaneous) propagator, there is enhancement in the $q^+ \longrightarrow 0$ limit. A simple calculation shows that they do have non-zero contributions.

We can do a bit better by replacing the chain with the exchange of the (approximately) complete set of states obtained by the valence approximation. We finally find that the net effect of them (which comes from the zero-mode sector) is just to renormalize the external field θ,

$$\theta \rightarrow \theta_e ff \equiv \theta \left[1 - \sum_n \frac{|g_V(n)|^2}{M_n^2} \right] \tag{9}$$

where $g_V(n) = \int_0^1 dx \Phi_n(x)$ and $\Phi_n(x)$ is the n-th eigenstate (with eigenvalue M_n) of eq.(8). We end up with the following equation for the meson,

$$M^2 \Phi(x) = \frac{m^2 - 1}{x(1-x)} \Phi(x) + \int_0^1 dy \left(1 - \frac{1}{(x-y)^2} \right) \Phi(y) + i\theta_{eff} \frac{\partial}{\partial x} \Phi(x) \tag{10}$$

2.3 variational calculation in the chiral limit

We consider the following variational ansatz for the meson state,

$$\Phi(x) = x^{\beta}(1-x)^{1-\beta} + x^{1-\beta^*}(1-x)^{\beta^*}, \tag{11}$$

leading to

$$\frac{\langle\Phi|H|\Phi\rangle}{\langle\Phi|\Phi\rangle} = 1 + \frac{m^2}{a} + \frac{a^2+b^2}{3a}\pi^2 - \theta_{eff}\frac{b}{a} + \mathcal{O}(\beta^2), \tag{12}$$

where $\beta = a + ib$. By minimizing it by varying a and b, we find

$$\beta_{min} = \frac{\sqrt{3}}{\pi}\sqrt{m^2 - \frac{3}{4\pi^2}\theta_{eff}^2} + i\frac{3}{2\pi^2}\theta_{eff} \tag{13}$$

and

$$M^2 \equiv \left.\frac{\langle\Phi|H|\Phi\rangle}{\langle\Phi|\Phi\rangle}\right|_{min} = 1 + \frac{2\pi}{\sqrt{3}}\sqrt{m^2 - \frac{3}{4\pi^2}\theta_{eff}^2} \tag{14}$$

By noting that $g_V(n=0) = 1 - \mathcal{O}(\beta^2)$ and $\theta_{eff}/\theta = M^2 - 1 + \mathcal{O}(\beta^2)$, we find $\theta_{eff} = 2\pi m\theta/\sqrt{3} + \mathcal{O}(\theta^3)$ and

$$M^2(\theta, m) = 1 + \frac{2\pi m}{\sqrt{3}}\left(1 - \frac{\theta^2}{2}\right) + \mathcal{O}(m^2) + \mathcal{O}(\theta^4). \tag{15}$$

This is consistent with the known result (4) up to a 2% discrepancy[6].

3 Numerical analysis

Finally we present some preliminary numerical results. We discretize the interval $[0,1]$ of x in eq.(10) into $N+1$ pieces and impose the boundary conditions $\Phi(0) = \Phi(1) = 0$. Then we diagonalize the $N \times N$ Hermitian (complex!) matrix (Hamiltonian) numerically. In the following, $N = 2000$. Fig. 2 shows how theta parameter is renormalized by the vacuum polarization. The screening effect is larger for smaller fermion masses. In the massless limit, $\theta_{eff} \to 0$ irrespective to θ_{true}. This is consistent with the fact that there is no effect of theta on the meson in the massless Schwinger model. In Fig. 3 and Fig. 4, we show a typical behavior of the wave function for a small fermion mass and for a large fermion mass, respectively.

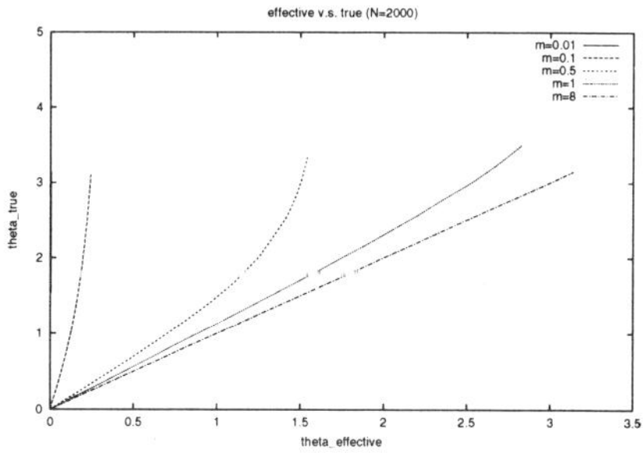

Fig. 2 Effective v.s. True theta.

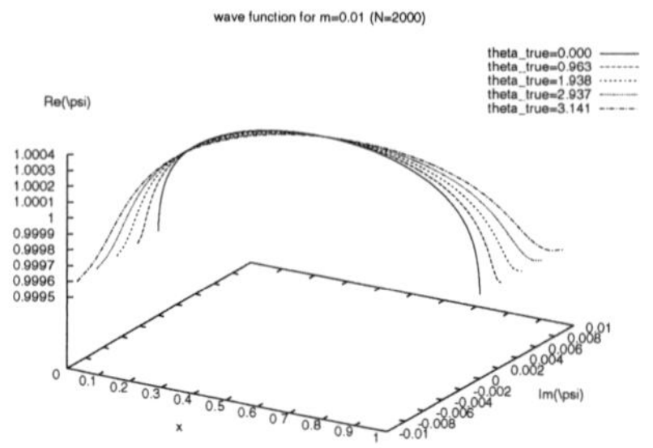

Fig. 3. wave function for a small mass.

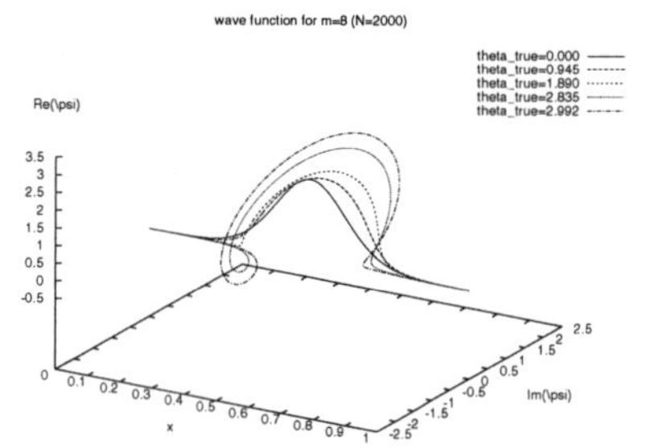

Fig. 4. wave function for a large mass.

It is easy to see that the wave function behaves differently. It implies that there is a phase transition in between. A more direct indication is obtained by looking at how the meson mass changes as a function of θ_{true}. Fig. 5 shows that the meson mass appears to vanish discontinuously at $\theta_{true} = \pi$. We would like to emphasize that we are able to represent a phase transition without explicilty including zero modes.

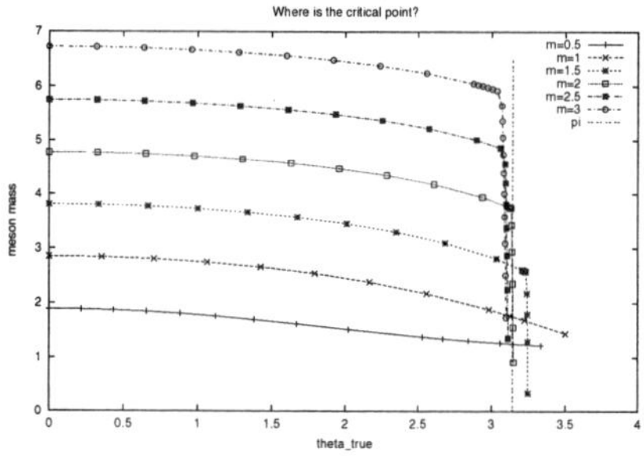

Fig. 5. meson mass v.s. θ_{eff}

Acknowledgments

The author would like to thank Matthias Burkardt for the collaboration on which this contribution is based. He would also like to thank the organizers, especially Dr. Yabu, for their efforts to make the school/workshop very successful.

References

1. See R. Perry's lectures at the school as an introduction.
2. T. Maskawa and K. Yamawaki, *Prog. Theor. Phys.* **56**, 1649 (1976), H. C. Pauli and S. Brodsky, *Phys. Rev.* D **32**, 193 (1985); *ibid* **32**, 2001 (1985).
3. K. Harada, A. Okazaki, and M .Taniguchi, PRD **55**, 4910 (1997).
4. M. Burkardt and K. Harada, *Phys. Rev.* D **57**, 5950 (1998), work in preparation.
5. S. Coleman, *Ann. Phys. (N.Y.)* **101**, 239 (1976).
6. K. Harada, T. Heinzl, and C. Stern, *Phys. Rev.* D **57**, 2460 (1998).

Light-Cone Approach for Exclusive Reactions and Decays in QCD

Chueng-Ryong Ji and Ho-Meoyng Choi
Department of Physics, North Carolina State University, Raleigh, NC 27695-8202, USA

We discuss the usefulness of light-cone degrees of freedom in the exclusive processes. We investigate the electroweak form factors and semileptonic decay rates of pseudoscalar mesons using the constituent quark model based on the light-cone degrees of freedom. Our results demonstrate the broader applicability of light-cone approach including the time-like region of exclusive processes.

1 Introduction

While the well-known successful examples of light-cone approach are often referred to inclusive processes such as the parton model and the PQCD corrections in deep inelastic lepton hadron scattering, etc., even the quark counting rule in exclusive processes rigorously predicted from the light-cone formulation of PQCD is still under severe debate[1]. "How high is really high?" is the main debate; $Q^2 \sim 100?$, 10? or 1? $\mathrm{GeV}^2/\mathrm{c}^2$ as the relevant minimum momentum scale for the PQCD applicability to exclusive processes. While the debate is still going on, we recently applied our light-cone quark model[2] to exclusive meson reactions and decay processes at small Q^2 region($\leq$ few $\mathrm{GeV}^2/\mathrm{c}^2$) and found that the model works fairly well for various mass spectra and wavefunction-related observables. In this presentation, we mainly discuss our light-cone constituent quark model predictions for exclusive reactions and decays with small momentum transfers(i.e. $\mathrm{Q}^2 \leq$ few GeV^2/c^2). Especially, we focus on the calculations of observables in time-like $|Q^2|$ region overcoming the diffculties associated with the quark-antiquark pair creation("Z-graph") in the light-cone approach. As an explicit example of application to time-like $|Q^2|$ region, we present our recent analysis of semileptonic weak decay processes[3]. In the next section(Section 2), we discuss the distinguished features of light-cone degrees of freedom. In Section 3, we describe our QCD motivated light-cone quark model. In Section 4, we present our model calculations of semileptonic pseudoscalar meson decay processes as an example of application to the time-like $|Q^2|$ region. Conclusions and discussions follow in Section 5.

2 Light-Cone Degrees of Freedom

In the light-cone approach, a hadron is characterized by a set of Fock state wave functions, the probability amplitudes for finding different combinations of bare quarks and gluons in the hadron at a given light-cone time $\tau = t + z/c$. These wave functions provide the essential link between hadronic phenomena at short distances(perturbative) and at long distances(non-perturbative)[4]. The distinguished features in the light-cone approach are the simplicity of the vacuum except the zero-modes and the dynamical property of rotation operators. The vacuum at equal τ presents a dramatic difference compared to the vacuum at equal t. For a particle which has the mass m and the four-momentum $k = (k^0, k^1, k^2, k^3)$, the relativistic energy-momentum relation of the particle at equal-τ is given by

$$k^- = \frac{\vec{k}_\perp^2 + m^2}{k^+}, \tag{1}$$

where the light-cone energy conjugate to τ is given by $k^- = k^0 - k^3$ and the light-cone momenta $k^+ = k^0 + k^3$ and $\vec{k}_\perp = (k^1, k^2)$ are orthogonal to k^- and form the light-cone three-momentum $\underline{k} = (k^+, \vec{k}_\perp)$. The rational relation given by Eq.(1) is drastically different from the irrational energy-momentum relation at equal-t, $k^0 = \sqrt{\vec{k}^2 + m^2}$, where the energy k^0 is conjugate to t and the three-momentum vector $\vec{k}$ is given by $\vec{k} = (k^1, k^2, k^3)$. The important point here is that the signs of k^+ and k^- are correlated while at equal t the signs of k^0 and $\mathbf{k}$ are not correlated. Thus the momentum k^+ is always positive because only the positive energy k^- makes the system evolve to the future direction (*i.e.* positive τ), while the momentum k^3 can be either positive or negative even though k^0 is positive to evolve the system in the furture direction (*i.e.* positive t). This provides a remarkable feature to the light-cone vacuum; *i.e.*, the Fock state vacuum is an eigenstate of the full Hamiltonian. Consequently, all bare quanta in hadronic Fock states are associated with the hadron and none are disconnected elements of the vacuum. Furthermore, the problem of boost operators at equal t changing particle numbers can be cured by this framework since the quantization surface $\tau = 0$ is invariant under both longitudinal and transverse boosts defined at equal τ. However, the quantization surface $\tau = 0$ is not invariant under the transverse rotation whose direction is perpendicular to the direction of the quantization axis z at equal τ. Thus, the transverse angular momentum operator involves the interaction that changes the particle number and it is not easy to specify the total angular momentum of a particular hadronic state. Also τ is not invariant under parity. We avoid these problems by using the Melosh transformation of each constituents from equal t to equal τ. The Melosh transformation uniquely determines the assignment of angular momentum, parity and charge conjugation for the light-cone wave function of hadrons.

Table 1: Optimized quark masses m_q[GeV] and the gaussian parameters β[GeV] for both harmonic oscillator and linear potentials obtained from the variational principle. q=u and d.

Potential	m_q	m_s	m_c	m_b	$\beta_{q\bar{q}}$	$\beta_{s\bar{s}}$	$\beta_{q\bar{s}}$	$\beta_{q\bar{c}}$	$\beta_{q\bar{b}}$
H.O.	0.25	0.48	1.8	5.2	0.32	0.37	0.34	0.42	0.50
Linear	0.22	0.45	1.8	5.2	0.37	0.41	0.39	0.47	0.53

3 QCD Motivated Light-Cone Quark Model

The key idea in our light-cone quark model for mesons is to saturate the Fock state expansion by the constituent quark and anti-quark and use the variational principle for a QCD motivated Hamiltonian to fix all the model parameters by comparing with the meson mass spectra. The QCD motivated effective Hamiltonian for the description of the meson mass spectra is given by

$$H_{q\bar{q}} = H_0 + V_{q\bar{q}} = \sqrt{m_q^2 + k^2} + \sqrt{m_{\bar{q}}^2 + k^2} + V_{q\bar{q}}. \tag{2}$$

We use the usual confining interaction potential $V_{q\bar{q}} = V_0(r) + V_{\rm hyp}(r)$ given by

$$V_{q\bar{q}} = a + b\mathcal{V}_{\rm conf.}(r) - \frac{4\kappa}{3r} + \frac{2\vec{S}_q \cdot \vec{S}_{\bar{q}}}{3m_q m_{\bar{q}}} \nabla^2 V_{\rm Coul}, \tag{3}$$

where $\mathcal{V}_{\rm conf.}(r) = r[r^2]$ is linear[harmonic oscillator] type potential. Our variational method first evaluates $< \phi|[H_0 + V_0]|\phi >$ with a trial function $\phi(k^2) = Ne^{-k^2/2\beta^2}$ that depends on the parameters (m, β) and varies these parameters until the expectation value of $H_0 + V_0$ is a minumum. Once these model parameters are fixed, then, the mass eigenvalue of each meson is obtained by $M_{q\bar{q}} =< \phi|[H_0 + V_0]|\phi > + < \phi|H_{\rm hyp}|\phi >$. Our model parameters are summarized in Table 1. Both the mass spectra and the wave functions of the light pseudoscalar (π, K, η, η') and vector$(\rho, K^*, \omega, \phi)$ mesons were analyzed using our light-cone constituent quark model[2]. The mixing angles of $\omega - \phi$ and $\eta - \eta'$ were predicted and various physical observables such as decay constants, charge radii, and radiative decay rates *etc.* were calculated in ref.[2]. Our numerical results in ref.[2] had a good agreement with the available experimental data. In Table 2, the decay constants and charge radii of heavy mesons are summarized.

Table 2: Decay constants[MeV] and charge radii[fm^2] for various heavy pseudoscalar and vector mesons.

Potential	f_D	f_{D^*}	f_B	f_{B^*}	$r^2_{D^+}$	$r^2_{D^0}$	$r^2_{B^+}$	$r^2_{B^0}$
H.O.	127.1	149.6	113.8	122.3	0.182	−0.309	0.420	−0.208
Linear	139.2	168.9	121.2	131.4	0.176	−0.301	0.438	−0.217

4 Semileptonic Pseudoscalar Meson Decays

The semileptonic decays of a pseudoscalar meson $Q_1\bar{q}$ into another pseudoscalar meson $Q_2\bar{q}$ are governed by the weak vector current as follows

$$< P_2|\bar{Q_2}\gamma^\mu Q_1|P_1 > = f_+(q^2)(P_1+P_2)^\mu + f_-(q^2)(P_1-P_2)^\mu, \qquad (4)$$

where $q^\mu = (P_1 - P_2)^\mu$ is the four-momentum transfer to the leptons. In the LFQM calculations presented in ref.[5-7], $q^+ \neq 0$ frame has been used to calculate the weak decays in the time-like region $m_l^2 \leq q^2 \leq (M_1 - M_2)^2$, with $M_{1[2]}$ and m_l being the initial[final] meson mass and the lepton(l) mass, respectively. However, when the $q^+ \neq 0$ frame is used, the inclusion of the non-valence contributions arising from quark-antiquark pair creation("Z-graph") is inevitable and this inclusion may be very important for heavy-to-light and light-to-light decays. Nevertheless, the previous analyses[5,6,7] in $q^+ \neq 0$ frame considered only valence contributions neglecting non-valence contributions. In this work, we circumvent this problem by calculating the processes in $q^+ = 0$ frame and analytically continuing to the time-like region. The $q^+ = 0$ frame is useful because only valence contributions are needed. However, one needs to calculate the component of the current other than J^+ to obtain the form factor $f_-(q^2)$. Since J^- is not free from the zero-mode contributions even in $q^+ = 0$ frame[8], we use $J_\perp$ instead of J^- to obtain f_-. In the standard $q^+ = 0$ frame, we obain the form factors $f_+(q^2)$ and $f_-(q^2)$ using the matrix element of the "+" and "$\perp$"-components of the current, J^μ, respectively, and then continued to the time-like $q^2 > 0$ region by changing $q_\perp$ to $iq_\perp$ in the form facors. We note that our analytic continuation method is equivalent to that of ref.[9] where the form factors are obtained by the dispersion representations through the (Gaussian) wave functions of the initial and final mesons. The more detailed formulations were presented in ref.[3]. The rates for the semileptonic decays $K \to \pi$, $B \to \pi(D)$, and $D \to \pi(K)$ are presented in Table 3 and compared with the available experimental data[10]. Our numerical results are in a good agreement with the available experimental data.

Table 3: Branching ratio $Br(A \to Xl\nu_l)$ for various semileptonic decays.

Processes		$Br^{\text{Th.}}$	$Br^{\text{exp.}}$
$K \to \pi$	H.O.	$(37.60 \pm 0.60)\%$	$(38.78 \pm 0.27)\%$
	Linear	$(37.90 \pm 0.61)\%$	
$B \to \pi$	H.O.	$(1.20 \pm 0.14^{+0.24+0.63}_{-0.25-0.36}) \times 10^{-4}$	$(1.8 \pm 0.4 \pm 0.3 \pm 0.2)$ $\times 10^{-4}$
	Linear	$(1.33 \pm 0.16^{+0.27+0.71}_{-0.28-0.40}) \times 10^{-4}$	
$D \to K$	H.O.	$(3.43 \pm 1.33)\%$	$(3.64 \pm 0.20)\%$
	Linear	$(3.46 \pm 1.34)\%$	
$D \to \pi$	H.O.	$(2.24 \pm 0.33) \times 10^{-3}$	$(3.8^{+1.2}_{-1.0}) \times 10^{-3}$
	Linear	$(2.28 \pm 0.34) \times 10^{-3}$	
$B \to D$	H.O.	$(2.36 \pm 0.36)\%$	$(2.35 \pm 0.2 \pm 0.44)\%$
	Linear	$(2.47 \pm 0.37)\%$	

5 Conclusions and Discussions

The space-like form factors[11] are well described by our light-cone quark model. Analyzing the semileptonic decays of a pseudoscalar meson into another pseudoscalar meson using the same model, we demonstrated that our model can also be applied to the time-like exclusive processes. The form factors $f_\pm$ are obtained in $q^+ = 0$ frame and then analytically continued to the time-like region by changing $q_\perp$ to $iq_\perp$ in the form factors. The matrix element of the "$\perp$" component of the current J^μ is used to obtain the form factor f_-. Our numerical results are in a good agreement with the available experimental data. We also confirmed that our anlaytic continuation method is equivalent to that of ref.[9] where the form factors are obtained by the dispersion representations through the (Gaussian) wave functions of the initial and final mesons. For the estimation of the zero-mode contribution, we calculated the "$-$" component of the current. To the extent that the zero-modes have a significant contribution to some physical observables[3], it seems conceivable that the condensation of zero-modes could lead to the nontrivial realization of chiral symmetry breaking in the light-cone quantization approach. However, the zero-mode contribution seems highly suppressed as the quark mass increases[3,8]. The detailed analysis of heavy-to-heavy and heavy-to-light semileptonic decays will be presented in another communication. In any case, the success of our calculations seems to reveal the effectiveness of light-cone degrees of freedom in exclusive processes.

Acknowledgments

We would like to thank the organizers and all others who helped make this meeting such an enjoyable success. This work was supported by the U.S. Department of Energy(DE-FG-02-96ER40947). The North Carolina Supercomputing Center and the National Energy Research Scientific Computing Center are also acknowledged for the grant of computing time allocation.

References

1. S. J. Brodsky, C. -R. Ji, A. Pang and D. Robertson, *Phys. Rev.* D **57**, 245 (1998); A. Szczepaniak, C. -R. Ji and A. Radyushkin, *Phys. Rev.* D **57**, 2813 (1998); N. Isgur and C. H. Llewellyn Smith, *Phys. Rev. Lett.* **52**, 1080 (1984); *Phys. Lett.* B **217**, 535 (1989); *Nucl. Phys.* B **317**, 526 (1989).
2. H.-M. Choi and C.-R. Ji: hep-ph/9711450.
3. H.-M. Choi and C.-R. Ji, to appear in Phys. Rev. D(hep-ph/9807500).
4. G. P. Lepage, S. J. Brodsky, T. Huang, and P. Mackenzie: in *Particle and Fields 2*, edited by A.Z. Capri and A.N. Kamal (Plenum, New York, 1983), p.83.
5. N.B. Demchuk et al., Phys. At. Nuclei **59**, 2152 (1996).
6. I. L. Grach et al., *Phys. Lett.* B **385**, 317 (1996).
7. H.-Y. Cheng et al., *Phys. Rev.* D **55**, 1559 (1997).
8. H.-M. Choi and C.-R. Ji, *Phys. Rev.* D **58**, 071901 (1998).
9. D. Melikhov, *Phys. Lett.* B **394**, 385 (1997); *Phys. Rev.* D **53**, 2460 (1996).
10. Particle Data Group, R. M. Barnett et al., *Phys. Rev.* D **54**, 1 (1996).
11. H.-M. Choi and C.-R. Ji, *Nucl. Phys.* A **618**, 291 (1997); *Phys. Rev.* D **56**, 6010 (1997).

PROTON SPIN PROBLEM IN QCD

Nikolai Kochelev
Bogoliubov Laboratory of Theoretical Physics,
Joint Institute for Nuclear Research, Dubna, Moscow region, 141980 Russia

Vicente Vento
Departament de Física Teòrica and Institut de Física Corpuscular, Universitat de València-CSIC E-46100 Burjassot (Valencia), Spain

It is shown that a consistent treatment of the axial anomaly leads to an explicit cancellation between the the infrared and ultraviolet contributions to the flavor singlet axial charge. This result is a consequence of the simultaneous crossing of the zero point energy and the ultraviolet cut-off by quark levels of defined chirality from the vacuum in the presence of gluon fields. Based on this consideration we give the arguments in favor of a large violation of the Ellis-Jaffe sum rule by predicting the vanishing of the flavor singlet axial charge g_A^0 .

1 Introduction

During the last ten years there have been many attempts, both theoretical and experimental (see review [1]) to understand how the spin is distributed among the different components of the proton. This investigations have been labeled generically as *the proton spin problem.*

There are two lines of thought, among the many theoretical developments providing an explanation of the small portion of nucleon spin carried by the quarks, which merit our attention here. One of them attributes this fact to the contribution of the polarized gluons via the anomaly to the first moment of $g_1(x, Q^2)$ [2,3,4], which results from the analysis of the triangle diagram contribution to the singlet axial-vector current. In this case only the large $k_\perp \to \infty$ region of integration over transfer momentum of quark in the triangle diagram is involved, which is associated with the *ultraviolet* description of the axial anomaly.

The other line suggests that a large negative quark polarization, leading to a non-perturbative contribution to the flavor singlet axial charge (FSAC), is responsible for the proton spin. In QCD, this polarization can be generated by the presence of non-perturbative gluon fields called instantons, which lead to the famous t' Hooft interaction. Microscopically this interaction arises, because in the presence of the instanton field, quark-antiquark pairs are created by the zero energy quark modes. This approach therefore is connected with the *infrared* treatment of the non conservation of the axial charge (see review [5]).

The aim of this presentation is to proof that one should consider both contributions, the infrared and the ultraviolet, to the singlet axial-vector charge together, in order to be consistent with the properties of the QCD vacuum. In order to do so we have shown that these contributions are related to non-perturbative properties of the QCD vacuum, i.e., the behavior of the Dirac level spectrum. Moreover, due to Dirac level number conservation in the vacuum, they are equal in magnitude and of opposite sign cancelling each other. Our message is therefore that the small value of g_A^0 can be understood as a consequence of non-perturbative effects in QCD, namely the level motion of the Dirac spectrum in the vacuum.

2 Two faces of axial anomaly and their contribution to flavor singlet axial charge

Present wisdom tells us that the key to resolve the *spin crisis* is in the non conservation of the flavor singlet axial-vector current

$$\partial_\mu J^0_{\mu 5}(x) = 2i \sum_i m_i \bar{q}\gamma_5 q + 2N_f \frac{\alpha_s}{8\pi} G^a_{\mu\nu} \widetilde{G^a_{\mu\nu}}, \tag{1}$$

due to axial anomaly term, which is the last in Eq.(1). A very detailed account of the various ways to get Eq.(1) was presented in the review by Shifman [6]. In there, it is discussed how the anomaly presents itself with two faces, the infrared and the ultraviolet one. The reason for this duality being that there are two ways of getting the adequate result: the infrared one, by looking at the motion of the low lying levels of well defined chirality in the Dirac vacuum; the ultraviolet one, by looking to the corresponding high levels after introducing a gauge invariant cut-off. We will make use of this description to show that one should take into account both, the infrared and the ultraviolet, contributions in our process and that, for the FSAC, they tend to cancel.

Our basic argument goes as follows. The number of Dirac levels with definite chirality should be conserved. Fig. 1 of ref. [6] is very clarifying for the discussion that follows. The crossing of the levels by the zero energy point, which leads to a valence quark depolarization, should be accompanied by a simultaneous crossing by the levels of the ultraviolet boundary. As shown by Shifman, both contributions are equal in magnitude, but the flow is opposite and therefore their sign opposite, therefore their simultaneous consideration produces a cancellation. In the simplified model of ref. [6] this result is immediate. In terms of the more familiar language in four dimensions, the crossing of the ultraviolet cut-off determines the perturbative contribution of the gluons to the non conservation of the FSAC. The conservation of the number of

Dirac level implies therefore, that the two contributions, infrared and *perturbative* cancel each other and that the FSAC is conserved. This conservation implies that a Goldstone boson is not needed in this channel to insure axial-vector current conservation. This statement reflects the solution to the $U_A(1)$ problem.

A similar analysis can be done following the description of Mueller [7] for the massive case. The paralelism is complete, since the intermediate non chiral states do not contribute, and the cancellation is explicitly shown. In this case it appears as the cancellation between the charge coming from the anomaly, the infrared term, and the charge coming from the mass term, which arises from the ultraviolet region, and corresponds to the perturbative term. It is necessary to stress that the mass term in Eq. (1) plays a crucial role in the cancellation and therefore it can neither be neglected nor treated as a small perturbation. We must stress that up to now we are dealing with the vacuum, thus no valence quarks have been considered in the argument, only levels in a Dirac sea.

For the proton, the same analysis is valid, but one has to take care of the additional valence quark terms. In the massless case, chirality is a good quantum number and the analysis of number level conservation follows through without problem. In the massive case, there are additional contributions coming from the region where chirality is not a good quantum number. One expects these contributions, which arise from the mass terms, to be small [8]. In the case of the instanton vacuum they have been shown to be zero since, besides the negative sea quark polarization mentioned above, the zero modes flip the spin of the valence quarks [11]. This is the main difference one encounters between QED and QCD, the chiral limit is good in QCD but not in QED. In the former the spin flips, due to the presence of the instantons, and the valence contribution vanishes. In the latter it does not and the valence contribution remains.

The authors of ref. [8] have arrived to similar conclusion, however we must stress, that it is the cancellation between the anomaly (ultraviolet ΔG), and the screaned (due to depolarization) valence contributions which make the FSAC vanish. In particular within instanton model for $N_f = 1$, and only in this case, the two terms vanish independently.

It should be emphasised that this cancellation has been found in several papers on the axial anomaly contribution to polarized DIS [1]. For example, in the the perturbative approach to axial anomaly the contribution to the matrix element of the singlet axial-vector current comes from triangle diagram.

The total result has two terms arising from two different regions of integration over $k_\perp$ [a]

$$< g, s|J^0_{\mu 5}|g, s >_{pert} \sim -\frac{2N_f\alpha_s}{4\pi}\left(1 - \int_0^1 \frac{2m^2(1-x)dx}{m^2 + P^2x(1-x)}\right), \qquad (2)$$

where P^2 is the virtuality of gluon. There is an *explicit* cancellation in Eq.(2) between these contributions from the two different regions of momentum transfer of the quarks in the box diagramm, the small $k_\perp$ region, $k_\perp^2 \approx m^2$, and the large $k_\perp$ region, $k_\perp^2 \to \infty$ when $P^2/m^2 \to 0$, i.e. the limit of small current quark masses.

The two terms in Eq.(2) are directly related to the two different terms in Eq.(1). The last term in Eq.(2) represents the matrix element $m\bar{q}\gamma_5 q$ in Eq.(1) and therefore, in the usual treatment of axial anomaly contribution to DIS, it can be absorbed in the redefinition of the quark polarization densities. The first term in Eq.(2), which comes from ultraviolet region, is usually considered as the gluon contribution to proton spin [b].

The widespread argument used to eliminate the second term in Eq.(2) is that in the region of small $k_\perp$ perturbative QCD does not apply and therefore in this region this term should be substituted by non-perturbative QCD effects. This argument does not apply to our non-perturbative calculation where, we should stress once more, it is precisely the quark mass term which is resposible for the vanishing.

One approach to take into account non-perturbative QCD effects in DIS is the instanton model for the QCD vacuum [10]. In the framework of this model the matrix element of the flavor singlet axial current between gluon states with different momenta $Q = p' - p$ is [12]

$$< g', s|J^0_{\mu 5}|g, s >_{inst} \sim (K_2(\rho Q) - \frac{2K_1(\rho Q)}{Q\rho}), \qquad (3)$$

where ρ is the instanton size. In Eq.(3) the first term comes from *non-zero modes* in the instanton field and corresponds to the anomaly term in Eq.(1). The second term has its origin in the *zero modes* and is related to the mass term in Eq.(1). Again, as in the perturbative case, at $Q^2 \to 0$ by using expansion of the functions $K_2(Q\rho) \to 2/(Q\rho)^2$, $K_1(Q\rho) \to 1/Q\rho$, we observe an *explicit* cancellation between the two contributions in Eq.(3). This cancellation is the fundamental reason behind the resolution of the $U_A(1)$ problem. At $Q^2 \neq 0$ there are two massless poles in Eq.(3). One of them is the usual Goldstone pole which is related to the mass term in Eq.(1), the other comes from the

---

[a] For simplicity we put below the same value for all quark masses.
[b] See also ref. [9] for QED.

anomaly. The consequence of the explicit cancellation between these two poles is the appearence of the massive η' meson.

The calculation just described is the conventional one in the framework of the instanton model for the QCD vacuum (see for example [11], [10]), and as such purely non-perturbative. However, we must point out, that in many papers where this type of calculations were performed only zero modes were taken into account. We have shown the importance of non-zero modes in the calculation of the matrix element of the flavor singlet axial current.

One could also calculate the matrix element of divergence of the flavor singlet axial-vector current between proton states in the framework of the same instanton model for the QCD vacuum. It has been done in the particular case of $N_f = 1$ in a paper by Forte and Shuryak [11]. Again the explicit cancellation has been found. This is however a model dependent statement.

It is evident that the cancellation phenomenon is a general statement of the QCD vacuum and can be explained by quotating from Shifman's review [6] : *Both phenomena, though - the crossing of the zero-energy point and the departure (arrival) of the levels via the ultraviolet cut-off - occur simultaneously and represent, actually, two different facets of the one and the same anomaly, which admits both, the infrared and ultraviolet interpretation.*

Therefore we come to the conclusion that, due to the properties of the QCD vacuum, a cancellation between the infrared and the ultraviolet contributions takes place, and therefore, a small matrix element between proton states of the right side of divergence of the singlet axial-vector current is to be expected, which implies an almost conserved FSAC.

The matrix element between nucleon states of the flavor singlet axial-vector current is given by

$$< p'|J^0_{\mu 5}|p >= g^0_A(Q^2)\bar{P}\gamma_\mu\gamma_5 P + g^0_P(Q^2)q_\mu\bar{P}\gamma_5 P \tag{4}$$

and by recovering the matrix element of Eq.(1) we immediately arrive to the conclusion that

$$g^0_A = 0, \tag{5}$$

which is the main result of this paper. Therefore, we predict very large violation of the Ellis-Jaffe sum rule [13].

The modern experimental data on spin-dependent structure function g_1 [1] give the value of singlet axial-vector current $g^0_A(5\div 10GeV^2) \approx 0.2\div 0.28$. From our point of view, the difference between Eq.(5) and data comes from the inaccessible low x region where the *negative* contribution from quark zero-modes should occur [14]. Naive Regge extrapolation at $x \to 0$, used in experimental papers should be used with utmost precaution.

3 Conclusion

We have shown that there is a cancellation between the contributions from the ultraviolet and the infrared regions to the FSAC and that it appears as a consequence of the general properties of the QCD vacuum. Therefore it should be respected by all models.

A main observation of our work is the relevance that the quark mass term in the equation for divergence of singlet axial-vector current plays in the cancellation due to existence of zero quark modes in the QCD vacuum.

In spite of the fact that these two contributions combined explicitly cancel in the FSAC, their contributions to the polarized structure functions does not, because each of them has a different x dependence. So in the large x region the contribution from gluons should be large while at low x a large contribution from zero modes is expected. These two contributions have also a different structure for the particles in the final state. The ultraviolet contribution leads to quark and antiquark jets with high $k_\perp$ and different helicities for the quark and antiquark. On the contrary, the infrared contribution is associated with the production of a $q\bar{q}$ pairs with small $k_\perp$ and the same helicity for quark and antiquark [9].

We have presented a new point of view which emphasizes, how the non-perturbative behavior of the QCD vacuum, leads to a better understanding of the proton spin problem.

References

1. M.Anselmino, A.Efremov, and E.Leader, *Phys. Rep.* **261**, 1 (1995).
2. A.V.Efremov and O.A.Teryaev, JINR preprint E2-88-287 (1988).
3. G.Altarelli and G.G.Ross, *Phys. Lett.* B **212**, 391 (1988).
4. R.D.Carlitz,J.C.Collins and A.H.Muller,*Phys. Lett.* B **214**, 229 (1988).
5. A.E.Dorokhov, N.I.Kochelev, and Yu.A. Zubov *Int. J. of Mod. Phys.* A **8**, 603 (1993) 603.
6. M.A. Shifman, *Phys. Rep.* **209**, 341 (1991).
7. A.H. Mueller, *Phys. Lett.* B **234**, 517 (1990).
8. Igor Halperin and Ariel Zhitnitsky, hep-ph/9706251.
9. A. Freund and L.M. Sehgal, *Phys. Lett.* B **41**, 90 (1994).
10. T.Schäfer, E.V.Shuryak, hep-ph/9610451.
11. S.Forte and E.V.Shuryak,*Nucl. Phys.* B **357**, 153 (1991).
12. B.V.Geshkenbein and B.L.Ioffe,*Nucl. Phys.* B **166**, 340 (1980).
13. J.Ellis and R.L.Jaffe, *Phys. Rev.* D **9**, 1444 (1974).
14. N.I.Kochelev, *Phys. Rev.* D **57**, 5539 (1998).

INDUCED PSEUDOSCALAR COUPLING CONSTANT

IL-TONG CHEON
Department of Physics, Yonsei University, Seoul, 120-749, Korea
E-mail: itcheon@phya.yonsei.ac.kr

MYUNG KI CHEOUN
Department of Physics, Yonsei University, Seoul, 120-749, Korea
E-mail: cheoun@phya.yonsei.ac.kr

The recent TRIUMF experiment for $\mu^- p \to n\nu_\mu \gamma$ gave a surprising result that the induced pseudoscalar coupling constant g_P was larger than the value obtained from $\mu^- p \to n\nu_\mu$ experiment as much as 44 %. With the axial vector current derived by extension of the Gell-Mann-Levy equation, we found an additional term to the matrix element of Beder and Fearing. This additional term plays a key role to restore the reliability of $g_P(-0.88m_\mu^2) = 6.77g_A(0)$.

Recently, the TRIUMF group measured the radiative muon capture (RMC) on a proton, $\mu^- p \to n\nu_\mu \gamma$, and extracted a surprising result [2] for the induced pseudoscalar coupling constant

$$\hat{g}_P \equiv g_P(-0.88m_\mu^2)/g_A(0) = 9.8 \pm 0.7 \pm 0.3 \ . \tag{1}$$

The PCAC (Partially Conserved Axial Current) predicts

$$g_P(-0.88m_\mu^2) = \frac{2m\ m_\mu}{m_\pi^2 + 0.88m_\mu^2} g_A(0) = 6.77g_A(0). \tag{2}$$

This 44% discrepancy is serious because the theoretical value of g_P is predicted in a fundamental manner based on the PCAC imposed on the axial vector current and agrees with the ordinary muon capture (OMC) value. As long as the PCAC is assumed to be creditable, a doubt may be cast on the result of TRIUMF experiment. However, the measured photon energy spectrum seems to be reliable in view point of their enough experimental experiences in TRIUMF. In order to solve this puzzle, one has to reexamine carefully the Beder-Fearing formula [3,4] used to extract the g_P value from the measured spectrum.

The chiral perturbation calculation (ChPT) was recently carried out for OMC [5] and well reproduced the PCAC prediction, i.e. $\hat{g}_P = 6.77$. It is also consistent with the result of the heavy baryon chiral perturbation theory [6]. Therefore, we believe that the calculation based on PCAC is reliable. Since the RMC amplitude is generated merely by the minimal coupling procedures from the OMC amplitude [3], the calculation method based on PCAC might preserve the confidence even for the RMC.

The axial current coupled to the electromagnetic field can be derived in an elegant manner, i.e. a standard gauge transformation. We start from the ordinary linear σ-model given by the Lagrangian

$$\mathcal{L}_0 = \bar{\Psi}[i\gamma^\mu\partial_\mu + g(\sigma + i\vec{\tau}\cdot\vec{\pi}\gamma_5)]\Psi + \frac{1}{2}[(\partial_\mu\vec{\pi})^2 + (\partial_\mu\sigma)^2] - \frac{1}{2}\mu^2(\vec{\pi}^2 + \sigma^2) - \frac{\lambda}{4}(\vec{\pi}^2 + \sigma^2)^2 \tag{3}$$

Although this Lagrangian is invariant under the SU(2) isovector infinitesimal chiral and gauge transformations, $\Psi \to \Psi' = (1 + i\gamma_5\vec{\eta}\cdot\frac{\vec{\tau}}{2})\Psi, \bar{\Psi} \to \bar{\Psi}' = \bar{\Psi}(1 + i\gamma_5\vec{\eta}\cdot\frac{\vec{\tau}}{2})$, $\sigma \to \sigma' = \sigma + \vec{\eta}\cdot\vec{\pi}$ and $\vec{\pi} \to \vec{\pi}' = \vec{\pi} - \vec{\eta}\sigma$, it describes only a massless fermion. In order to create the pion mass, the chiral symmetry breaking term $\zeta\sigma$ should be included into $\mathcal{L}_0$. Since the vacuum expectation value of the σ field does not vanish, i.e. $\langle 0|\sigma|0\rangle = f_\pi$, the σ field is shifted as $\sigma \to \tilde{\sigma} = \sigma - f_\pi$ and, then, $\langle 0|\tilde{\sigma}|0\rangle = 0$ is fulfilled. Then, the pion and sigma meson's masses can be given as $m_\pi^2 = \mu^2 + \lambda f_\pi^2$, $m_\sigma^2 = \mu^2 + 3\lambda f_\pi^2$ and $\zeta = f_\pi m_\pi^2$, and the nucleon mass is $m = -gf_\pi$.

The resulting Lagrangian without the symmetry breaking term $\zeta\sigma$ becomes

$$\begin{aligned}\mathcal{L}_0 = \bar{\Psi}[i\gamma^\mu\partial_\mu - m + g(\tilde{\sigma} + i\vec{\tau}\cdot\vec{\pi}\gamma_5)]\Psi + \frac{1}{2}[(\partial_\mu\vec{\pi})^2 + (\partial_\mu\tilde{\sigma})^2] \\ -\frac{m_\pi^2}{2}\vec{\pi}^2 - \frac{m_\sigma^2}{2}\tilde{\sigma}^2 - f_\pi[\lambda(\vec{\pi}^2 + \tilde{\sigma}^2) + m_\pi^2]\tilde{\sigma} - \frac{\lambda}{4}(\vec{\pi}^2 + \tilde{\sigma}^2)^2 + const\end{aligned} \tag{4}$$

By the relation

$$exp(\frac{i}{f_\pi}\gamma_5\vec{\tau}\cdot\vec{\phi}) = cos(\frac{\phi}{f_\pi}) + i\gamma_5\vec{\tau}\cdot\hat{\phi}sin(\frac{\phi}{f_\pi}) \equiv \frac{1}{f_\pi}[\tilde{\sigma} + i\gamma_5\vec{\tau}\cdot\vec{\pi}] \; , \tag{5}$$

where $\hat{\phi} = \vec{\phi}/\phi$, and replacement of $\tilde{\sigma} \to \tilde{\sigma}' = \frac{m}{g} + f_\pi cos(\frac{\phi}{f_\pi})$, the Lagrangian can be rewritten for $g = -m/f_\pi$ as

$$\mathcal{L}_0' = \bar{\Psi}[i\gamma^\mu\partial_\mu + gf_\pi exp(\frac{i}{f_\pi}\vec{\tau}\cdot\vec{\phi}\gamma_5)]\Psi + \frac{1}{2}(\partial_\mu\vec{\phi})^2 \quad , \tag{6}$$

where $\mathcal{L}_0' = \mathcal{L}_0 - const.$

With this Lagrangian, the axial coupling constant appears to be unity. However, it is well known as $g_A = 1.25$. To cure this defect, the following chiral invariant Lagrangian is added to $\mathcal{L}_0'$ [7],

$$\mathcal{L}_1 = C_1\bar{\Psi}\gamma^\mu\gamma_5\frac{\vec{\tau}}{2}\Psi\cdot f_\pi\partial_\mu\vec{\phi} \; , \tag{7}$$

where C_1 is determined so as to give $g_A = 1.25$. Our Lagrangian is

$$\mathcal{L}_0'' = \mathcal{L}_0' + \mathcal{L}_1, \tag{8}$$

with $g_A = 1 + C_1 f_\pi^2$.

It is easy to see that the Lagrangian $\mathcal{L}_0''$ holds a global chiral symmetry under the infinitesimal phase transformations $\Psi \to \Psi' = (1 + i\gamma_5 \frac{\vec{\tau}}{2} \cdot \vec{\eta})\Psi$, $\vec{\phi} \to \vec{\phi}' = \vec{\phi} - f_\pi \vec{\eta}$. Even when the derivative is replaced by a covariant derivative, i.e. $\partial_\mu \to D_\mu = \partial_\mu - ie\epsilon_\mu$ where ϵ_μ is the photon polarization vector, $\mathcal{L}_0'$ remains in the global chiral symmetry, since higher order terms than e^2 and $\vec{\eta}^2$ are ignored and thus, $e\vec{\eta}$ yields negligible small quantities.

For deriving the axial current from the Lagrangian, let us first obtain the extended Euler equation by defining the action,

$$\mathcal{S} = \int d^4x \mathcal{L}(\phi_i(x), D_\mu^\pm \phi_i(x)) \ , \tag{9}$$

where $D_\mu^\pm = \partial_\mu \pm \kappa_\mu$. Under an infinitesimal variation $\delta\phi_i(x)$, the stationary condition yields

$$0 = \delta\mathcal{S} = \int d^4x [\frac{\partial \mathcal{L}}{\partial \phi_i} \delta\phi_i + \frac{\partial \mathcal{L}}{\partial (D_\mu^\pm \phi_i)} \delta(D_\mu^\pm \phi_i)] \ . \tag{10}$$

Since $\delta(D_\mu^\pm \phi_i) = \delta(\partial_\mu \phi_i \pm \kappa_\mu \phi_i) = \partial_\mu(\delta\phi_i) \pm \kappa_\mu \delta\phi_i$, the integrand becomes

$$\begin{aligned} &\frac{\partial \mathcal{L}}{\partial \phi_i} \delta\phi_i + \frac{\partial \mathcal{L}}{\partial (D_\mu^\pm \phi_i)} (\partial_\mu \delta\phi_i) \pm \kappa_\mu \frac{\partial \mathcal{L}}{\partial (D_\mu^\pm \phi_i)} \delta\phi_i \\ &= [\frac{\partial \mathcal{L}}{\partial \phi_i} - (\partial_\mu \mp \kappa_\mu) \frac{\partial \mathcal{L}}{\partial (D_\mu^\pm \phi_i)}] \delta\phi_i + \partial_\mu [\frac{\partial \mathcal{L}}{\partial (D_\mu^\pm \phi_i)} \delta\phi_i] \ . \end{aligned} \tag{11}$$

Thus, the extended Euler equation is found as

$$\frac{\partial \mathcal{L}}{\partial \phi_i} - D_\mu^\mp \frac{\partial \mathcal{L}}{\partial (D_\mu^\pm \phi_i)} = 0 \ . \tag{12}$$

Next, we derive the extended Gell-Mann-Levy equations. The variation $\delta\phi_i = -\eta_a(x) F_i^a$ changes the Lagrangian $\mathcal{L}$ as $\mathcal{L} + \delta\mathcal{L}$, where $\delta\mathcal{L}(\phi_i, D_\mu^\pm \phi_i)$. Then, we have

$$\begin{aligned} \delta\mathcal{L} &= \frac{\partial \mathcal{L}}{\partial \phi_i} \delta\phi_i + \frac{\partial \mathcal{L}}{\partial (D_\mu^{(\pm)} \phi_i)} \delta(D_\mu^\pm \phi_i) \\ &= -\frac{\partial \mathcal{L}}{\partial \phi_i} \eta_a(x) F_i^a - \frac{\partial \mathcal{L}}{\partial (D_\mu^{(\pm)} \phi_i)} (D_\mu^\pm \eta_a(x)) F_i^a \\ &= -[D_\mu^\mp (\frac{\partial \mathcal{L}}{\partial (D_\mu^{(\pm)} \phi_i)} F_i^a)] \eta_a(x) - [\frac{\partial \mathcal{L}}{\partial (D_\mu^{(\pm)} \phi_i)} F_i^a](D_\mu^\pm \eta_a(x)) \ , \end{aligned} \tag{13}$$

where we used the extended Euler equation in (12). Defining $(\frac{\partial \mathcal{L}}{\partial(D_\mu^\pm \phi_i)} F_i^a) = -A_a^\mu$, we obtain the extended Gell-Mann-Levy equations,

$$A_a^\mu = -\frac{\partial \delta\mathcal{L}}{\partial(D_\mu^\pm \eta_a)} \quad , \quad D_\mu^\mp A_a^\mu = -\frac{\partial \delta\mathcal{L}}{\partial \eta_a} \; . \tag{14}$$

If $\partial_\mu \eta_a = 0$, i.e. η_a is independent of x, eq.(13) reduces to $\delta\mathcal{L} = (\partial_\mu A_a^\mu)\eta_a$. Since $\delta\mathcal{L} = 0$ if the lagrangian holds a global symmetry under the infinitesimal variation, we have $\partial_\mu A_a^\mu = 0$. This statement is synonymous with the Noether's theorem.

Under the local transformations, $\bar{\Psi} \to \bar{\Psi}' = \bar{\Psi}(1 + i\gamma_5 \frac{\vec{\tau}}{2} \cdot \vec{\eta}(x)), \Psi \to \Psi' = (1 + i\gamma_5 \frac{\vec{\tau}}{2} \cdot \vec{\eta}(x))\Psi, \vec{\phi} \to \vec{\phi}' = \vec{\phi} - f_\pi \vec{\eta}(x)$, and $\partial_\mu \to D_\mu = \partial_\mu - ie\epsilon_\mu$, the Lagrangian, $\mathcal{L}_0''$ reads as

$$\begin{aligned} \tilde{\mathcal{L}}_0 = {} & \bar{\Psi}[i\gamma^\mu D_\mu + g f_\pi exp(\frac{i}{f_\pi}\gamma_5 \vec{\tau} \cdot \vec{\phi})]\Psi + C_1 \bar{\Psi}\gamma^\mu \gamma_5 \frac{\vec{\tau}}{2}\Psi \cdot f_\pi D_\mu \vec{\phi} \\ & - g_A \bar{\Psi}\gamma^\mu\gamma_5 \frac{\vec{\tau}}{2}\Psi \cdot (D_\mu \vec{\eta}) + \frac{1}{2}(D_\mu \vec{\phi})^2 - (D^\mu \vec{\phi}) \cdot f_\pi (D_\mu \vec{\eta}). \end{aligned} \tag{15}$$

For this case, eq.(13) yield

$$\delta\tilde{\mathcal{L}}_0 = -g_A \bar{\Psi}\gamma^\mu \gamma_5 \frac{\vec{\tau}}{2}\Psi \cdot (D_\mu \vec{\eta}) - (D^\mu \vec{\phi}) \cdot f_\pi (D_\mu \vec{\eta}) \; . \tag{16}$$

Thus, by eq.(14), the axial current and its divergence are given as

$$A_a^\mu = g_A \bar{\Psi}\gamma^\mu\gamma_5 \frac{\tau_a}{2}\Psi + f_\pi D^\mu \phi_a \quad , \quad D_\mu^{(+)} A_a^\mu = 0 \; . \tag{17}$$

Here, notice that ${D_\mu}^{(+)} = \partial_\mu + ie\epsilon_\mu$ because ${D_\mu}^{(-)} = D_\mu = \partial_\mu - ie\epsilon_\mu$. If we add the chiral symmetry breaking term $-\frac{m_\pi^2}{2}\vec{\phi}^2$ to $\mathcal{L}_0''$, the second equation in (17) yields

$$D_\mu^{(+)} A_a^\mu = -m_\pi^2 f_\pi \phi_a \; , \tag{18}$$

which is the extended form of the PCAC. The same equation was also given by Adler [8].

Operating a covariant derivative $D_\mu^{(+)}$ on eq.(17) and equating it to eq.(18), we find

$$-f_\pi (D_\mu^{(+)} D^\mu + m_\pi^2)\phi_a = g_A D_\mu^{(+)} \bar{\Psi}\gamma^\mu\gamma_5 \frac{\tau_a}{2}\Psi \; , \tag{19}$$

i.e., in ignoring e^2-order term,

$$f_\pi (q_\mu^2 - m_\pi^2)\phi_a = g_A [2mi\bar{\Psi}\gamma_5 \frac{\tau_a}{2}\Psi + ie\bar{\Psi}\epsilon_\mu \gamma^\mu \gamma_5 \frac{\tau_a}{2}\Psi] \; . \tag{20}$$

When the solution of this equation ϕ_a with a definition, $g_A/(q^2 - m_\pi^2) = -g_P/2mm_\mu$, is substituted into the first equation of (17), we obtain

$$A_a^\mu(x) = \bar{\Psi}(x)[g_A\gamma^\mu\gamma_5 + \frac{g_P(q^2)}{m_\mu}q^\mu\gamma_5 - \frac{eg_P(q^2)}{m_\mu}\epsilon^\mu\gamma_5]\frac{\tau_a}{2}\Psi(x) + \frac{eg_P(q^2)}{2mm_\mu}q^\mu[\bar{\Psi}(x)\epsilon_\alpha\gamma^\alpha\gamma_5\frac{\tau_a}{2}\Psi(x)] \ , \tag{21}$$

where e^2 term is ignored. The fourth term is missing [9] in the previous calculations [3,4]. As will be shown below, it is actually a very important term corresponding to the seagull term in the pion photoproduction [10].

Following the Fearing's formulation, one can evaluate the relativistic amplitude of RMC on a proton as

$$M_{fi} = M_a + M_b + M_c + M_d + M_e + \Delta M_e \tag{22}$$

with

$$\begin{aligned} M_a &= -\epsilon_\alpha\bar{u}_n\Gamma^\delta(Q)u_p \cdot \bar{u}_\nu\gamma_\delta(1-\gamma_5)\frac{\not{\mu} - \not{k} + m_\mu}{-2k\cdot\mu}\gamma^\alpha u_\mu \ , \\ M_b &= \epsilon_\alpha[AL_\delta\bar{u}_n\Gamma^\delta(K)\frac{\not{p} - \not{k} + m_p}{-2k\cdot p}(\gamma^\alpha - i\kappa_p\frac{\sigma^{\alpha\beta}}{2m_p}k_\beta)u_p \ , \\ M_c &= \epsilon_\alpha L_\delta\bar{u}_n(-i\kappa_n\frac{\sigma^{\alpha\beta}}{2m_n}k_\beta)\frac{\not{n} + \not{k} + m_n}{2k\cdot n}\Gamma^\delta(K)u_p \ , \\ M_d &= +\epsilon_\alpha L_\delta\bar{u}_n(\frac{2Q^\alpha + k^\alpha}{Q^2 - m_\pi^2}\frac{g_P(K^2)}{m_\mu}K^\delta\gamma_5)u_p \ , \\ M_e &= \epsilon_\alpha L_\delta\bar{u}_n(\frac{ig_M}{2m}\sigma^{\delta\alpha} + \frac{g_P(Q^2)}{m_\mu}\gamma_5 g^{\delta\alpha})u_p \ , \\ \Delta M_e &= +\epsilon_\alpha L_\delta\bar{u}_n(\frac{g_P(K^2)}{2mm_\mu}Q^\delta\gamma_5\gamma^\alpha)u_p \ , \end{aligned} \tag{23}$$

where

$$\Gamma^\delta(q) = g_V\gamma^\delta + \frac{ig_M}{2m}\sigma^{\delta\beta}q_\beta + g_A\gamma^\delta\gamma_5 + \frac{g_P(q^2)}{m_\mu}q^\delta\gamma_5 \ , \tag{24}$$

$L_\delta = \bar{u}_\nu\gamma_\delta(1-\gamma_5)u_\mu, K = n - p + k$ and $Q = n - p$ with momenta of neutron, proton and photon, n, p and k, respectively. And $m \sim m_p \sim m_n$. Other constants are taken as $g_V = 1.0, g_A = -1.25, g_M = 3.71, \kappa_p = 1.79$ and $\kappa_n = -1.91$ [3]. M_e term is originated from the third term in eq.(21) and ΔM_e term comes from the fourth term. But the latter, ΔM_e, is missing in the paper by

Fearing[3,4]. Accordingly, this term was not included in the previous procedure of extracting g_P value from the experimental photon energy spectrum[2]. The transition rate is given by

$$\frac{d\Gamma_{RMC}}{dk} = \frac{\alpha G^2 |\phi_\mu|^2 m_N}{(2\pi)^2} \int_{-1}^{1} dy \frac{k E_\nu^2}{W_0 - k(1-y)} \frac{1}{4} \sum_{spins} |M_{fi}|^2 , \quad (25)$$

where α is the fine structure constant, G is the standard weak coupling constant, $y = \hat{k} \cdot \hat{\nu}$, $k_{max} = (W_0^2 - m_n^2)/2W_0$, $E_\nu = W_0(k_{max} - k)/[W_0 - k(1 - \nu)]$, $W_0 = m_p + m_n -$ (muon binding energy) and $|\phi_\mu|^2$ is the absolute square of muon wave function averaged over the proton which is taken as a point Coulomb.

In order to compare to the experimental results, we take the following steps. For liquid hydrogen target, muon capture is dominated through the ortho and para $p\mu p$ molecular states[2,11]. Since these molecular states can be attributed to the combinations of hyperfine states of μp atomic states[11] i.e. single and triplet states, we decompose the statistical spin mixture $\frac{1}{4}\sum_{spins}|M_{fi}|^2$ into such hyperfine states by reducing 4×4 matrix elements to 2×2 spin matrix elements. At this step, we confirmed that when the ΔM_e term was not included, eq.(25) reproduced the curves given in ref. [4]. Finally, by exploiting the mixture of muonic states relevant in experiments[2], we calculate the photon energy spectrum. The count number of the photons is now expressed as

$$N = Z \frac{d\Gamma_{RMC}}{dk} . \quad (26)$$

Here Z is determined by adjusting the value of $d\Gamma_{RMC}/dk$ without ΔM_e term for $\hat{g}_P \equiv g_P(-0.88 m_\mu^2)/g_A(0) = 9.8$ so as to agree with the best fit curve shown in Fig.2 of ref. [2]. With this value of Z, we have to examine the case of ΔM_e included.

Our results show that the curve in ref. [2] can be obtained without ΔM_e term for $\hat{g}_P = 9.8$, or with ΔM_e term for $\hat{g}_P = 6.77$. On the other hand, the result calculated without ΔM_e term for $\hat{g}_P = 6.77$ is obviously much lower than the measured spectrum. Although there is a minor discrepancy due to the neglect of higher order contribution and other degree of freedom such as Δ, our result shows that ΔM_e term restores the credit of $\hat{g}_P = 6.77$.

Since the contribution of Δ degree of freedom is known to be a few percent [4], it is not included in the present calculation. Vector mesons such as ρ and ω make also very small contributions. Higher order terms are pointed out to be insignificant[3].

The pion field actually interacts in virtual state with the nucleon and therefore the πNN form factor may be taken into account as an off-shell effect. However, the standard πNN form factor $f_{\pi NN}(q^2) = (\Lambda^2 - m_\pi^2)/(\Lambda^2 - q^2)$ with $\Lambda^2 = m_\rho^2 + m_\pi^2$ participates through an effective $g_P(q^2)$, i.e. $\tilde{g}_P(q^2) = g_P(q^2) f_{\pi NN}(q^2)$ but gives only $4 \sim 5\%$ contribution, because the process occurs at low momentum transfer, $q^2 = -0.88 m_\mu^2$.

As another attempt, the ChPT calculations of RMC have also been carried out [12,13], but the $\hat{g}_P = 6.77$ value could not be extracted. Their results are, more or less, the same as those of Fearing's calculation [3,4], in which the important "Seagull" term was missing. Thereby, these calculations may have to be reexamined, taking higher order terms into account. It should be noted that the ChPT can satisfy the gauge invariance but it becomes obscure if the Δ degree of freedom is taken into account.

In the present framework, our calculation shows that $\hat{g}_P = 6.77$ is reasonable for both OMC and RMC on a proton.

Acknowledgment

This work was supported by the KOSEF (961-0204-018-2) and the Korean Ministry of Education (BSRI-97-2425). We thank F.Khanna for having called our attention to this problem.

References

1. G.Bardin *et al.*, Phys. Lett. **104B**, 320 (1981) and T.P.Gorringe *et al.*, Phys. Rev. Lett. **72**, 3472 (1994).
2. G.Jonkmans *et al.*, Phys. Rev. Lett. **77**, 4512 (1996).
3. H.W.Fearing, Phys. Rev. **21**, 1951(1980).
4. D.S.Beder and H.W.Fearing, Phys. Rev. **D35**, 2130 (1987).
5. H.W.Fearing, R.Lewis, N.Mobed and S.Schrer, preprint, hep-ph/9702394.
6. V.Bernard, N.Kaiser and Ulf-G. Meissner, Phys. Rev. **D50**, 6899(1994).
7. E.Kh.Akhmedov, Nucl. Phys. **A500**, 596(1989).
8. S.L.Adler, Phys. Rev., **139**, B1638 (1965).
9. It was firstly pointed out by the present author. See preprints, nucl-th/9811009, nucl-th/9811005, and nuclt-th/9811025.
10. F.A.Berends, A.Donnache and D.L.Weaver, Nucl. Phys. **B4**, 54(1967).
11. D.D.Bakalov, M.P.Faifman, L.I.Ponomarev and S.I.Vinitsky, Nucl. Phys. **A384**, 302(1982).
12. T.Meissner, F.Myhrer and K.Kubodera, preprint, nucl-th/9707019.
13. S.Ando and D.P.Min, preprint, hep-th/9707504.

MASSIVE ϕ^4 MODEL AT FINITE TEMPERATURE – RESUMMATION PROCEDURE *A LA* RG IMPROVEMENT –

HISAO NAKKAGAWA* AND HIROSHI YOKOTA**
Institute for Natural Science, Nara University, 1500 Misasagi-cho, Nara 631-8502, JAPAN
* *nakk@daibutsu.nara-u.ac.jp*, ** *yokotah@daibutsu.nara-u.ac.jp*

In this paper the phase structure of the massive $\lambda\phi^4$ model at finite temperature ($T \neq 0$) is investigated by applying a resummation method inspired by the renormalization-group (RG) improvement to the one-loop effective potential. The resummation method *a la* RG-improvement is shown to work quite successfully by resumming up systematically large correction-terms of $O(\lambda T/\mu)$ and of $O(\lambda(T/\mu)^2)$. The temperature-dependent phase transition of the model is shown to proceed through the second order transition. The critical exponents are determined analytically and are compared with those in other analyses.

1 Introduction

Understanding the phase structure and the mechanism of phase transition of quantum field theories at finite temperature/density is important to understand the evolution of Universe and the physics to be searched by the ultrarelativistic heavy-ion experiments planned at the BNL-RHIC and CERN-LHC [1].

To investigate analytically the phase structure of relativistic quantum field theory, the effective potential (EP) is used as a powerful tool. Perturbative calculation of the EP at finite temperature, however, suffers from various troubles: the poor convergence or the breakdown of the loop expansion [2], and the strong dependence on the renormalization-scheme (RS). These troubles have essentially the same origin, i.e., they come about with the emergence of large perturbative corrections depending explicitly on the RS. Thus to break a way out of these troubles we must carry out the systematic resummation [3,4,5,6] of dominant large correction terms, and at the same time we must also solve the problem of the RS-dependence.

Recently simple but very efficient renormalization group (RG) improvement procedures for resumming dominant large correction terms are proposed in vacuum [7] and in thermal [8] field theories. This procedure was originally proposed to solve the problem of the strong RS-dependence of the EP calculated through the loop-expansion method.

It is worth noticing here that in the massive scalar $\lambda\phi^4$ model at high temperature the large correction terms appearing in the L-loop EP have the structures as follows; i) terms proportional to powers of the temperature T:

i-a) $(\lambda(T/\mu)^2)^L$, i-b) $(\lambda(T/\mu))^L$, and ii) terms proportional to powers of the *logs*: ii-a) $(\lambda \ln(T/\mu))^L$, ii-b) $(\lambda \ln(M/\mu))^L$, where M is the large mass scale appearing in the theory.

In this paper we present the result of application of the resummation procedure *a la* RG [8,9] to the massive scalar ϕ^4 model renormalized at the temperature of the environment T. We have found that the proposed resummation procedure *a la* RG works efficiently, not only by resolving the problem of the RS-dependence, but also by properly as well as systematically resumming terms having the structures i-a) and i-b) above. As for the details of the analyses, see Ref. 9 and the paper to appear [10].

2 Improving the effective potential through resummation and the phase structure of massive ϕ^4 model at $T \neq 0$

Let us consider the massive scalar ϕ^4 model at finite temperature renormalized at an arbitrary mass-scale μ and at the temperature of the environment T (hereafter we call this scheme as the T-renomalization). The key idea to resolve the RS-ambiguity is to use correctly and efficiently the fact that the exact EP satisfies a homogeneous renormalization group equation (RGE) with respect to change of the arbitrary parameter $\mu \to \bar{\mu} = \mu e^t$.

In the scalar ϕ^4 model the dominant large corrections appear as a power function of the effective variable τ (for more details, see Refs. 9 and 10)

$$\begin{aligned}\tau/\lambda &\equiv \Delta_1 \\ &= \frac{T^2}{2\pi^2 M^2}\left\{L_1\left(\frac{T^2}{M^2}\right) - \frac{\pi^2}{12}\right\} - \frac{1}{2\pi^2}L_2\left(\frac{T^2}{\mu^2}\right) + \frac{b}{2}\left(\ln\frac{M^2}{\mu^2} - 1\right), \quad (1)\end{aligned}$$

where $b = 1/16\pi^2$, $M^2 = m^2 + \lambda\phi^2/2$ and $M^2\Delta_1$ is nothing but (a part of) the renormalized one-loop self-energy correction,

$$M^2\Delta_1 \equiv \frac{1}{2}\sum\!\!\!\!\!\!\int \frac{1}{k^2 - M^2} + (\text{one-loop counter term}), \quad (2)$$

$$L_i\left(\frac{1}{a^2}\right) \equiv \frac{\partial^i}{\partial(a^2)^i}L_0\left(\frac{1}{a^2}\right), \quad (i \geq 1),$$

$$L_0\left(\frac{1}{a^2}\right) \equiv \int_0^\infty k^2\, dk \ln[1 - \exp\{-\sqrt{k^2 + a^2}\}] .$$

The resummation of dominant $O(\lambda(T/\mu)^2)$ terms in the T-renomalization can be automatically performed through renormalization, giving the remormalized mass-squared $m^2 \simeq m_0^2 + \frac{1}{24}\lambda T^2$, appearing as a mass-term in the propagator with which the perturbative calculation is performed, where m_0 denotes the renormalized mass in the vacuum theory.

At the one-loop level the RGE's satisfied by the renormalized coupling and mass-squared can be solved exactly, giving solutions to the running parameters $\bar{m}^2$ and $\bar{\lambda}$ as

$$\bar{M}^2 = \bar{m}^2 + \frac{1}{2}\bar{\lambda}\phi^2, \quad \bar{m}^2 = m^2 f^{-1/3}, \quad \bar{\lambda} = \lambda f^{-1}, \tag{3}$$

$$f = 1 - 3\lambda \left[bt + \frac{1}{2\pi^2} \left\{ L_2\left(\frac{T^2}{\bar{\mu}^2}\right) - L_2\left(\frac{T^2}{\mu^2}\right) \right\} \right].$$

Up to now $\bar{\mu}$ in the above eqations (3) can be arbitrary, with μ being fixed at the initial value of renormalization. *Our RG-improvement procedure, i.e., the resummation procedure a la RG, can be carried out by choosing the RS-fixing parameter $\bar{\mu}$ so as to satisfy $\bar{\tau}(t) = 0$, namely to make the one-loop radiative correction to the mass fully vanish.*

The RS-fixing equation $\bar{\tau}(t) = 0$ gives the mass-gap equation [6,8,9]

$$M^2 = m^2 + f(\bar{M}^2)\bar{M}^2 - f(\bar{M}^2)^{2/3} m^2 , \tag{4}$$

which determines, in the HT regime where $T/\mu \gg 1$, the RS-parameter $\bar{\mu}$ being exact up to T-independent constant as

$$\bar{\mu} = \frac{\bar{M}}{2} . \tag{5}$$

Now we can study the consequences of the RG-improvement in the T-renormalization, with solutions $\bar{\lambda}$, $\bar{m}^2$, and $\bar{\mu}$, Eqs. (3) and (5). The RG improvement can then be performed analytically [8,9,10], obtaining the improved EP as

$$\bar{V}_1 = \frac{1}{2}\bar{m}^2\phi^2 + \frac{1}{4!}\bar{\lambda}\phi^4 + \bar{h}\bar{m}^4$$
$$+ \frac{\bar{M}^4}{2}\left[-\frac{b}{4} + \frac{T^4}{\pi^2 \bar{M}^4} L_0\left(\frac{T^2}{\bar{M}^2}\right) - \frac{T^2}{2\pi^2 \bar{M}^2} L_1\left(\frac{T^2}{\bar{M}^2}\right) - \frac{T^2}{24\bar{M}^2} \right] \tag{6}$$

$$= \frac{1}{2}\bar{m}^2\phi^2 + \frac{1}{4!}\bar{\lambda}\phi^4 - \frac{\bar{m}^4}{2\bar{\lambda}} - \frac{T\bar{M}^3}{48\pi} + \cdots . \tag{7}$$

With the RG-improved EP, $\bar{V}_1$, Eqs. (6) and (7), we can see the nature of the temperaure-dependent phase-transition of the model; i) At low temperature below $T_c \sim \sqrt{24|m^2|/\lambda}$ the EP has twofold structure showing the existence of two phases, Fig. 1a, the ordinary mass phase and the

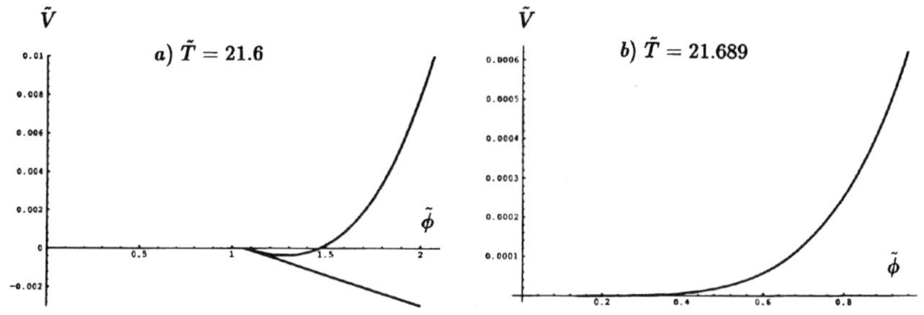

Figure 1: The RG improved effective potential of the massive ϕ^4 model at two temperatures in the T-renormalization: a) $\tilde{T}_1 = 21.6$ and b) $\tilde{T}_2 = \tilde{T}_c = 21.689$ with $\tilde{T} \equiv T/|m_0|$. $\tilde{V} \equiv [\bar{V}_1(\tilde{\phi}) - \bar{V}_1(\tilde{\phi}_{min})]/|m_0|^4$, $\tilde{\phi} \equiv \phi/|m_0|$ and $\lambda = 1/20$.

small mass phase. The ordinary mass phase, with its counterpart in the tree-level potential, is the symmetry-broken phase which develops its minimum at $\phi = \phi_0 \sim \{T_c^4|m^2|^3/\lambda\}^{1/10}$. The small mass phase is a new "symmetric" phase, without having any counterpart in the tree-level potential, with a linearly decreasing potential unbounded from below, indicating the simple ϕ^4 model becoming an unstable theory at low temperature. As the temperature becomes higher the the minimum of the ordinary mass phase eventually diminishes, and ii) at the critical temperature T_c the minimum of the potential at non-zero ϕ completely disappears. The EP shows a symmetric structure in ϕ with the minimum at $\phi = 0$, $V(\phi) - V(0) \propto \phi^{\delta+1}$, $\delta \sim 5.0$, Fig. 1b, and iii) at high temperature above T_c the EP remains symmetric in ϕ with its minimum at $\phi = 0$. Transition between the symmetry-broken phase at low temperature and the symmetric phase at high temperature proceeds through the second order transition.

3 Critical exponents

Here we present the critical exponents determined from the RG-improved one-loop EP in the T-renormalization, Eqs. (6) and (7). In this case we can calculate the critical exponents through analytic manipulation.

The definition of the critical exponents are as follows; 1) On the behavior at $\phi = \phi_0$ around $T \simeq T_c$: $\phi_0 \propto (T_c - T)^{\beta}$, $d^2V/d\phi^2|_{\phi=\phi_0} \propto |T_c - T|^{\gamma}$, $V(\phi_0) - V(0) \propto |T_c - T|^{2-\alpha}$. 2) On the behavior at $T = T_c$: $V(\phi) - V(\phi_0 = 0) \propto \phi^{\delta+1}$ or $dV/d\phi \propto \phi^{\delta}$. Here ϕ_0 denotes the position of the true minimum, and T_C denotes the critical temperature, which are determined as

Table 1: Critical exponents obtained from various methods.

	β	γ	δ	α
Our result	0.3	1.2	5.0	0.2
mean-field	0.5	1.0	3.0	0.0
lattice [12]	0.324	1.24	4.83	0.113
experimental [11]	0.325	1.24		0.112

$\phi_0 \simeq \{54^3 T_c^4 |m^2|^3/\lambda(8\pi)^4\}^{1/10}$, $T_c \simeq \sqrt{24|m_0^2|/\lambda} - 3|m_0^2|/4\pi\mu$.

Results are summarized in Table 1, showing that our result deviates significantly from the mean-field values and agrees reasonably with the experimental data [11,12].

4 Summary and discussion

In this paper we proposed a new resummation method inspired by the renormalization-group improvement. Applying this resummation procedure *a la* RG-improvement to the one-loop effective potential in the massive scalar ϕ^4 model renormalized at the temperature of the environment T, we found important observations; The $O(\lambda(T/\mu)^2)$-term resummation, thus the so-called hard-thermal-loop resummation [3] in this model, can be simply done through the T-renormalization itself. With the lack of freedom we can set only one condition to choose the RS-fixing parameter, which actually ensures to absorb the large terms of $O(\lambda T/\mu)$, thus only the partial resummation of these terms can be carried out.

It is to be noted that all the results obtained are essentially the same as those in the T_0-renormalization case [9,10]: the second order phase transition between the ordinary mass broken phase at low temperature and the symmetric phase at high temperature, and the existence of the unstable small mass phase at low temperature. In this sense our resummation method gives stable results so long as the terms of $O(\lambda T/\mu)$ are systematically resummed. The critical exponents are determined by analytic manipulation, showing the significant deviation from the mean-field values and the reasonable agreement with the experimental data [11,12]. For details, see Refs. 9 and 10.

As noticed above, the RG-improved EP in the simple massive ϕ^4 model has, below the critical temperature T_c, an unstable small-mass phase in addition to the ordinary symmetry-broken phase. This unstable phase also appears in the same model at exact zero-temperature, indicating its appearence being not the artifact due to the crudeness of approximation on the temperature-dependent correction terms. Though the origin of its appearence is not fully understood, it may have a relation with the triviality of the model, which is an interesting problem for further studies. The $O(N)$ symmetric model in the large-N limit exists as a stable theory without having such an unstable phase[8].

Acknowledgments

This work is partly supported by the Special Research Grant-in-Aid of the Nara University.

References

1. See, e.g., Quark Matter '97, *Nucl. Phys.* A **638** (1998).
2. A. D. Linde, *Rep. Prog. Phys.* **42**, 389 (1979); *Phys. Lett.* B **96**, 289 (1980); D. J. Gross, R. D. Pisarski and L. G. Yaffe, *Rev. Mod. Phys.* **53**, 43 (1981).
3. E. Braaten and R. D. Pisarski, *Nucl. Phys.* B **337**, 569 (1990); ibid. **339**, 310 (1990); F. Frenkel and J. C. Taylor, ibid. **334**, 199 (1990).
4. L. Dolan and R. Jackiw, *Phys. Rev.* D **9**, 3320 (1974).
5. E. Braaten, *Phys. Rev. Lett.* **74**, 2164 (1995); E. Braaten and A. Nieto, *Phys. Rev.* D **51**, 6990 (1995); ibid. **53**, 3421 (1996); F. Karsch, A. Patkos and P. Petreczky, *Phys. Lett.* B **401**, 69 (1997).
6. G. Amelino-Camelia and S.-Y. Pi, *Phys. Rev.* D **47**, 2356 (1993).
7. M. Bando, T. Kugo, N. Maekawa and H. Nakano, *Phys. Lett.* B **301**, 83 (1993).
8. H. Nakkagawa and H. Yokota, *Mod. Phys. Lett.* A **11**, 2259 (1996).
9. H. Nakkagawa and H. Yokota, in *TFT98* [http://xxx.lanl.gov/html/hep-ph/9811469], proceedings of the 5th International Workshop on Thermal Field Theories and Their Applications, Regensburg, Germany, 10–14 August 1998, edited by U. Heinz.
10. H. Nakkagawa and H. Yokota, papers to appear.
11. J. Zinn-Justin, *Quantum Field Theory and Critical Phenomena* (Claredon, Oxford, 1989).
12. H.W.J. Blöte *et al.*, *Physica* A **161**, 1 (1989).

EFFECTS OF MESON MASS DECREASE ON SUPERFLUIDITY IN NUCLEAR MATTER

TOMONORI TANIGAWA
Department of Physics, Kyushu University,
Fukuoka 812-8581, Japan
E-mail: tomo2scp@mbox.nc.kyushu-u.ac.jp

MASAYUKI MATSUZAKI
Department of Physics, Fukuoka University of Education,
Munakata, Fukuoka 811-4192, Japan
E-mail: matsuza@fukuoka-edu.ac.jp

$^1S_0$ pairing in nuclear matter is investigated by taking the hadron mass decrease into account via the "In-Medium Bonn potential" which was recently proposed by Rapp *et al.* The resulting gap is significantly reduced in comparison with the one obtained with the original Bonn-B potential and we ascertain that the meson mass decrease is mainly responsible for this reduction.

Superfluidity in nuclear matter is one of the important issues in nuclear physics since it forms the foundation for nuclear structure theory and has close relationship to neutron-star physics. To describe a nuclear many-body system, relativistic models are attracting attention as well as non-relativistic ones. This is due to the success of Quantum Hadrodynamics (QHD) originated with the study by Chin and Walecka in 1970's.[1]

We studied $^1S_0$ pairing in nuclear matter by means of QHD, which has been succeeded in describing the various phenomena of nuclei, that is, not only the saturation property of nuclear matter, but spherical, deformed and rotating nuclei. Therefore improvement on the description of pairing which is essential in open shell nuclei would make QHD more reliable. To speak of neutron-star physics, relativistic effect might be important due to high density which is a few times larger than the normal nuclear matter density.

As is well known, QHD is an effective field theory of hadronic degrees of freedom. The Lagrangian density for the present model is like this:

$$\begin{aligned}\mathcal{L} = &\bar{\psi}(i\gamma_\mu\partial^\mu - M)\psi + \frac{1}{2}(\partial_\mu\sigma)(\partial^\mu\sigma) - \frac{1}{2}m_\sigma^2\sigma^2 \\ &- \frac{1}{4}\Omega_{\mu\nu}\Omega^{\mu\nu} + \frac{1}{2}m_\omega^2\omega_\mu\omega^\mu - \frac{1}{4}\boldsymbol{R}_{\mu\nu}\cdot\boldsymbol{R}^{\mu\nu} + \frac{1}{2}m_\rho^2\boldsymbol{\rho}_\mu\cdot\boldsymbol{\rho}^\mu \\ &- g_\sigma\bar{\psi}\sigma\psi - g_\omega\bar{\psi}\gamma_\mu\omega^\mu\psi - g_\rho\bar{\psi}\gamma_\mu\boldsymbol{\tau}\cdot\boldsymbol{\rho}^\mu\psi - \frac{1}{3}g_2\sigma^3 - \frac{1}{4}g_3\sigma^4,\end{aligned} \quad (1)$$

$$\Omega_{\mu\nu} = \partial_\mu \omega_\nu - \partial_\nu \omega_\mu, \quad \boldsymbol{R}_{\mu\nu} = \partial_\mu \boldsymbol{\rho}_\nu - \partial_\nu \boldsymbol{\rho}_\mu. \tag{2}$$

Here ψ is the nucleon field, σ stands for σ-boson field, ω for ω-meson field and ρ for ρ-meson field. Non-linear self-coupling terms for σ-boson, which are crucial for a realistic description of nuclear properties within Relativistic Mean Field Theory (RMFT), are also included. Deriving the particle-particle (p-p) channel interaction in the relativistic models is one of the current topics though practical approach, namely, the use of a non-relativistic force (*e.g.* Gogny force) in p-p channel, has been adopted in the finite nuclei calculations.

The first study of pairing by means of QHD was done by Kucharek and Ring in 1991. [2] In order to incorporate the pairing field, meson fields must be quantized and the anomalous Green's function is defined by the Gor'kov factorization. The customary manipulation gives Relativistic-Hartree-Bogoliubov (RHB) equation. Hence a transcendental equation of effective nucleon mass for the mean field and an ordinary non-linear gap equation for the pairing field organize the non-linear system of equations.

The resulting gap obtained by adopting the one-boson-exchange (OBE) interaction with RMFT parameter set such as NL1 is too large to achieve a consensus. Aforementioned "practical approach" for finite nuclei is a remedy for this excessive gap problem. In nuclear matter, alternatively, a realistic force can be used as p-p interaction and produces a consistent result with the non-relativistic studies, where the maximum pairing gap ranges from 2-4 MeV. [3]

A possible extension in alignment with the above policy is to take the change of the hadron properties into consideration. In the pioneering work done by Brown and Rho, it was pointed out that the change of hadron masses may occur in conformity with the chiral symmetry arguments. [4] Although there is still controversy on this subject, some experiments seem to support the vector meson mass decrease. The change is expressed by the linear relation between the masses and the density:

$$\frac{M^*}{M} = \frac{m^*_{\rho,\omega}}{m_{\rho,\omega}} = \frac{\Lambda^*_{\rho,\omega}}{\Lambda_{\rho,\omega}} = 1 - C\frac{\rho}{\rho_0}, \quad C = 0.15, \tag{3}$$

where M is the nucleon mass, $m_{\rho,\omega}$ are the masses of the ρ- and ω-meson and $\Lambda_{\rho,\omega}$ are the cutoff masses in the form factors applied to each nucleon-meson vertex. The scaling factor C is taken to be 0.15, which is almost in line with the one obtained from QCD sum rules. This relation is often referred to as "Brown-Rho (BR) scaling."

Then, Rapp *et al.* showed that hadron mass decrease conforming to this scaling was compatible with the saturation property of nuclear matter.[5] They constructed the OBE potential just by adding two extra σ-bosons to the original Bonn-B potential with slightly modified parameters. This "In-Medium Bonn potential" makes the investigation of the medium effects on superfluidity quite tractable. Accordingly we adopt this potential as the p-p interaction in the gap equation.

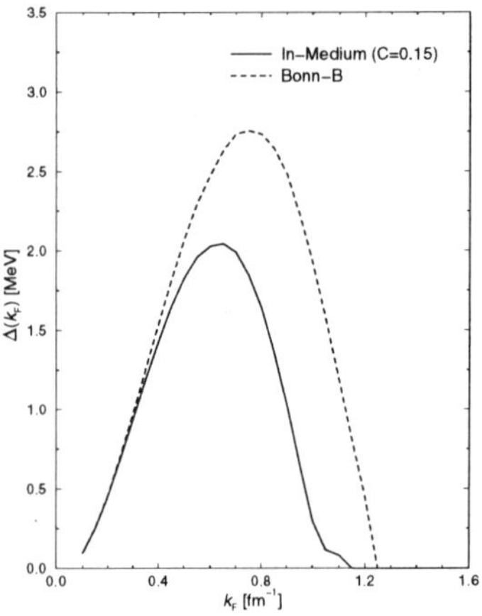

Figure 1: The pairing gaps at the Fermi surface as a function of Fermi momentum k_F. Details are in the text.

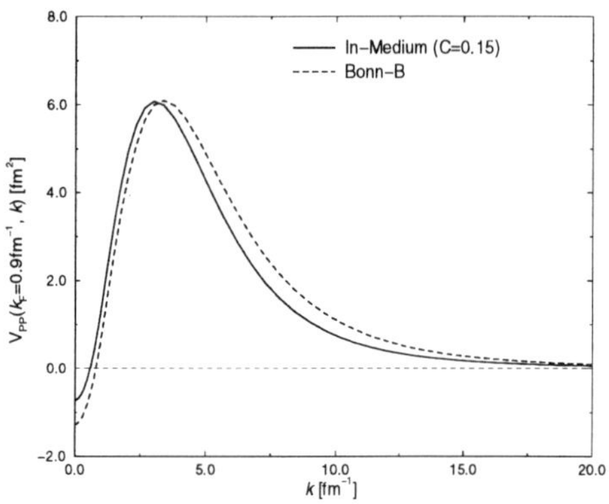

Figure 2: The particle-particle channel interactions as a function of momentum, with a Fermi momentum $k_F = 0.9$ fm^{-1}. Details are in the text.

The resulting pairing gap at the Fermi surface is shown in Figure 1, which is drawn as a function of Fermi momentum with the solid line. The dashed line corresponds to the gap obtained by using the original Bonn-B potential. Comparison shows that the inclusion of hadron mass decrease in concert with BR scaling reduces the gap significantly, however, the values stay in the physically acceptable range.

Before going into the detailed discussion of the gap, the structure of the gap equation has to be reviewed to clarify how the contributions come from each momentum region. As pointed out by Rummel and Ring,[3] on the one hand, positive contributions come from the low- and high-momentum region where the gap has the opposite sign to the potential, which is due to the minus sign in the integrand of the gap equation. On the other hand, negative contributions come from the intermediate-momentum region where the both potentials have a repulsive peak.

Next, to see the mass decrease effects on the p-p interaction, In-Medium Bonn potential and the original Bonn-B potential are given in Figure 2, which represents the shift of the In-Medium Bonn potential to the lower-momentum region. In other words, the hadron mass decrease leads to reduction of magnitude in the region giving the positive contributions to the gap as mentioned above. This is the main reason why reduction of the gap occurs in the case of In-Medium Bonn potential.

But a question may arise: which hadron is responsible for this reduction, nucleon or meson? In order to answer this, we also calculate the gap applying BR scaling only to either hadron. The result is shown in Figure 3, where the solid, dashed and long-dashed line correspond to the case of decreasing only the nucleon mass, only the meson masses and the both, respectively. This figure ascertains that the reduction of the gap is mainly due to the meson mass decrease though the nucleon is responsible for it to some extent.

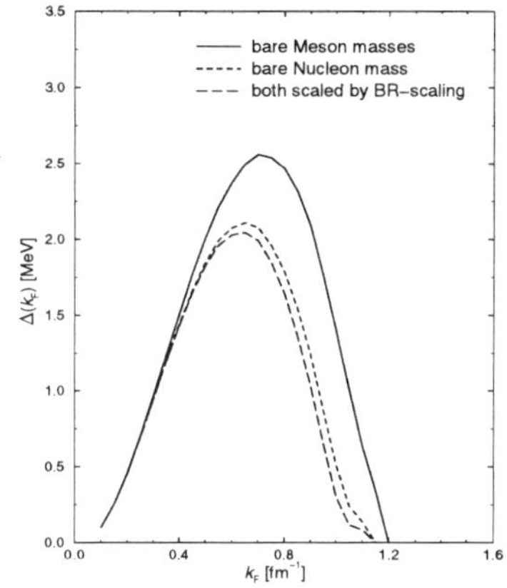

Figure 3: The pairing gaps obtained with the partially BR scaled p-p interactions. Details are in the text.

In summary, we performed the numerical calculation of the pairing gap using In-Medium Bonn potential as the p-p interaction. The resulting gap is considerably reduced in comparison with the one obtained with the original Bonn-B potential and is consistent with the non-relativistic studies. The use of the meson theoretic potential reveals that the vector meson mass decrease, or lengthening the range of repulsive forces accounts for the reduction. Whether the polarization of Dirac sea as well as Fermi sea is effective for it remains a conjecture left for further investigation. [6]

References

1. B. D. Serot and J. D. Walecka, *Adv. Nucl. Phys.* **16**, 1 (1986).
2. H. Kucharek and P. Ring, *Z. Phys.* A **339**, 23 (1991).
3. A. Rummel and P. Ring, *preprint* (1996).
4. G. E. Brown and M. Rho, *Phys. Rev. Lett.* **66**, 2720 (1991).
5. R. Rapp *et al.*, *LANL e-Print archive* nucl-th/9706006.
6. M. Matsuzaki and P. Ring, *ibid.* nucl-th/9712060; M. Matsuzaki, *Phys. Rev.* C **58**, 3407 (1998).

EOS for Kaon Condensed Matter at Finite Temperature

M. Yasuhira and T. Tatsumi

Department of Physics, Kyoto University,
Kitashirakawa, Kyoto 606-8502 JAPAN
E-mail: yasuhira@ruby.scphys.kyoto-u.ac.jp,
tatsumi@ruby.scphys.kyoto-u.ac.jp

A new formulation is presented to treat thermal and quantum fluctuations around the kaon condensate on the basis of chiral symmetry. With the idea of chiral rotation, we can formulate the inclusion of fluctuations in a transparent way. Goldstone-mode corresponding to K^- appears in the kaon condensed phase and thermal and quantum fluctuations around the condensate give sizable effects on the thermal properties. Nucleon and kaon loops are properly renormalized to give finite contribution to equation of state(EOS).

1 Introduction

In high density matter physics, kaon condensatation(at zero temperature) has been disscussed in relation to physics of neutron stars, for example, through the large softening of EOS, the kaon condensed phase considerably reduces the maximum mass of neutron stars.

Recently the finite temperature EOS seems to be very important with respect to protoneutron stars (PNS) and their delayed collapse. PNS are formed after supernovae explosions and some of them may collapse to the low mass black holes during the stage of the liberation of neutrinos and/or the initial cooling.

Here we discuss the kaon condensate at finite temperature($T \neq 0$). In the recent works, Prakash et al [1] discussed the finite-temperature kaon condensate in the mean field theory, while it was not based on chiral symmetry. Thorsson and Ellis[2] derived the thermodynamic function in the kaon-condensed phase but their argument was limitted to the zero-temperature case. We have presented a new formulation to treat the thermal and quantum fluctuations around the condensate based on chiral symmetry[7,8]. By the extension of the idea of chiral rotation on the chiral manifold G/H with $G = SU(3)_L \times SU(3)_R$, $H = SU(3)_V$ and using the imaginary-time path-integral method, we can get the thermodynamic function in a transparent way.

2 Formulation

2.1 Chiral Rotation

We start from the Kaplan-Nelson Lagrangian[3] in which the meson fields correspond to the coordinates spanned on the chiral manifold $G/H \simeq SU(3)$: Goldstone field can be written as $U(\phi_a) = \exp(2iT_a\phi_a)$ with the generators of $SU(3)$ Lie algebra T_a and the meson fields ϕ_a. In the mean-field approximation the condensed state(**K**) is generated by the chiral rotation[4] ζ, $\zeta = \exp\{i(T_+\langle K^+\rangle + T_-\langle K^-\rangle)/\sqrt{2}f\}$, with $T_\pm = T_4 \pm iT_5$.

In the usual treatment of fluctuations meson field is deviated into two parts, $\phi_a = \langle\phi_a\rangle + \tilde{\phi}_a$. In terms of chiral rotation on the manifold this corresponds to the path: **O→FK** in Fig.1. We take another path:**O→F→FK**.

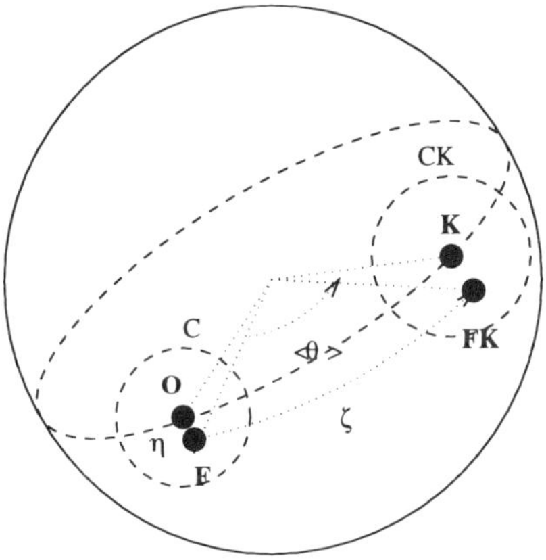

Fig.1: Chiral rotations on the $SU(3)$ manifold. Circles C, CK indicate the fluctuation "areas" around the meson vacuum (**O**) and kaon condensed state (**K**), respectively. $\langle\theta\rangle$ represents the thermal average of the condensate.

Then U field is parametrized as $U(\langle\theta\rangle = |\langle K^\pm\rangle|, \tilde{\phi}_a) = \zeta\eta^2\zeta$ with fluctuation field $\eta = \exp(iT_a\tilde{\phi}_a)$. This also corresponds to the separation of zero mode[5]. In other words, we introduce the *local* coordinate around the condensate, so that we can use the approximately flat curvature and don't worry about the Lee-Yang term.

2.2 Dispersion Relation

Substituting the above parametrization for U, we perform the imaginary time path integral, and thereby get the thermodynamic function including both effects of the fluctuations and condensate. Since the symmetric part of the

Lagrangian is invariant under the grobal rotation ζ, the dynamics is governed only by the symmetry-breaking part.

From the pole of the thermal Green function, we can get the dispersion relation of kaonic modes in the kaon condensed phase. There appear two modes; one is the very massive mode and the other is the soft mode, $E_{-}(p=0)=0$, corresponding to K^{-}. We can see this soft mode is the Goldstone-mode by the spontaneous breaking of V-spin symmetry (SSB). These massive and soft modes affects the thermal properties through the Bose-Einstein distribution function, and in particular the soft mode is very important as it should be.

2.3 Zero-point fluctuations

Since our thermodynamic function includes the zero-point divergences through nucleon and kaon loops, we should properly renormalize it. By the renormalizations of the self-energies for kaons and nucleons at $T=0$, we can get the finite contribution. As a result we find kaons' zero-point energies (ZP) are tiny but nucleons' ZP have the sizable effects and reduce the condensate.

3 Numerical Results and Discussions

In Fig.2 we show the pressure to see thermal effects on EOS and energy gain per baryon to estimate the ZP effects.

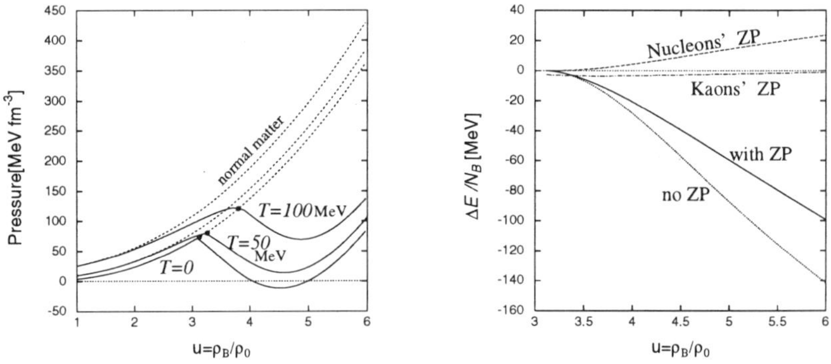

Fig.2: [left]Thermal effect for EOS. Pressure at $T=0,50,100$MeV for kaon existing matter with solid lines and for normal matter with dotted lines. Stiffer pressure lines correspond to higher temperature. The dots mean the critical points in each

case. [right]ZP effect $\Delta E/N_B \equiv (E_{total} - E_{total}^{normal\ matter})/N_B$, and ZP contributions at $T = 0$. Solid line shows the results with the zero-point fluctuation by nucleon, while dotted line those in the tree level. Dashed line shows the ZP contribution. Kaon ZP contributions are estimated by the use of optimum values(dash-dotted one).

Critical density lies around $(3.08 - 3.87)\rho_0$ ($\rho_0 = 0.16 fm^{-3}$) in this case. As for the pressure, we can see the thermal effects are 2 or 3 times stronger in the kaon condensed phase than those in the normal phase. It is to be noted that even in the non-condensed phase, the existence of thermal kaons considerably soften EOS. Hence it is clear that the thermal kaons play an important role as the thermal effects. As for the ZP contributions, we can see that the kaons' ZP is tiny but nucleons' ZP gives rise to a sizable effect. Nucleons' ZP affects the thermal properties of kaon condensed matter through the change of optimum values of the parameters like $\langle\theta\rangle$.

By the extension of the idea of chiral rotation we have succeeded in the treatment of the fluctuations around the condensate based on chiral symmetry. In the condensed phase, Goldstone-mode appears as a result of SSB and strongly affects the thermal properties of the system.

With this formulation we intend to calculate EOS including the ZP and neutrino trapping effects and study the static properties of protoneutron stars. Finally we would simulate the delayed collapse dynamically.

References

1. M. Prakash *et al*, Phys. Reports. **280** (1997) 1.
2. V. Thorsson and P.J. Ellis, Phys. Rev. **D55** (1997) 5177.
3. D.B. Kaplan and A.E. Nelson, Phys. Lett. **B175** (1986) 57; **B179** (1986) 409(E).
4. T. Tatsumi, Prog. Theor. Phys. Suppl. **120** (1995) 111 and references cited therein.
5. J. Gasser and H. Leutwyler, Phys. Lett. **B188** (1987) 477.
6. T.W. Baumgarte *et al*, Astrophys. J. **443** (1996) 680.
7. T. Tatsumi and M. Yasuhira, *nucl-th/9811054*, Phys. Lett. **B** in press,
8. T. Tatsumi and M. Yasuhira, *nucl-th/9811067*.

KAON CONDENSATION IN AN EXCLUDED VOLUME MEAN FIELD MODEL WITH MESON EXCHANGE INTERACTION

NIRVIKAR PRASAD, V. K. TIWARI AND C. P. SINGH
Department of Physics, Banaras Hindu University,
Varanasi 221 005, INDIA.
E-mail: cpsingh@banaras.ernet.in

We develop an equation of state for a multicompoment hadron gas in the mean field formalism by incorporating excluded-volume correction arising due to a hard-core repulsion between a pair of baryons and meson exchange correction in the kaon- sector. We use it to study the kaon-condensation problem and find that kaon-condensation phenomenon is a model-dependent one. It occures in such a description of a hot and dence hadron gas when hard-core radius of baryons $r_N \leq 0.6$ fm.

1 Introduction

Currently there is much interest on the problem of kaon-condensation in a baryon-rich hadron gas at a low temperature [**1**–**3**]. Such a situation occurs in the core of a neutron star [**4**]. The idea that, above some critical density, the ground state of baryonic matter might contain a Bose-Einstein condensate of kaons was first given by Kaplan and Nelson [**1**–**2**].

Several efforts were made, recently, for constructing a consistent approach to kaon condensation based on the chiral models. Here we use a more phenomenological approach based on the one-boson exchange model. Recently we have developed a mean-field formalism which includes the excluded-volume correction for the hard-core size of the baryons in a multicomponent hadron gas [**5**]. Our purpose here is to compare the predictions of traditional meson exchange picture describing the kaon-nucleon intercation with those of the chiral model.

2 Method of Calculation

In this calculation, we have included the following ingredients:

(1) An attractive interactions between baryons due to the exchange of a scalar ρ–meson.

(2) A repulsive interaction arising due to ω–exchange interaction.

(3) The hard core repulsion between baryons represented by a new thermodynamic mean-field potential $U_{VdW}(n,T)$.

(4) An attractive interaction between K–mesons due to a scalar σ–meson exchange.

(5) A repulsive intercation between K–mesons due to a vector ω–meson exchange.

In our model [**5**], the hard-core of a baryon is an additional parameter. We consider the multicomponent hadron gas consisting of low-lying baryons N, Λ, σ antibaryons $\bar{N}$, $\bar{\Lambda}$, $\bar{\sigma}$ and mesons π, k, k^*, ρ, ω, ϕ, η respectively.

The expression for the total pressure of the hadron gas in our model is

$$P = \frac{1}{3}\sum_j g_j \int \frac{d^3k}{(2\pi)^3}\frac{k^2}{(M_j^{*2}+k^2)^{1/2}}[f_j + f_{\bar{j}}] + P_{VdW}(n,T) + P_B(n_B) + P_\sigma(\sigma) + \sum_m p_m(T) \quad (1)$$

where $f_{j(\bar{j})} = [\exp(\frac{(M_j^{*2}+k^2)^{1/2}+U_{VdW}(n,T)\pm U_{Bj}(n_B)\mp\mu_j}{T}) + 1]^{-1}$.

Here $j = N$, Λ, Σ and $\bar{j}$ stands for their antiparticles. $n = \sum_j(n_j + n_{\bar{j}})$ and $n_B = \sum_j n_{Bj}(\mu_j, T)$ with $N_{Bj} = n_j - n_{\bar{j}}$

Futhermore, we have Van-der Waals' hard-core repulsion terms: $P_{VdW}(n,T) = nT\frac{V_0 n}{1-V_0 n}$ and $U_{VdW}(n,T) = T\frac{V_0 n}{1-V_0 n} - T\ln(1-V_0 n)$

P_B is the baryon density dependent mean-field pressure $P_B = \frac{1}{2}{m_\omega}^2{\omega_0}^2$ where ω_0 is the time-component of the ω-exchange field:

$\omega_0 = \frac{1}{{m_\omega}^2}[g_{\omega NN}n_{BN} + g_{\omega\Lambda\Lambda}n_{B\Lambda} + g_{\omega\Sigma\Sigma}n_{B\Sigma}]$ and $U_{Bj}(n_B) = g_{\omega jj}\omega_0$

The vector interaction of ω-meson with nucleons and hyperons modifies their chemical potential as $\nu_j = \mu_j - U_{Bj}$. Similarly the attractive σ scalar interaction modifies the masses of the baryons as ${M_j}^* = M_j - g_{\sigma jj}\sigma$ and the corresponding pressure is $P_\sigma(\sigma) = -\frac{1}{2}{m_\sigma}^2\sigma^2$. In order to know ${M_j}^*$ of the jth baryon, one should know the scalar field σ which can be obtained from the following thermodynamic consistency conditions: $n\frac{\partial U_{VdW}(n,T)}{\partial n} = \frac{\partial P_{VdW}(n,T)}{\partial n}$ and $\sum_i n_{Bi}\frac{\partial U_{Bi}(n,T)}{\partial n_B} = \frac{\partial P_B(n,T)}{\partial n_B}$. Moreover, the dependence of the effective masses M_j on T and μ_j can be obtained from the following equation: $(\frac{\partial p}{\partial {M_j}^*})_{T,\mu_j} = 0$ Therefore the expression for the scalar field is

$$\sigma = (\frac{1}{{m_\sigma}^2})\sum_j g_{\sigma jj}(n_{\sigma j} + n_{\sigma\bar{j}}) \quad (2)$$

with $n_{\sigma j(\bar{j})} = g_j \int \frac{d^3k}{(2\pi)^3}\frac{{M_j}^*}{({M_j}^{*2}+k^2)^{\frac{1}{2}}} f_{j(\bar{j})}$

To obtain a solution for M^* is a difficult task since it involves a set of coupled and self-consistent equations. We use the following values of the parameters: ${c_s}^2 = \frac{g^2_{\sigma NN}}{m^2_\sigma} = 374.84 GeV^{-2}$; ${c_v}^2 = \frac{g^2_{\omega NN}}{m^2_\omega} = 283.59 GeV^{-2}$ and $g_{\sigma YY} = (\frac{2}{3})g_{\sigma NN}$; $g_{\omega YY} = (\frac{2}{3})g_{\omega NN}$

In the one boson exchange model for the kaon–sector, we use the following scalar and vector meson interaction Lagrangians[6]:

$L_{\sigma k} = -g_{\sigma k} m_k \sigma k\bar{k}$ and $L_{\omega k} = i g_{\omega k}(-\bar{k}\partial_\mu k V^\mu + k\partial_\mu \bar{k} V^\mu)$

where V_μ is the vector ω–meson field. We can simplify notation by introducing an effective kaon mass: ${m_k^*}^2 = {m_k}^2 - g_{\sigma k} m_k \sigma$. Thus the number densities of kaons (antikaons) can be written as:

$$n_{K(\bar{K})} = g_K \int \frac{d^3k}{(2\pi)^3} \frac{1}{\exp\left[\frac{({m_K^*}^2 + k^2)^{1/2} \pm X \mp \mu_K}{T}\right] - 1} \quad (3)$$

where g_k is the degeneracy factor for kaons, μ_k is the strangeness chemical potential and $X = g_{\omega k}\omega_0$. In Eq.(3) we have ignored X^2 term. In terms of μ_k , we can write the chemical potential for a hyperon as: $\mu_{\Lambda(\sigma)} = \mu_N - \mu_K$. In order to solve the above set of equations, we require another constraint equation which we get by invoking the strangeness neutrality condition: $n_{B\Lambda} + n_{B\bar{\Sigma}} + (n_K - n_{\bar{K}} + n_K{}^* - n_{\bar{K}}{}^*) = 0$ Moreover, we have used $\frac{g_{\sigma k} g_{\sigma NN}}{g_{\sigma NN}} = 62.8\, GeV^{-2}$ [6] which ensures that the potential felt by kaons is same for any parametrization of the KN interactions. We further use $g_{\omega k} = \frac{g_{\rho\pi\pi}}{2} = 3.02$ which is supported by the $SU(3)$ ideal mixing scheme.

3 Results and Conclusions

In **Fig.1,** we find that the kaon-condensation occurs in the Walecka model near μ_B=2000 MeV and T=100 MeV. In comparison, our generalised excluded volume model reveals that the Kaon-condensation occurs near $\mu_B = 2800$ MeV and T = 100 MeV when $r_N = 0.6$ fm. However, for $r_N = 0.8$ fm, we find that the possibility for the occurence of Kaon-condensation is remote.
In **Fig.2**, we have shown the same variations at T = 40 MeV.
In **Fig.3**, Curve A represents the calculation in the Walecka model and shows a rapid increasing behaviour. The excluded volume correlations with r_N =0.6fm decreases the value of n_B/n_0 slightly (Curve B). However, for r_N=0.8 fm the repulsion term becomes a prominent one and modifies the calculation significantly. The trend of saturation is clearly seen in the values of n_B/n_0 for $\mu_B \geq$ 1500 MeV.

In conclusion, we have shown that the kaon-condensation is a model dependent phenomenon and its occurence in a cold baryon-rich hadron gas depends how we model the attractive and repulsive interactions existing between baryons and Kaons. In a excluded volume model, formulated in the mean-field framework, it sensitively depends on the hard-core size of the baryons. However, the meson-exchange model of incorporating interactions between Kaons and baryons gives similar results as obtained in the framework of chiral model.

Acknowledgments

N. Prasad and V.K.T are thankful to University Grants Commission (UGC), Govt. of India for financial assistance.

References

1. D.B. Kaplan and A.E. Nelson, *Phys. Lett.* B175, 57 (1986); *Phys. Lett.* B179, 409 (1986).
2. A.E. Nelson and D.B. Kaplan, *Phys. Lett.* B192, 193 (1987).
3. G.E. Brown, K. Kubodera and M. Rho, *Phys. Lett.* B192, 273 (1987).
4. M. Prakash *et al*, *Phys. Rep.* 280, 1 (1997); C.H. Lee, *Phys. Rep.* 275, 255 (1996).
5. V.K. Tiwari, Nirvikar Prasad and C.P. Singh, *Phys. Rev.* C58, 439 (1998); V.K. Tiwari, K.K. Singh, Nirvikar Prasad and C.P. Singh, *Nucl. Phys.* A637, 159 (1998).
6. A. Muller-Groeling, K. Holinde and J. Speth, *Nucl. Phys.* A513, 557 (1990).

FIGURE CAPTIONS

FIG.1:- The variations of the effective mass m_K^* and the strangeness chemical potential μ_K with respect to the baryon chemical potential μ_B at T=100 MeV Curves A, B and C represent the values of m_K^* as calculated in 1) Walecka model, 2) our model with hard-core radius as 0.6 fm and 3) our model with hard-core radius as 0.8 fm, respectively. Curves A" ,B" and C" show the values of μ_K as calculated in the above three cases, respectively.

FIG.2:- The variation of the same quantities as shown in **Fig.1** at a lower temperture T=40 MeV.

FIG.3:- The variations of the baryon density n_B/n_0 with respect to the baryon chemical potenttial μ_B at T=100 MeV. Curves A, B, and C show the calculations in Walecka model and in our model with r_N=0.6 and 0.8fm, respectively.

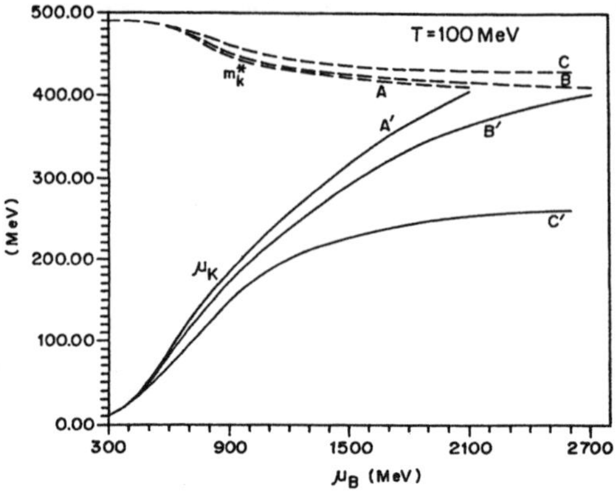

FIG. 1

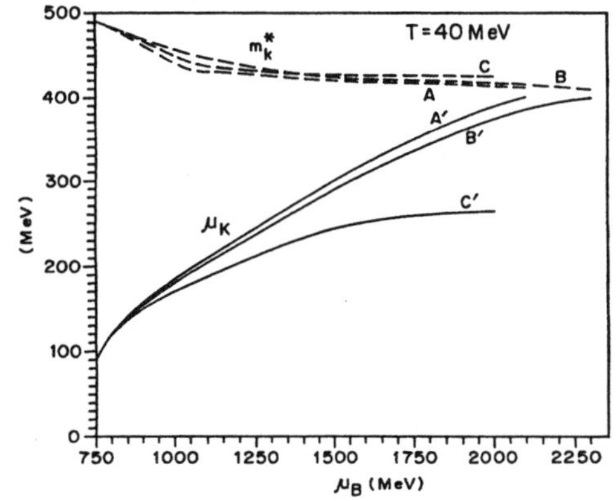

FIG. 2

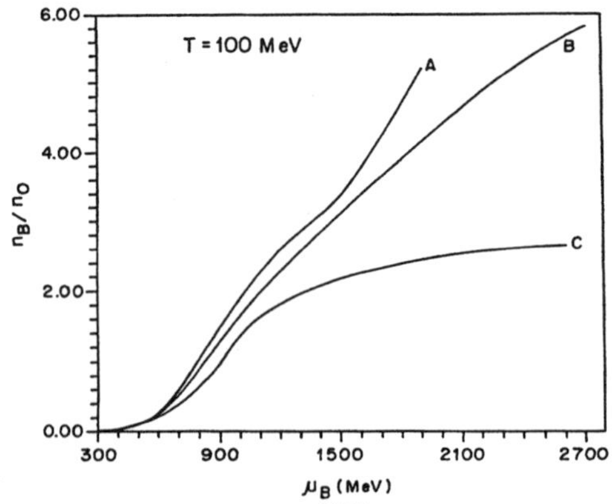

FIG. 3

APPROXIMATE DISCRETE γ_5 SYMMETRY IN NUCLEAR MATTER

H. KOUNO
Department of Physics, Saga University, 1 Honjo-machi,
Saga 840-8502, Japan
E-mail: kounoh@cc.saga-u.ac.jp

Using the quantum hadrodynamics, the discrete γ_5-symmetry (the discrete chiral symmetry), which is the partial symmetry of the continuous chiral symmetry, is studied in nuclear matter, both phenomenologically and theoretically. The theoretical analyses have been done by the non-perturbative renormalization group equations. It is shown that this symmetry may be restored approximately in nuclear matter.

1 Introduction

In the framework of the quantum hadrodynamics (QHD),[1] the effective nucleon mass M^* becomes smaller in nuclear matter. The nucleon mass term breaks the discrete γ_5-symmetry (the discrete chiral symmetry) which is the partial symmetry of the continuous chiral symmetry. If this symmetry is restored in nuclear matter, the odd-terms of σ-meson self-interactions is suppressed as in the case of the effective nucleon mass. We study this problems both phenomenologically and theoretically. The theoretical analyses have been done by the non-perturbative renormalization group equations (NPRGE). [2] This report is based on the collaboration works in Ref. [3], where the details of our results are shown.

2 Phenomenological Analyses

Our phenomenological model[4,5] is based on the following "improved" (or "effective") Lagrangian of σ-ω model,

$$L = \bar{\psi}(i\gamma_\mu\partial^\mu - M + g_s\phi - g_v V_\mu\gamma^\mu)\psi$$

$$+\frac{1}{2}\partial_\mu\phi\partial^\mu\phi - \frac{1}{2}m_s^2\phi^2 - \frac{1}{2}F_{\mu\nu}F^{\mu\nu} + \frac{1}{2}m_v^2V_\mu V^\mu - \sum_{n=0}^{6} c_n\phi^n, \qquad (1)$$

where ψ, ϕ, V_μ, M, m_s, m_v, g_s, g_v and c_n are the nucleon field, the σ-meson field, the ω-meson field, the nucleon mass, the σ-meson mass, the ω-meson mass, the σ-nucleon coupling, the ω-nucleon coupling and the couplings of n-th order σ-meson self-interaction, respectively and $F_{\mu\nu} = \partial_\mu V_\nu - \partial_\nu V_\mu$.

Using the Lagrangian (1), we have done the Hartree calculations for nuclear matter. Different from the usual calculations, the Lagrangian (1) can be used only below some finite cutoff Λ_0 and the exact value of Λ_0 is *unknown*. However, according to the NPRGE, we can use the arbitrary cutoff Λ which is smaller than Λ_0, if we re-define the couplings at Λ. Such a re-definition is done automatically, if we determine the couplings in Eq. (1) phenomenologically. The phenomenological determinations of c_5 and c_6, in addition to the other couplings, make the Λ-dependence of the physical results very small. Although we only show the results with the parameter set PS11 in Ref. [5], we get similar results with the other favorable parameter sets. [3]

The discrete γ_5-symmetry is defined by imposing the action is invariant under the following transformation,

$$\phi' = -\phi, \quad \psi' = \pm\gamma_5\psi, \qquad \bar{\psi}' = \mp\bar{\psi}\gamma_5, \tag{2}$$

where the plus (minus) sign in the second equation and the minus (plus) sign in the third equation are for proton (neutron). Therefore, the odd terms of σ-meson self-interactions are forbidden as well as the fermion mass term.

To see the discrete γ_5-symmetry, we expand the tolal energy density ε around $M^* = M$ at zero baryon density $\rho = 0$.

$$\varepsilon(\rho = 0, \phi) = \sum_{n=0} \frac{1}{n!}\frac{\partial^n}{\partial\phi^n}\varepsilon(\rho = 0, \phi)|_{M^*=M}\phi^n \equiv \sum_{n=0} c_n^*\phi^n. \tag{3}$$

We show the absolute values of the effective coupling c_n^* with the normalization factor $1/g_s^n/M^{4-n}$ in the first line (a) of Table 1. It seems that there is no suppressions of odd terms. Next, we expand the energy density around $M^* = M_{st}^*$ at the saturation density ρ_{st} where M_{st}^* is the effective nucleon mass at ρ_{st}. The results are shown in the second line (b) of Table 1. There may be approximate discrete γ_5-symmetry at ρ_{st}. Finally, we expand $\varepsilon(\rho, \phi)$ around $M^* = M_{st}^*$ but put $\rho = 0$. The results are shown in the third line (c) in Table 1. We see that the odd term couplings are suppressed except for $n = 1$. Naturally, nonvanishing c_1^* corresponds to the fact that $M^* = M_{st}^*$ is not the solution of the equation of motion of σ-meson. To analyze these results, we use the NPRGE, since in our model above, the couplings $c_n^*(l = 0...6)$ are determined phenomenologically and are *not* be calculated.

3 Theoretical Analyses

Since there is not enough space to describe our method of NPRGE, we only summarize the main results. For the details of the formalism and the results, see Ref. [3].

Table 1: The absolute values of normalized effective couplings $c_n^*/g_s^n/M^{4-n}$ of σ-meson self-interactions

n	1	2	3	4	5	6
(a)	0.0	0.0014	0.0	0.0	0.00094	0.0018
(b)	0.0	0.0015	0.00038	0.0010	0.00011	0.0024
(c)	0.0013	0.0013	0.00062	0.0014	0.00062	0.0017
(d)	0.092	0.059	0.022	0.11	0.015	0.053

For example, the NPRGE for the cubic σ-meson self-interaction coupling

$$\frac{dC_3}{dt} = \frac{1}{(3!)(16\pi^2)}\left(120\frac{C_5}{1+2C_2} - 432\frac{C_3C_4}{(1+2C_2)^2} + \cdots\right) \qquad (4)$$

where C_n is the effective coupling of the n-th σ-meson self-interaction in the Wilsonian effective action and $t = \log(\Lambda_0/\Lambda)$. Each term in the r. h. s. of the equation (4) has at least *one* odd coupling in the numerator. Therefore, Eq. (4) vanishes if we put all odd couplings zero at the initial. This is true for other odd couplings. Therefore, the odd couplings remain zero through the whole evolution, if we put them zero at the initial. This mechanism may raise the possibility that approximate discrete γ_5-symmetry remains through the evolution as well as the exact symmetry, i.e., if we put the odd couplings to be small, they remains small through the whole evolution. In the last line (d) of Table 1, we show an example of the absolute value of the σ-meson self-interaction couplings C_n, with the normalization factor $1/g_s^n/M^{4-n}$, at $t = 4$ where almost all quantum corrections have been included into the effective potential. (The initial conditions are the same as in Ref. [3].) It is seen that the odd couplings of the σ-meson self-interactions suppressed except for C_1 as in the third line (c) of Table 1. However, nonvanishing value of the coupling C_1 means that the effective potential does not have minimum at $\phi = 0$ and the σ-meson field ϕ has non-zero vacuum expectation value $<\phi>$. Therefore, we must shift the ϕ-field to the minimum to consider the stable σ-meson. This shift enlarges the other odd couplings and make the nucleon mass larger.

Our interpretation on the phenomenological results in the previous section is very simple. At the initial, there are approximate discrete γ_5-symmetry and the nucleon mass is not so large at the bare Lagrangian. However, quantum corrections yield the large nonvanishing c_1^* and the field ϕ has a non-zero vacuum expectation value which causes the re-definition of ϕ and breaks the approximate symmetry. Finally, the medium effects cancels the vacuum fluctuations effects and the approximate discrete γ_5-symmetry is restored in nuclear matter.

We remark that the low-order couplings $c_n^*(n = 1 \sim 6)$ of the σ-meson self-interactions are produced mainly by the quantum effects in high-energy region and be almost constant in low-energy region. [3] Therefore, they are not much affected by the fermion mass term which becomes important in low-energy region. As a result, the approximate discrete γ_5-symmetry is well-preserved for these couplings in NPRGE evolutions. These properties of the low-order couplings also justify the one-loop approximations, i.e., the Hartree approximation, at low-energy. [3]

4 Summary

In this paper, the approximate discrete γ_5-symmetry is studied phenomenologically. There may be the restoration of this symmetry at finite density. The results are analyzed by NPRGE. It is shown that the discrete γ_5-symmetry is preserved explicitly at each step in NPRGE evolutions. It is also shown that the medium effects cancels the large linear ϕ-term and makes the nucleon mass smaller. There are some interesting extensions of this work. For example, it is interesting to study the tensor coupling between the ω-meson and the nucleon, since it is forbidden by the discrete γ_5-symmetry.

Acknowledgments

The author would like to thank K.-I. Aoki, T. Kohomura and H. Yoneyama for useful discussions and suggestions. This work is supported in part by the Japanese Scientific Foundation (Grant No. 09740208).

References

1. B. D. Serot and J. D. Walecka, "The Relativistic Nuclear Many-Body Problem" in *Advances in nuclear physics*, vol. **16**, ed. by J .W. Negele and E. Vogt (Plenum Press, New York, 1986).
2. K. G. Wilson, Rev. Mod. Phys. **47** (1975), 773; See also the references in Ref [3].
3. H. Kouno, M. Nakai, A.Hasegawa and M. Nakano, Preprint SAGA-HE-137(nucl-th/9808058).
4. H. Kouno, T. Mitsumori, Y. Iwasaki, K. Sakamoto, N. Noda, K. Koide, A. Hasegawa and M. Nakano, Prog. Theor. Phys. **97** (1997), 91.
5. H. Kouno, K. Sakamoto, Y. Iwasaki, N. Noda, T . Mitsumori, K. Koide, A. Hasegawa and M. Nakano, Prog. Theor. Phys. **98** (1997), 1123.

SPONTANEOUS SYMMETRY BREAKING OF THE LIGHT-FRONT ϕ^4_{1+1} THEORY

MASA-AKI TANIGUCHI
Department of Physics, Nagoya University, Nagoya 464-8602, Japan
E-mail: mass@eken.phys.nagoya-u.ac.jp

We study $(1+1)$-dimensional ϕ^4 theory quantized on the light cone. We pay much attention to the light-front zero mode, and find the second-order phase transition in the strong coupling region by solving the zero-mode constraint with a new method. We also obtain the critical coupling constant, $\lambda_c \simeq 31.8\mu_R^2$, which is consistent with the equal-time calculations, and the critical exponent, $\beta = 1/2$.

The light-front Tamm-Dancoff (LFTD) approximation[1] is a new promising numerical approach to solve the nonperturbative problems. It has been successfully applied to two-dimensional models,[2,3,4] and there is a possibility to make new progress in the study of the relativistic bound-state problems once it is applied to QCD. Recently Brisudova, Perry, and Wilson have derived the confining potential[5] of the heavy quark systems directly from QCD by combining this approximation with the similarity transformation on the Hamiltonian. It is important to note that this new approach not only reproduces known results correctly, but also brings us new results.[1-5]

The success of the LFTD approximation is based on the vacuum simplicity of the light-front field theory (LFFT). In LFFT the vacuum is almost trivial and therefore the particle picture matches with LFFT very well. However there is a problem in LFFT how we can derive the spontaneous symmetry breaking with this simple vacuum. It is widely believed[6] that the light-front zero mode plays an important role for the spontaneous symmetry breaking. In order to treat it properly, it is convenient to put the system into a finite box ($x^- \in [-L, L]$) and impose the periodic boundary condition.[6] We can treat the zero mode separately from the other modes under this condition.

In this article, we study $(1+1)$-dimensional ϕ^4 theory to clarify that we can derive the spontaneous symmetry breaking by treating the light-front zero mode properly. There are several papers on this subject.[7,8] They solve the zero mode constraint by applying the approximations which are based on the truncation of the particle number. They showed that there is the broken phase in the strong coupling region and obtained the critical coupling constant, which is consistent with the equal time calculations. However they both failed to obtain the correct critical exponent of the order parameter $\langle\phi\rangle$ in their works. They obtained $\beta = 1/2$, which is the same as one in the mean field approximation. In addition, it is technically difficult to extend their methods to higher orders. We investigate the zero mode more precisely with a new method. We show that there is the second-order phase transition in the strong coupling region and obtain the critical coupling constant and critical exponent in ϕ^4_{1+1} theory.

The Lagrangian density of ϕ^4_{1+1} theory is given by $\mathcal{L} = \partial_+\phi\partial_-\phi - \mu^2\phi^2/2 - \lambda\phi^4/4!$. We put the system into a finite light-like box ($x^- \in [-L, L]$) and impose the periodic boundary condition, $\phi(x^- = L) = \phi(x^- = -L)$, to separate the zero mode from the non-zero mode,

$$\begin{aligned}\phi(x) &= \phi_0(x^+) + \tilde{\phi}(x), \\ &= \phi_0(x^+) + \sum_{n>0}[\phi_n(x^+)e^{-ip_n^+x^-} + \phi_{-n}(x^+)e^{ip_n^+x^-}], \quad (p_n^+ \equiv \frac{n\pi}{L}) \qquad (1)\end{aligned}$$

where the zero mode is defined as $\phi_0(x^+) \equiv (1/2L)\int_{-L}^{L} dx^-\phi(x)$, and ϕ_n's are the Fourier components of the non-zero mode $\tilde{\phi}$.

This system has constraints because the Lagrangian density is the first order in the light-cone time derivative. Non-trivial one is, so called, the zero mode constraint,

$$\frac{\lambda}{3!}{\phi_0}^3 + \mu^2\phi_0 + \frac{\lambda}{3!}([\tilde{\phi}^2]_0\phi_0 + [\tilde{\phi}\phi_0\tilde{\phi}]_0 + \phi_0[\tilde{\phi}^2]_0) + \frac{\lambda}{3!}[\tilde{\phi}^3]_0 \approx 0, \qquad (2)$$

where $[O]_0$ is the zero-mode of an (composite) operator O and defined by $[O]_0 \equiv \frac{1}{2L}\int_{-L}^{L} dx^- O(x)$. The composite operators in Eq. (2) have been made to symmetric ordering, since it is natural in quantizing the theory.

The equal-x^+ commutation relations are as follows:

$$[\phi_m, \phi_n] = \frac{1}{4\pi m}\delta_{m+n,0}, \quad [\phi_0, \phi_0] = 0, \qquad (3)$$

$$[\phi_0, \phi_n] = \frac{\lambda}{8\pi n}(\phi_0\phi_n + \phi_n\phi_0 + \sum_{m\neq 0,n}\phi_m\phi_{n-m})\alpha_0^{-1}, \qquad (4)$$

where $\alpha_0 \equiv \mu^2 + \lambda/2({\phi_0}^2 + [\tilde{\phi}^2]_0)$. Eq. (3) are the same as the canonical ones but Eq. (4) is not. Although the right-hand side of Eq. (4) is q-number, we just write it down formally here, neglecting the ordering of the operators. In this work, we will show that the operator ordering in Eq. (2) and (4) is essential to lead to the spontaneous symmetry breakdown in this system.

The zero mode is not a dynamical degree of freedom in this system. It is clear that the zero mode should be written as a function of the non-zero modes according to Eq. (2). The Fock space should be constructed only with the non-zero modes, and therefore the vacuum is trivial, which is defined by

$\phi_n|0\rangle = 0$ $(n > 0)$, where the ϕ_n $(n > 0)$ is the annihilation operator as usual. The creation operator is defined as $\phi_n^\dagger \equiv \phi_{-n}$ $(n > 0)$ which creates a particle with the momentum $n\pi/L$. The order parameter of Z_2-symmetry is $\langle\phi\rangle = \langle\phi_0\rangle = v$ in this model. From now on, we denote the zero mode as $\phi_0 \equiv v + w$, where w is the normal-ordered operator part of the zero mode.

We are now going to obtain the vacuum expectation value v by solving the zero mode constraint (2). It is clear that the vacuum expectation value of Eq. (2) is written as follows:

$$\frac{\lambda}{3!}v^3 + \mu^2 v + \frac{\lambda}{2!}\langle[\tilde{\phi}^2]_0\rangle v + \frac{\lambda}{3!}\langle[\tilde{\phi}w\tilde{\phi}]_0\rangle = 0 \tag{5}$$

where a relation, $\langle\phi_0^3\rangle = \langle\phi_0\rangle^3 = v^3$, has been used. It holds because there is no independent state with zero momentum except the vacuum state in this system. We also used $\langle[\tilde{\phi}^3]_0\rangle = 0$ according to the particle number conservation. Here, we define the renormalized mass as $\mu_R^2 \equiv \mu^2 + \frac{\lambda}{2!}\langle[\tilde{\phi}^2]_0\rangle$, because the second term corresponds to the one loop self-energy diagram in the two point function which is the only divergent diagram in ϕ_{1+1}^4 theory.

The last term in Eq. (5) can be evaluated easily by using the commutation relation (4). It has an expression of $\langle[\tilde{\phi}w\tilde{\phi}]_0\rangle = \sum_{n<0}\langle\phi_{-n}w\phi_n\rangle$ in the momentum expansion, and $\langle\phi_{-n}w\phi_n\rangle$ can be calculated by using Eq. (4),

$$\langle\phi_{-n}w\phi_n\rangle = \langle\phi_{-n}\phi_n\rangle v\frac{\frac{\lambda}{4n\pi}\langle\alpha_0^{-1}\rangle}{1 - \frac{\lambda}{8n\pi}\langle\alpha_0^{-1}\rangle} = \frac{-vx}{2\pi n(n-x)}, \tag{6}$$

where we defined the new variable $x \equiv \frac{\lambda}{8\pi}\langle\alpha_0^{-1}\rangle = \frac{\lambda}{8\pi(\mu_R^2 + \lambda v^2/2)}$ instead of the coupling constant λ, and used $\langle\phi_{-n}\phi_n\rangle = -1/4n\pi$ for $n < 0$. Note that $\langle\alpha_0^{-1}\rangle = \langle\alpha_0\rangle^{-1}$ holds because of the momentum conservation.

By Combining Eq. (5) and (6), we obtain the following equation for v,

$$v[v^2 + \frac{1}{4\pi}(\psi(x) + \gamma - \frac{1}{2x})] = 0, \tag{7}$$

where $\psi(x)$ is the digamma function and γ is the Euler's constant. This equation can be solved numerically. For the small coupling constant, $\lambda < \lambda_c \simeq 31.8\mu_R^2$, We find that the only solution for Eq. (7) is $v = 0$, and therefore this phase is the symmetric phase. When the coupling constant is strong enough, $\lambda > \lambda_c$, there are two solutions other than $v = 0$. These solutions behave as $v \propto \pm(\lambda - \lambda_c)^\beta$ and give a same energy smaller than that of $v = 0$. Therefore, this phase is called broken and this phase transition is the second-order, because the order parameter, $v = \langle\phi\rangle$, is continuous to $v = 0$ at the critical coupling constant λ_c. We also obtain the critical exponent β and its value is 1/2, which is known as one in the mean field approximation.

It is well-established that there exists the second-order phase transition[9] between weak and strong coupling regions in ϕ^4_{1+1} theory, and the critical coupling is given as $22\mu_R^2 < \lambda_c < 55\mu_R^2$ in the usual equal-time calculations.[10] Therefore our result is consistent with the equal-time one. However, our critical exponent β is the same as the mean field value. The correct value is known as $\beta = 1/8$.[11] Here we should explain this discrepancy. In Eq. (4) we have fixed the position of the operator α_0^{-1} at the end of other operators. Therefore, we have been able to obtain Eq. (6) easily by using the trivial relation, $\langle 0|\alpha_0^{-1}|0\rangle = \langle 0|\alpha_0|0\rangle^{-1}$, in the light-front quantization. However, α_0^{-1} enters also between other operators in general. What we did in Eq. (4) and (6) is to neglect the commutation relation between α_0^{-1} and other operators, and it corresponds to the lowest order estimation in the semi-classical approximation. To obtain the correct critical exponent, we must do higher order calculations in this approximation and add the quantum corrections.

The author is grateful for T. Sugihara, M. Yahiro and K. Yamawaki for fruitful discussions. I also thank K. Harada for comments. This work has been partially supported by a Grant-in-Aid for Scientific Research from the Ministry of Education, Science and Culture of Japan (No.90001951).

References

1. R. J. Perry, A. Harindranath, and K. G. Wilson, Phys. Rev. Lett. **65**, 2959 (1990); S. J. Brodsky, G. McCartor, H. C. Pauli and S. S. Pinsky, Particle World **3**, 109 (1993); K. G. Wilson, T. S. Walhout, A. Harindranath, W. M. Zhang, R. J. Perry, and S .D. Glazek, Phys. Rev **D49**, 6720 (1994).
2. H. Bergknoff, Nucl.Phys. **B122**, 215 (1977); Y. Mo and R. J. Perry, J. Comp. Phys. **108**, 159 (1993); K. Harada, M. Sugihara, M. Taniguchi, M. Yahiro, Phys. Rev. **D49**, 4226 (1994).
3. T. Sugihara, M.Matsuzaki, and M. Yahiro, Phys. Rev. **D50**, 5274 (1994).
4. K. Harada, A. Okazaki, and M. Taniguchi, Phys. Rev. **D55**, 4910 (1997).
5. M. Brisudova, R.J. Perry, K.G. Wilson, Phys.Rev.Lett.**78**, 1227 (1997).
6. T. Maskawa and K. Yamawaki, Prog. Theo. Phys. **56**, 270 (1976).
7. S. S. Pinsky, and B. van de Sande, Phys. Rev. **D49**, 2001, (1994)
8. T.Heinzl, C.Stern, E.Werner, and B.Zellermann, Z.Phys.**C72**, 353 (1996)
9. S. Chang, Phys. Rev. **D13**, 2778 (1976).
10. M. Funke, U. Kaulfuss, and H. Kummel, Phys. Rev. **D35**, 621 (1987); H. Kroger, R.Girard, and G. Dufour, Phys. Rev. **D35**, 3944 (1987).
11. G.A. Baker Jr., B.G. Nickel, and D.I. Meiron, Phys.Rev.**B17**, 1365, (1978); J.C. Le Guillou, and J. Zinn-Justin, Phys.Rev.**B21**, 3976 (1980).

LIGHT-CONE WAVEFUNCTION REPRESENTATION OF ELECTROWEAK MATRIX ELEMENTS

DAE SUNG HWANG

Department of Physics, Sejong University, Seoul 143-747, Korea
E-mail: dshwang@kunja.sejong.ac.kr

In general, each semileptonic exclusive decay amplitude receives two contributions, a diagonal $\Delta n = 0$ parton-number-conserving amplitude and a $\Delta n = -2$ contribution in which a quark and an antiquark from the initial hadron Fock state annihilate to the leptonic current. The general formalism can be used as a basis for systematic approximations to heavy hadron decay amplitudes such as hard perturbative QCD contributions. We illustrate the general formalism using a simple perturbative model of composite hadrons. Our analysis demonstrates the occurrence of "zero-mode" endpoint contributions to matrix elements of the "bad" j^- currents in the Drell-Yan frame when $q^+ \to 0$.

1 Introduction

In this paper we shall give formulas for the current matrix elements $\langle A|J^\mu|B\rangle$ describing general transition between hadrons B and A. The formulas are in principle exact, given the light-cone wavefunctions of hadrons. Our results generalize the expressions for the elastic form factors obtained by Drell and Yan [1,2] and West [3]. The underlying formalism is the light-cone Hamiltonian Fock expansion in which hadron wavefunctions are decomposed on the free Fock basis of QCD [4]. In this formalism, the full Heisenberg current J^μ can be equated to the current j^μ of the non-interacting theory which in turn has simple matrix elements on the free Fock basis. The entire electroweak current matrix element is in general given by the sum of the diagonal $n \to n$ and off-diagonal $n+1 \to n-1$ transitions. As we shall see, an important feature of a general analysis is the emergence of singular $\delta(x)$ "zero-mode" contributions in the "bad" – current $j^- = j^0 - j^z$ ¿from the off-diagonal matrix elements if the choice of frame dictates $q^+ = 0$. The contents of this paper is based on the work done in collaboration with Brodsky. It is advised to refer Ref. [5] for the details.

2 Matrix Elements of Electroweak Currents

The light-cone Fock expansion is defined as the projection of an exact eigensolution of the full light-cone quantized Hamiltonian on the solutions of the free Hamiltonian with the same global quantum numbers. The coefficients of the Fock expansion are the complete set of n-particle light-cone wavefunctions, $\{\psi_n(x_i, k_{\perp i}, \lambda_i)\}$. The coordinates $x_i, k_{\perp i}$ are internal relative coordinates, independent of the total momentum of the bound state, and satisfy $0 < x_i < 1$, $\sum_i^n x_i = 1$ and $\sum_i^n k_{\perp i} = 0_\perp$. Here $x = \frac{k^+}{P^+} = \frac{k^0+k^3}{P^0+P^3}$ and we use the metric convention $a \cdot b = \frac{1}{2}(a^+b^- + a^-b^+) - \vec{a}_\perp \cdot \vec{b}_\perp$. The evaluation of the semileptonic decay amplitude $B \to A\ell\bar{\nu}$ requires the matrix element of the weak current between hadron states $\langle A|j^\mu(0)|B\rangle$. The outgoing leptonic current carries $q^\mu = (q^+, q_\perp, q^-) = \left(\Delta P^+, q_\perp, \frac{q^2+q_\perp^2}{\Delta P^+}\right)$, and the value of $\Delta = q^+/P^+$ is determined from four-momentum conservation $\frac{q^2+q_\perp^2}{\Delta} + \frac{m_A^2+q_\perp^2}{1-\Delta} = m_B^2$.

For the $n \to n$ diagonal term ($\Delta n = 0$), the final-state hadron wavefunction has arguments $\frac{x_1-\Delta}{1-\Delta}$, $\vec{k}_{\perp 1} - \frac{1-x_1}{1-\Delta}\vec{q}_\perp$ for the struck quark and $\frac{x_i}{1-\Delta}$, $\vec{k}_{\perp i} + \frac{x_i}{1-\Delta}\vec{q}_\perp$ for the $n-1$ spectators. We thus have a formula for the diagonal (parton-number-conserving) matrix element of the form [5]:

$$\langle A|J^\mu|B\rangle_{\Delta n=0} = \sum_{n,\ \lambda}\prod_{i=1}^{n}\int_\Delta^1 dx_1 \int_0^1 dx_{i(i\neq 1)} \int \frac{d^2\vec{k}_{\perp i}}{2(2\pi)^3}\, \delta\left(1-\sum_{j=1}^n x_j\right)$$
$$\times\delta^{(2)}\left(\sum_{j=1}^n \vec{k}_{\perp j}\right) \psi^\dagger_{A(n)}(x_i', \vec{k}_{\perp i}', \lambda_i)\ j^\mu\ \psi_{B(n)}(x_i, \vec{k}_{\perp i}, \lambda_i) \quad (1)$$

where

$$\begin{cases} x_1' = \frac{x_1-\Delta}{1-\Delta}, & \vec{k}_{\perp 1}' = \vec{k}_{\perp 1} - \frac{1-x_1}{1-\Delta}\vec{q}_\perp \text{ for the struck quark} \\ x_i' = \frac{x_i}{1-\Delta}, & \vec{k}_{\perp i}' = \vec{k}_{\perp i} + \frac{x_i}{1-\Delta}\vec{q}_\perp \text{ for the } (n-1) \text{ spectators.} \end{cases} \quad (2)$$

A sum over all possible helicities λ_i is understood. If quark masses are neglected the vector and axial currents conserve helicity. We also can check that $\sum_i^n x_i' = 1$, $\sum_i^n \vec{k}_{\perp i}' = \vec{0}_\perp$.

For the $n+1 \to n-1$ off-diagonal term ($\Delta n = -2$), let us consider the case where partons 1 and $n+1$ of the initial wavefunction annihilate into the leptonic current leaving $n-1$ spectators. Then $x_{n+1} = \Delta - x_1$, $\vec{k}_{\perp n+1} = \vec{q}_\perp - \vec{k}_{\perp 1}$. The remaining $n-1$ partons have total momentum $((1-\Delta)P^+, -\vec{q}_\perp)$. The final wavefunction then has arguments $x_i' = \frac{x_i}{(1-\Delta)}$ and $\vec{k}_{\perp i}' = \vec{k}_{\perp i} + \frac{x_i}{1-\Delta}\vec{q}_\perp$. We thus obtain the formula for the off-diagonal matrix element [5]:

$$\langle A|J^\mu|B\rangle_{\Delta n=-2} = \sum_{n\ \lambda}\int_0^\Delta dx_1 \int_0^1 dx_{n+1} \int \frac{d^2\vec{k}_{\perp 1}}{2(2\pi)^3} \int \frac{d^2\vec{k}_{\perp n+1}}{2(2\pi)^3}$$

$$\times \prod_{i=2}^{n} \int_0^1 dx_i \int \frac{d^2 \vec{k}_{\perp i}}{2(2\pi)^3} \delta\left(1 - \sum_{j=1}^{n+1} x_j\right) \delta^{(2)}\left(\sum_{j=1}^{n+1} \vec{k}_{\perp j}\right)$$

$$\times \psi^{\dagger}_{A(n-1)}(x'_i, \vec{k}'_{\perp i}, \lambda_i)\ j^{\mu}\ \psi_{B(n+1)}(\{x_1, x_i, x_{n+1} = \Delta - x_1\},$$

$$\{\vec{k}_{\perp 1}, \vec{k}_{\perp i}, \vec{k}_{\perp n+1} = \vec{q}_{\perp} - \vec{k}_{\perp 1}\}, \{\lambda_1, \lambda_i, \lambda_{n+1} = -\lambda_1\}). \quad (3)$$

Here $i = 2, 3, \cdots, n$ with

$$x'_i = \frac{x_i}{1-\Delta}, \qquad \vec{k}'_{\perp i} = \vec{k}_{\perp i} + \frac{x_i}{1-\Delta}\vec{q}_{\perp} \quad (4)$$

label the $n-1$ spectator partons which appear in the final-state hadron wavefunction. We can again check that the arguments of the final-state wavefunction satisfy $\sum_{i=2}^{n} x'_i = 1$, $\sum_{i=2}^{n} \vec{k}'_{\perp i} = \vec{0}_{\perp}$.

3 Example—ϕ^3 Perturbation Theory

As an explicit example and check on the above formalism, we shall consider the electromagnetic vector current matrix element of a neutral composite system composed of two charged scalars ($e_a + e_b = 0$) where the light-cone wavefunctions are known explicitly from perturbation theory. To construct the model, we consider a 3+1 dimensional system represented by the Lagrangian:

$$\begin{aligned} \mathcal{L} = &(\partial_\mu \phi_a + i e_a A_\mu \phi_a)^\dagger (\partial^\mu \phi_a + i e_a A^\mu \phi_a) - m_a^2 \phi_a^\dagger \phi_a \quad (5) \\ &+(\partial_\mu \phi_b - i e_b A_\mu \phi_b)^\dagger (\partial^\mu \phi_b - i e_b A^\mu \phi_b) - m_b^2 \phi_b^\dagger \phi_b \\ &+\frac{1}{2}\partial_\mu \Phi \partial^\mu \Phi - \frac{1}{2} M^2 \Phi\Phi + g\Phi(\phi_a^\dagger \phi_b + \phi_b^\dagger \phi_a). \end{aligned}$$

We can derive the light-cone amplitudes from the covariant amplitude by integrating over the k^- variable. The amplitude of the transition from Φ to Φ in one loop is given as follows ¿from the Feynman rules:

$$\mathcal{M}^\mu = i e_a g^2 \int \frac{d^4 k}{(2\pi)^4} \frac{(2k-q)^\mu}{(k^2 - m_a^2 + i\epsilon)\,((k-q)^2 - m_a^2 + i\epsilon)((k-P)^2 - m_b^2 + i\epsilon)}$$

$$+ \left(a \leftrightarrow b\right). \quad (6)$$

When we perform the integration over k^-, the integral does not vanish only for $0 \le x \le 1$ where $k^+ = xP^+$.

After the k^- integration and some arrangements, we obtain [5]

$$\mathcal{M}^\mu = (2P-q)^\mu F(q^2) \tag{7}$$

$$= \int_\Delta^1 dx \int \frac{d^2\vec{k}_\perp}{2(2\pi)^3}\,\psi_{(2)}(x'_a, x'_b, \vec{k'}_{a\perp}, \vec{k'}_{b\perp}; M, m_a, m_b) j^\mu_{(2\to2)a}$$

$$\times \psi_{(2)}(x_a, x_b, \vec{k}_{a\perp}, \vec{k}_{b\perp}; M, m_a, m_b)$$

$$+ \int_0^\Delta dx \int \frac{d^2\vec{k}_\perp}{2(2\pi)^3} \int_0^1 dy \int \frac{d^2\vec{k}_{y\perp}}{2(2\pi)^3}\,\psi_{(1)}\left(\frac{y}{1-\Delta}, \vec{k}_{y\perp} - \vec{P'}_\perp\right)$$

$$\times j^\mu_{(3-1)a}\psi_{(3)}(x, y, \Delta - x, \vec{k}_\perp, \vec{k}_{y\perp}, \vec{q}_\perp - \vec{k}_\perp; M, m_a, m_b)$$

$$+ \left(a \leftrightarrow b\right),$$

where

$$j_{(2\to2)a} = e_a \frac{P^+}{\sqrt{x(x-\Delta)}} \tag{8}$$

$$\times \left((2x-\Delta)\ ,\ \frac{1}{P^{+2}}\left(2\left(M^2 - \frac{(m_b^2 + \vec{k}_\perp^2)}{1-x}\right) - \frac{(q^2 + \vec{q}_\perp^2)}{\Delta}\right)\ ,\ (2\vec{k}_\perp - \vec{q}_\perp)\right),$$

$$\psi_{(2)}(x_a, x_b, \vec{k}_{a\perp}, \vec{k}_{b\perp}; M, m_a, m_b) = g\frac{1}{\sqrt{x_a x_b}}\,\frac{1}{\left(M^2 - \frac{m_a^2 + \vec{k}_{a\perp}^2}{x_a} - \frac{m_b^2 + \vec{k}_{b\perp}^2}{x_b}\right)},$$

$$j^\mu_{(3\to1)a} = e_a \frac{P^+}{\sqrt{x(\Delta - x)}}$$

$$\times \left((2x-\Delta)\ ,\ \frac{1}{P^{+2}}\left(2\frac{(m_a^2 + \vec{k}_\perp^2)}{x} - \frac{(q^2 + \vec{q}_\perp^2)}{\Delta}\right)\ ,\ (2\vec{k}_\perp - \vec{q}_\perp)\right),$$

$$\psi_{(3)}(x, y, \Delta - x, \vec{k}_\perp, \vec{k}_{y\perp}, \vec{q}_\perp - \vec{k}_\perp; M, m_a, m_b) = g^2 \frac{1}{\sqrt{x(1-x)^2(\Delta - x)y}}$$

$$\times \frac{1}{\left(M^2 - \frac{m_a^2 + \vec{k}_\perp^2}{x} - \frac{m_b^2 + (-\vec{k}_\perp)^2}{1-x}\right)\left(M^2 - \frac{m_a^2 + \vec{k}_\perp^2}{x} - \frac{m_a^2 + (\vec{q}_\perp - \vec{k}_\perp)^2}{\Delta - x} - \frac{M^2 + \vec{k}_{y\perp}^2}{y}\right)},$$

$$\psi_{(1)}(y, \vec{k}_{y\perp}) = \frac{2(2\pi)^3}{\sqrt{y}}\delta(y-1)\delta^2(\vec{k}_{y\perp}),$$

$$x_a = x, \quad x_b = 1 - x, \quad \vec{k}_{a\perp} = \vec{k}_\perp, \quad \vec{k}_{b\perp} = -\vec{k}_\perp,$$

$$x'_a = \frac{x_a - \Delta}{1-\Delta}, \quad x'_b = \frac{x_b}{1-\Delta}, \quad \vec{k'}_{a\perp} = \vec{k}_{a\perp} + (1 - x'_a)\vec{P'}_\perp,$$

$$\vec{k'}_{b\perp} = \vec{k}_{b\perp} - x'_b \vec{P'}_\perp, \quad \vec{P'}_\perp = -\vec{q}_\perp.$$

In (7) $\mathcal{M}^\mu = (2P - q)^\mu F(q^2)$ follows from $q_\mu \mathcal{M}^\mu = 0$.

For $q^2 \to 0$, $\vec{q}_\perp \to \vec{0}_\perp$ and $\Delta \to 0$, + component of (7) gives

$$F(0) = e_a \int \frac{d^2\vec{k}_\perp}{(2\pi)^3} \int_0^1 dx\ |\psi_{(2)}(x, 1-x, \vec{k}_\perp, -\vec{k}_\perp; M, m_a, m_b)|^2 \qquad (9)$$
$$+ e_b \int \frac{d^2\vec{k}_\perp}{(2\pi)^3} \int_0^1 dx\ |\psi_{(2)}(x, 1-x, \vec{k}_\perp, -\vec{k}_\perp; M, m_b, m_a)|^2 = 0,$$

where $\psi_{(2)}$ is given in (8). Each term can be normalized to unit charge, thus providing wavefunction renormalization in the model. Alternatively we can evaluate the $-$ component of (7) to obtain

$$F(0) = e_a \int \frac{d^2\vec{k}_\perp}{(2\pi)^3} \frac{1}{M^2} \left(\int_0^1 dx\ \frac{1}{x} \left(M^2 - \frac{m_b^2 + \vec{k}_\perp^2}{1-x} \right) \right. \qquad (10)$$
$$\left. \times |\psi_{(2)}(x, 1-x, \vec{k}_\perp, -\vec{k}_\perp; M, m_a, m_b)|^2 + \frac{1}{m_a^2 + \vec{k}_\perp^2} g^2 \right)$$
$$+ e_b \int \frac{d^2\vec{k}_\perp}{(2\pi)^3} \frac{1}{M^2} \left(\int_0^1 dx\ \frac{1}{x} \left(M^2 - \frac{m_a^2 + \vec{k}_\perp^2}{1-x} \right) \right.$$
$$\left. \times |\psi_{(2)}(x, 1-x, \vec{k}_\perp, -\vec{k}_\perp; M, m_b, m_a)|^2 + \frac{1}{m_b^2 + \vec{k}_\perp^2} g^2 \right),$$

where the $\frac{e_a}{m_a^2+\vec{k}_\perp^2} g^2$ and $\frac{e_b}{m_b^2+\vec{k}_\perp^2} g^2$ terms come from the singular contributions of the $\int_0^\Delta dx\, \psi_{(1)}\, j^-_{(3-1)}\, \psi_{(3)}$ terms in (7) when we take the limit $\Delta \to 0$. The $\perp$ components of (7) do not give more information since $(2\vec{P} - \vec{q})_\perp \to \vec{0}_\perp$ in the left hand side and the integrand of the right hand side is odd about $\vec{k}_\perp$.

The above analysis provides an explicit realization of the general formulas (1) and (3). In this simple model two transition matrix elements appear: $2 \to 2$ and $3 \to 1$. The equality of the formulas for (9) and (10) is a general condition which follows from gauge invariance and consistency of the light-cone formalism. We have verified the equality for the perturbative model by direct evaluation of the integrals. Note that the singular contributions in (10) includes the zero mode $\delta(x)$ contributions from the $n+1 \to n-1$ off-diagonal matrix element.

4 Conclusion

The necessity for this zero mode $\delta(x)$ terms were first noted in the pioneering work of Chang, Root and Yan [6], and Burkardt analyzed it in his studies of higher-twist parton distributions [7]. Glazek and Sawicki [8] have studied the relativistic two-body bound-state form factor in the 1+1 dimensional scalar field theory by considering both the spectator diagram and the Z diagram in the light-cone formalism. Here we see that the presence of such terms is a general feature of local operator matrix elements when one selects the simplified $q^+ = 0$ frame.

The off-diagonal $n+1 \to n-1$ contributions provide a new perspective on the physics of B-decays. A semileptonic decay involves not only matrix element where a quark changes flavor, but also a contribution where the leptonic pair is created from the annihilation of a $q\bar{q'}$ pair within the Fock states of the initial B wavefunction. The semileptonic decay thus can occur from the annihilation of a nonvalence quark and a valence antiquark in the initial hadron. This feature will carry over to exclusive hadronic B-decays, such as $B^0 \to \pi^- D^+$. In this case the pion can be produced from the coalescence of a $d\bar{u}$ pair emerging from the initial higher particle number Fock wavefunction of the B. The D meson is then formed from the remaining quarks after the internal exchange of a W boson.

Acknowledgments

The author wishes to thank Stanley J. Brodsky for the collaboration of this work. This work was supported in part by the Basic Science Research Institute Program, Ministry of Education, Project No. BSRI-97-2414, and in part by Non-Directed-Research-Fund, Korea Research Foundation 1997.

References

1. S.D. Drell and T.M. Yan, *Phys. Rev. Lett.* **24**, 181 (1970).
2. S.J. Brodsky and S.D. Drell, *Phys. Rev.* D **22**, 2236 (1980).
3. G.B. West, *Phys. Rev. Lett.* **24**, 1206 (1970).
4. G.P. Lepage and S.J. Brodsky, *Phys. Rev.* D **22**, 2157 (1980); *Phys. Lett.* B **87**, 359 (1979); *Phys. Rev. Lett.* **43**, 545, 1625(E) (1979).
5. S.J. Brodsky and D.S. Hwang, hep-ph/9806358, to appear in *Nucl. Phys.* B.
6. S.J. Chang, R.G. Root and T.M. Yan, *Phys. Rev.* D **7**, 1133 (1973).
7. M. Burkardt, *Nucl. Phys.* B **373**, 613 (1992).
8. St. Glazek and M. Sawicki, *Phys. Rev.* D **41**, 2563 (1990).

Form factor relations for B to light meson decay in Light Cone Approach

Bum-Hoon Lee
Department of Physics, Sogang University
Seoul 121-742, Korea
E-mail: bhl@ccs.sogang.ac.kr

We evaluate the form factors of $B \to \pi$ and $B \to \rho$ matrix elements by using the light cone formalism. With the reasonable approximation of $m_\pi/M_B = 0$, $m_\rho/M_B = 0$, $M_b/M_B = 1$ and $(1-x) << 1$, we derive the pole types of form factors and show also that there exists relations among form factors with 2 unknown functions.

1 Introduction

For the heavy to heavy quark decays like $b \to c$ transition, we can apply the heavy quark symmetry and it is possible to calculate the matrix elements through the heavy quark effective theory (HQET) [1]. However, for those involving $b \to u$ it is less likely that the heavy quark symmetry applies, and the determination of the matrix element has been heavily relied on the models for the form factors. The theoretical calculation of the form factors involving $b \to u$ transition is a difficult task because of the nonperturbative QCD physics. Recently there have been active investigations of the form factors of $B \to \pi$ and $B \to \rho$ by using quark model, QCD sum rule and lattice calculations [2].

In this work[3], we will calculate the form factors of $B \to \pi$ and $B \to \rho$ transitions by using the method [4] based on the meson theory of Brodsky and Lepage [5]. From the q^2 dependences we read the pole types of the form factors and also show the relations among the form factors in the limit of $m_\pi/M_B = 0$, $m_\rho/M_B = 0$, $M_b/M_B = 1$.

2 Form Factors and their relations

The hadronic matrix element for $B \to \pi$ transition is written in terms of hadronic form factors[6]:

$$< \pi^-(p_\pi)|V^\mu|B^0(p_B) >$$
$$= \left(r^\mu - \frac{m_B^2 - m_\pi^2}{q^2} q^\mu\right) F_1^{B\pi}(q^2) + \frac{m_B^2 - m_\pi^2}{q^2} q^\mu F_0^{B\pi}(q^2), \tag{1}$$

where $V^\mu = \bar{u}\gamma^\mu b$, $q^\mu = (p_B - p_\pi)^\mu$, $r^\mu = (p_B + p_\pi)^\mu$. In the similar way, the hadronic matrix element for $B \to \rho$ transition is written [6] in terms of hadronic

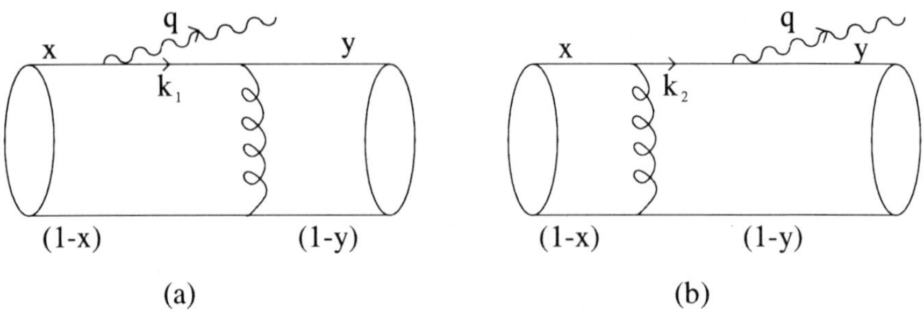

Figure 1: Feynman diagrams to the first order in α_s.

form factors V, A_0, A_1, and A_2.

We calculate the heavy to light transition matrix element by using the perturbative QCD factorization of exclusive amplitudes. To the first order in $\alpha_s = \alpha_s(Q^2)$, two Feynman diagrams in Fig. 1 give the following amplitude:

$$< \pi^-(p_\pi)|V^\mu|B^0(p_B) >= \frac{8\pi\alpha_s}{3}\int_0^1 dx \int_0^{1-\epsilon} dy\, \phi_B(x)$$
$$\times \Big[\frac{\mathrm{Tr}\{(\not p_\pi + m_\pi)\gamma_5\gamma^\nu \not k_1 \gamma^\mu(\not p_B + g(x)M_B)\gamma_5\gamma_\nu\}}{k_1^2 Q^2}$$
$$+\frac{\mathrm{Tr}\{(\not p_\pi + m_\pi)\gamma_5\gamma^\mu(\not k_2 + M_b)\gamma^\nu(\not p_B + g(x)M_B)\gamma_5\gamma_\nu\}}{(k_2^2 - M_b^2)Q^2}\Big]\phi_\pi(y), \quad (2)$$

and the similar expression for the decay into ρ. Here, $Q^\mu = (1-x)p_B^\mu - (1-y)p_\pi^\mu$, $k_1^\mu = -(1-x)p_B^\mu + p_\pi^\mu$, $k_2^\mu = p_B^\mu - (1-y)p_\pi^\mu$. The variables x and y represent the momentum fraction carried by the correspondig quarks. In Eq.(2) $g(x)$ is a phenomenological parameter for B meson wave function, and $\phi_\pi(y)$ and $\phi_B(x)$ are the distribution amplitudes for π and B mesons. We took the upper limit of the integration over momentum fraction y of a quark in the light meson as $1-\epsilon$. The integration in the interval $1-\epsilon \leq y \leq 1$ corresponds to the Drell-Yan-West [7] end-point region and is expected to be suppressed by a Sudakov form factor [4].

In the following, we take the approximations of $m_\pi/M_B = 0$, $m_\rho/M_B = 0$ and $M_b/M_B = 1$. For the B meson, it is expected that the momentum fraction of the light quark, $1-x$, is roughly given by m_{light}/M_B and small. Hence, we also take the approximation $(1-x) << 1$ with q^2 not too big.

After the calculation, we can organize the form factors in the following form:

$$F_i(q^2) = \frac{32\pi\alpha_s}{3M_B^2}\int_0^1 dx \int_0^{1-\epsilon} dy\, \phi_B(x)\ \phi_i(y)\frac{1}{(1-x)(1-y)^2} f_i, \tag{3}$$

where F_i=F_0,F_1,V,A_0,A_1,A_2, and $\phi_i(y) = \phi_\pi(y)$ for F_0 and F_1, and $\phi_i(y) = \phi_\rho(y)$ for V, A_0, A_1 and A_2. In (3) all f_i are given by

$$f_i = c1_i\frac{1}{z} + c2_i\frac{1}{z^2} \tag{4}$$

where ci_i and $c2_i$ are function of y only, and $z \equiv 1 - \frac{q^2}{M_B^2}$. The integration does not change the q^2 dependence and can be summarized as

$$F_0(q^2) = (a_\pi + b_\pi)\frac{1}{z}, \quad F_1(q^2) = 2b_\pi\frac{1}{z} + (a_\pi - b_\pi)\frac{1}{z^2}, \tag{5}$$
$$A_1(q^2) = a_\rho\frac{1}{z}, \quad A_2(q^2) = -V(q^2) = a_\rho\frac{1}{z^2}, \quad -A_0(q^2) = (a_\rho + b_\rho)\frac{1}{z^2},$$

where

$$a_i = \frac{32\pi\alpha_s}{3M_B^2}\int_0^1 dx \int_0^{1-\epsilon} dy\, \phi_B(x)\ \phi_i(y)\frac{2g-1}{(1-x)(1-y)^2}, \tag{6}$$
$$b_i = \frac{32\pi\alpha_s}{3M_B^2}\int_0^1 dx \int_0^{1-\epsilon} dy\, \phi_B(x)\ \phi_i(y)\frac{1}{(1-x)(1-y)}$$

with $i = \pi$ or ρ.

We emphasize that the q^2 dependences of the form factors given in (5) are independent of the shapes of the distribution amplitudes $\phi_B(x)$, $\phi_\pi(y)$, $\phi_\rho(y)$ and the value of the parameter ϵ. Their dependences appear only in the values of the constants a_i and b_i in (6), which affect the normalizations of the form factors. From (5) we find that $F_0(q^2)$ and $A_1(q^2)$ have the simple pole q^2 dependence, and $A_2(q^2)$, $V(q^2)$ and $A_0(q^2)$ have the dipole q^2 dependence. $F_1(q^2)$ has the mixture of the simple pole and dipole q^2 dependences, but the dipole q^2 dependence is dominant.

From (5) we find the relations among the form factors:

$$F_1(q^2) = F_0(q^2)(2 - \frac{1}{z}) + 2\frac{a_\pi}{a_\rho}A_1(q^2)(-1 + \frac{1}{z}) \tag{7}$$

$$F_0(q^2)\frac{1}{z} = -\frac{a_\pi + b_\pi}{a_\rho + b_\rho}A_0(q^2) \tag{8}$$

$$A_1(q^2)\frac{1}{z} = A_2(q^2) = -V(q^2). \tag{9}$$

This relation can be tested by measuring the differential branching ratios at $q^2 = 0$.

3 Conclusion

We calculated the form factors of $B \to \pi$ and $B \to \rho$ matrix elements by using the factorization formalism of perturbative QCD. The q^2 dependences of the form factors shows that $F_0(q^2)$ and $A_1(q^2)$ are the simple pole types and $A_2(q^2)$, $V(q^2)$ and $A_0(q^2)$ have the dipole types. $F_1(q^2)$ has the mixture of the simple pole and dipole q^2 dependences. We also derived the relations among the form factors. The pole types and the relations among the form factors are independent of the shapes of the distribution amplitudes $\phi_B(x)$, $\phi_\pi(y)$, $\phi_\rho(y)$ and the value of the parameter ϵ.

Acknowledgments

This work was supported in part by the Basic Science Research Institute Program, Ministry of Education, Project No. BSRI-98-2414, in part by Korea Science and Engineering Foundation through the SRC Program of SNU-CTP.

References

1. N. Isgur and M.B. Wise, Phys. Lett. B232 (1989) 113; B237 (1990) 527
2. H. Wittig, OUTP-97-59-P, hep-lat/9710088 (1997)
3. Dae Sung Hwang and Bum-Hoon Lee, hep-ph/9801421, To appear in Eur. Phys. J. C.
4. A. Szczepaniak, E.M. Henley and S.J. Brodsky, Phys. Lett. B243 (1990) 287; S.J. Brodsky, in *QCD 20 Years Later*, P.M. Zerwas and H.A. Kastrup, Ed. (World Scientific, 1993)
5. G.P. Lepage and S.J. Brodsky, *Phys. Lett.* B **87**, 359 (1979) ; Phys. Rev. D22 (1980) 2157; S.J. Brodsky and G.P. Lepage, Phys. Rev. D24 (1981) 1808; in *Perturbative Quantum Chromodynamics*, A.H. Mueller, Ed. (World Scientific, 1989)
6. M. Wirbel, B. Stech and M. Bauer, Z. Phys. C29 (1985) 637; C34 (1987) 103; M. Bauer and M. Wirbel, Z. Phys. C42 (1989) 671
7. S.D. Drell and T.M. Yan, Phys. Rev. Lett. 24 (1970) 181; G. West, Phys. Rev. Lett. 24 (1970) 1206
8. V.L. Chernyak and I.R. Zhitnitsky, Phys. Rept. 112 (1984) 173; V.L. Chernyak and A.R. Zhitnitsky, Nucl. Phys. B345 (1990) 137

TWIST THREE LIGHT-CONE DISTRIBUTION AMPLITUDES OF VECTOR MESONS IN QCD

K. TANAKA
Department of Physics, Juntendo University, Inba-gun, Chiba 270-1606, Japan
E-mail: tanakak@phys.sakura.juntendo.ac.jp

We present a systematic study of twist-3 light-cone distribution amplitudes of vector mesons in QCD, which is based on conformal expansion. A complete set of distribution amplitudes is constructed for ρ, ω, K^* and ϕ mesons, which satisfies all (exact) equations of motion and constraints from conformal expansion.

Amplitudes of hard (light-cone dominated) exclusive processes in QCD are expressed by the factorization formulae, which separate the short-distance dynamics from the long-distance one. For those processes producing a (light) vector meson V $(= \rho, \omega, K^*, \phi)$ in the final state, like exclusive semileptonic or radiative B decays ($B \to Ve\nu$, $B \to V + \gamma$) and the hard electroproduction of vector mesons ($\gamma^* + N \to V + N'$), a long-distance part involving the final vector meson is given by the vacuum-to-meson matrix element of the nonlocal light-cone operators $\langle 0|\bar{\psi}(z)\Gamma\lambda^i\psi(-z)|V\rangle$, corresponding to universal nonperturbative quantity called the light-cone distribution amplitudes (DAs).[1] (z_μ is a light-like vector $z^2 = 0$, and λ^i and Γ denote various flavor and Dirac matrices. We do not show the gauge phase factor connecting the quark fields.)

The DAs of higher twist are essential for systematic study of preasymptotic corrections to hard exclusive amplitudes. The higher twist DAs are also interesting theoretically because they contain new and direct information on hadron structure and the dynamics of QCD.

Here we will explicitly consider the "chiral-odd" DAs of the charged ρ meson with momentum P_μ and polarization vector ϵ_μ ($P^2 = m_\rho^2$, $\epsilon^2 = -1$, $P \cdot \epsilon = 0$); the "chiral-even" DAs can be treated similarly, and extension to other vector mesons is straightforward.[1] The relevant quark-antiquark DAs are defined with the chirality-violating Dirac matrices $\Gamma = \{\sigma^{\mu\nu}, 1\}$ as

$$\langle 0|\bar{u}(z)\sigma^{\mu\nu}d(-z)|\rho^-(P,\epsilon)\rangle = if_\rho^T \left[(\epsilon_\perp^\mu P^\nu - \epsilon_\perp^\nu P^\mu) \int_0^1 du\, e^{i\xi P\cdot z}\phi_\perp(u,\mu^2) \right.$$

$$\left. + (P^\mu z^\nu - P^\nu z^\mu) \frac{\epsilon \cdot z}{(P \cdot z)^2} m_\rho^2 \int_0^1 du\, e^{i\xi P \cdot z} h_\parallel^{(t)}(u,\mu^2) \right], \qquad (1)$$

$$\langle 0|\bar{u}(z)d(-z)|\rho^-(P,\epsilon)\rangle = -if_\rho^T m_\rho^2 (\epsilon \cdot z) \int_0^1 du\, e^{i\xi P\cdot z} h_\parallel^{(s)}(u,\mu^2), \qquad (2)$$

where we neglect the twist-4 terms and the quark mass corrections for simplicity (see Ref. [1] for the complete expressions). Our Lorentz frame is chosen as $P \cdot z = P^+ z^-$. The nonlocal operators on the l.h.s. is renormalized at scale μ. We set $\xi \equiv u - (1-u) = 2u - 1$, and f_ρ^T is the usual tensor decay constant as $\langle 0|\bar{u}(0)\sigma^{\mu\nu}d(0)|\rho^-(P,\epsilon)\rangle = i f_\rho^T(\epsilon^\mu P^\nu - \epsilon^\nu P^\mu)$. We wrote $\epsilon^\mu = \epsilon_\parallel^\mu + \epsilon_\perp^\mu$ with $\epsilon_\parallel^\mu = [P^\mu - z^\mu m_\rho^2/(P\cdot z)](\epsilon \cdot z)/(P \cdot z)$, so that the DAs with the subscripts $\parallel$ and $\perp$ describe longitudinally and transversely polarized ρ mesons, respectively. $\phi_\perp$ is of twist-2, while $h_\parallel^{(t)}$ and $h_\parallel^{(s)}$ are of twist-3. These DAs are dimensionless functions and describe the probability amplitudes to find the ρ in a state with quark and antiquark, which carry the light-cone momentum fractions u and $1-u$, respectively, and have a small transverse separation of order $1/\mu$.

The quark-antiquark-gluon DAs, which are of twist-3 and higher, can be defined similarly. There exists one chiral-odd DA of twist-3, given by

$$\langle 0|\bar{u}(z)\sigma^{\mu\nu} z_\nu g G_{\mu\lambda}(vz) z^\lambda d(-z)|\rho^-(P,\epsilon)\rangle$$
$$= (P\cdot z)(\epsilon \cdot z) f_{3\rho}^T m_\rho \int_0^1 \mathcal{D}\underline{\alpha}\, e^{-iP\cdot z(\alpha_u - \alpha_d + v\alpha_g)} \mathcal{T}(\underline{\alpha}, \mu^2), \qquad (3)$$

where $G_{\mu\nu}$ is the gluon field strength tensor, $\mathcal{D}\underline{\alpha} \equiv d\alpha_d d\alpha_u d\alpha_g \delta(1 - \alpha_d - \alpha_u - \alpha_g)$, and $\underline{\alpha}$ denotes the set of three light-cone momentum fractions: α_d (quark), α_u (antiquark), and α_g (gluon). $f_{3\rho}^T$ is the three-body tensor decay constant, [1] so that $\mathcal{T}$ is dimensionless. The three-particle DA $\mathcal{T}$ describes a higher component in the Fock-state wave function with additional gluon.

The basis of DAs defined above is overcomplete due to the constraints from the QCD equations of motion. Using $(i \not{D} - m_q)q(x) = 0$ $(q = u, d)$, we obtain

$$\frac{x^\nu}{2}\frac{\partial}{\partial x_\mu}\bar{u}(x)\left[\gamma_\mu, \gamma_\nu\right]_\pm d(-x) = \int_{-1}^{1} dv v^{(1\mp 1)/2} \bar{u}(x)\sigma^{\mu\nu} x_\nu g G_{\mu\lambda}(vx) x^\lambda d(-x)$$
$$- \frac{x^\nu}{2}\frac{\partial}{\partial y_\mu}\left\{\bar{u}(x+y)\left[\gamma_\mu,\gamma_\nu\right]_\mp d(-x+y)\right\}\Big|_{y\to 0} + i(m_u \pm m_d)\bar{u}(x)\not{x} d(-x), \qquad (4)$$

where $[\gamma_\mu, \gamma_\nu]_\pm \equiv \gamma_\mu\gamma_\nu \pm \gamma_\nu\gamma_\mu$. For $x^2 \to 0$, the vacuum-to-meson matrix elements of (4) yield a system of integral equations between two- and three-particle DAs. The total derivative term induces mixing between $h_\parallel^{(t)}$ and $h_\parallel^{(s)}$. We can solve [1] these coupled integral equations in a form $(i = t, s)$

$$h_\parallel^{(i)}(u,\mu^2) = \int_0^1 dv\, K_{WW}^{(i)}(u,v)\phi_\perp(v,\mu^2) + \int_0^1 \mathcal{D}\underline{\alpha}\, K_g^{(i)}(u,\underline{\alpha})\mathcal{T}(\underline{\alpha},\mu^2), \qquad (5)$$

where $K_X^{(i)}$ $(X = WW, g)$ are independent of μ. We omit the quark mass correction term for simplicity. The first term on the r.h.s. is the analogue of the Wandzura-Wilczek piece of the nucleon structure function $g_2(x, Q^2)$.

To analyze (5), we introduce the conformal partial wave expansion in the $m_q \to 0$ limit. [1] The conformal expansion of light-cone DAs is analogous to the partial wave expansion of wave functions in standard quantum mechanics (QM). In conformal expansion, the invariance of massless QCD under conformal transformations substitutes the rotational symmetry in QM. In QM, the purpose of partial wave decomposition is to separate angular degrees of freedom from radial ones (for spherically symmetric potentials). All dependence on the angular coordinates is included in spherical harmonics which form an irreducible representation of the group O(3), and the dependence on the single remaining radial coordinate is governed by a one-dimensional Schrödinger equation. Similarly, the conformal expansion of DAs in QCD aims to separate longitudinal degrees of freedom from transverse ones. All dependence on the longitudinal momentum fractions is included in terms of certain orthogonal polynomials which form irreducible representations of the so-called collinear subgroup of the conformal group, SL(2,R). The transverse-momentum dependence (the scale-dependence) is governed by simple renormalization group equations: the different partial waves, labeled by different "conformal spins", behave independently and do not mix with each other. Since the conformal invariance of QCD is broken by quantum corrections, mixing of different terms of the conformal expansion is only absent to leading logarithmic accuracy. Still, conformal spin is a good quantum number in hard processes, up to small corrections of $\mathcal{O}(\alpha_s^2)$.

The conformal expansion of the DAs on the r.h.s. of (5) reads

$$\phi_\perp(u, \mu^2) = 6u\bar{u} \sum_{n=0}^{\infty} a_n(\mu^2) C_n^{3/2}(\xi), \tag{6}$$

$$\mathcal{T}(\underline{\alpha}, \mu^2) = 360 \alpha_d \alpha_u \alpha_g^2 \sum_{k,l=0}^{\infty} \omega_{kl}(\mu^2) J_{kl}(\alpha_d, \alpha_u), \tag{7}$$

where $\bar{u} = 1 - u$, $\xi = 2u - 1$, $\alpha_g = 1 - \alpha_d - \alpha_u$, and $C_n^{3/2}$ and J_{kl} are particular Gegenbauer and Appell polynomials. [1] Using orthogonality relations of these polynomials, the expansion coefficients are expressed by matrix elements of local conformal operators: a_n and ω_{kl} are given by the operators of conformal spin $j = n + 2$ and $j = k + l + 7/2$, respectively (three-particle conformal representations are degenerate). Thanks to conformal symmetry, $K_X^{(i)}$ of (5) are resolved into the superposition of the terms of definite conformal spin. Thus $h_{\parallel}^{(i)}$ are given by the conformal expansion, where expansion coefficient of conformal spin j is expressed solely in terms of a_n and ω_{kl} with the same j.

Now all DAs are expressed, order by order in the conformal expansion, by independent matrix elements a_n and ω_{kl}. From (5)-(7), the μ^2-dependence of the DAs is governed by that of $a_n(\mu^2)$ and $\omega_{kl}(\mu^2)$, which is worked out in the leading logarithmic approximation utilizing conformal symmetry.[1,2] The results indicate that the relevant anomalous dimensions increase as $\sim \ln j$. This means that only the first few conformal partial waves contribute at sufficiently large scales. Therefore, the truncation of the conformal expansion at some low order provides a useful and consistent approximation of the full DAs.

We take into account the partial waves with $j \leq 9/2$, where the terms with $n \leq 2$ and $k+l \leq 1$ are retained in (6) and (7). Correspondingly, we get[1]

$$h_{\parallel}^{(s)}(u) = 6u\bar{u}\left[1 + a_1\xi + \left(\frac{1}{4}a_2 + \frac{35}{6}\zeta_3\right)(5\xi^2 - 1)\right] + 3\delta_+(3u\bar{u} + \bar{u}\ln\bar{u} + u\ln u) + 3\delta_-(\bar{u}\ln\bar{u} - u\ln u), \quad (8)$$

and similarly for $h_{\parallel}^{(t)}$. Here $\zeta_3 = f_{3\rho}^T/(f_\rho^T m_\rho)$, and we incorporated the quark mass corrections proportional to $\delta_\pm = f_\rho(m_u \pm m_d)/(f_\rho^T m_\rho)$ with f_ρ the usual vector decay constant. The results for other vector mesons are obtained by trivial substitutions. We emphasize that these results provide a consistent set of DAs involving a minimum number of independent nonperturbative parameters. These parameters are calculated from QCD sum rules taking into account SU(3)-breaking effects.[1] The results show that the three-particle contributions (the term with ζ_3 in (8)) are important and broaden the distributions.

In conclusion, we have developed a powerful framework for hard exclusive processes, which allows model building of the DAs consistent with all QCD constraints. Our formalism is applicable to arbitrary twist,[3] and our results are immediately applicable to a range of phenomenologically interesting processes.[4]

Acknowledgments

The author would like to thank P. Ball, V.M. Braun, Y. Koike, and N. Nishiyama for the collaboration. This work was supported in part by the Monbusho Grant-in-Aid for Encouragement of Young Scientists No. 09740215.

References

1. P. Ball, V.M. Braun, Y. Koike and K. Tanaka, *Nucl. Phys.* B **529**, 323 (1998), and references therein.
2. Y. Koike, N. Nishiyama and K. Tanaka, *Phys. Lett.* B **437**, 153 (1998).
3. P. Ball and V.M. Braun, hep-ph/9810475.
4. P. Ball and V.M. Braun, *Phys. Rev.* D **58**, 094016 (1998).

CORRELATION AND FINITE INTERACTION-RANGE EFFECTS IN HIGH ENERGY ELECTRON INCLUSIVE SCATTERING

A. KOHAMA
the Institute of Physical and Chemical Research (RIKEN),
2-1 Hirosawa, Wako-shi, Saitama 351-0198, JAPAN
E-mail: kohama@rikaxp.riken.go.jp

K. YAZAKI
Department of Physics, School of Science, University of Tokyo,
7-3-1 Hongo Bunkyo-ku Tokyo 113-0033, JAPAN

R. SEKI
Department of Physics and Astronomy,
California State University, Northridge, CA 91330, and
W. K. Kellogg Radiation Laboratory,
California Institute of Technology, Pasadena, CA 91125, USA

We calculate cross sections of high energy electron inclusive scattering off nuclear matter in a new and consistent formulation based on the Glauber approximation. We examine two kinds of zero-range approximations for the nucleon-nucleon interaction of the final-state interactions,

1 Introduction

Experimental data on high-energy electron-nucleus inclusive scattering have been accumulated for a decade[1]. The experiments were mostly done at SLAC, and will be done at the Jefferson Lab. Those data provide us with a way of studying where and how the quark-gluon degrees of freedom may come into sight.

The process that we are going to deal with is the (e, e') inclusive reaction in the nucleon quasi-elastic region, $e+A \rightarrow e'+N+(A-1)^*$. The incident electron has a four-momentum, $k_\mu = (\epsilon, \mathbf{k})$, and the outgoing electron, $k'_\mu = (\epsilon', \mathbf{k}')$. The momentum transfer is $q_\mu = k_\mu - k'_\mu = (\omega, \mathbf{q})$. We are interested in the quasi-elastic region, *i.e.*, $|\mathbf{q}| \sim 2$ [GeV/c], and $\omega \leq 2$ [GeV]. So the elementary process is considered to be mostly the electron-nucleon elastic scattering. We do not intend to explain the inelastic channel of the eN cross section, which takes part in the data in high-ω region.

In this work we calculate the effect of the final-state interaction (FSI) of the struck nucleon as precisely as possible, taking into account only the hadronic degrees of freedom. The theoretical treatment of this reaction is similar to that of the nuclear transparency in the $(e, e'p)$ reaction at large momentum transfers [2]. The present formulation allows us to describe the (e, e') and $(e, e'p)$ reactions in a unified way [3].

2 Formulation

In this section we explain our formulation of the electron-nucleus inclusive scattering based on the Green's function method [3]. We employ the Glauber approximation for the FSI, and try to treat the nuclear correlation properly in this approximation, which has not been done in the calculations of this process so far [4], [5].

We start with the (e, e') inclusive differential cross section on a target nucleus, A, which is assumed to be expressed as

$$\frac{d\sigma_{\mathrm{eA}}}{d\Omega d\omega} = \langle \frac{d\sigma_{\mathrm{eN}}}{d\Omega} \rangle_{\mathrm{el,on}} \, S(\omega, \mathbf{q}), \tag{1}$$

where the cross section on the r.h.s. is that of *on-shell* electron-nucleon elastic scattering for the same incident energy and scattering angle, and $\langle \ldots \rangle$ implies averaging over the spin, and the isospin, of the target nucleons. We avoid to bring about ambiguities due to the *off-shell* cross section, and concentrate on a consistent treatment of the FSI below. Therefore, the agreement of the numerical results with the experimental data is not necessarily our objective of this work.

$S(\omega, \mathbf{q})$ in eq. (1) is the response function defined by

$$\begin{aligned} S(\omega, \mathbf{q}) &\equiv \sum_X |\langle X | \hat{O}(\mathbf{q}) | A \rangle|^2 \, \delta(\omega - E_X) \\ &= -\frac{1}{\pi} \, \mathrm{Im} \, \langle A | \hat{O}^\dagger(\mathbf{q}) \, \frac{1}{\omega - \hat{H}_A + i\eta} \, \hat{O}(\mathbf{q}) \, | A \rangle, \quad (\eta > 0), \end{aligned} \tag{2}$$

where the initial state, A, is the ground state of the target nucleus, and the sum over the final states, X, is taken, because we are considering the inclusive process. $\hat{H}_A$ is the full hamiltonian of the target nucleus. The energy of the final state, E_X, is measured from that of the ground state of the target. $\hat{O}(\mathbf{q}) \equiv \exp\{i\mathbf{q} \cdot \mathbf{r}_1\}$ is a hard interaction operator, and we assume that the nucleon, 1, is struck.

Applying the fixed-scatterer approximation and the eikonal approximation to eq. (2), we obtain the expression for the response function in the following form;

$$S(\omega, \mathbf{q}) = \frac{1}{\pi v} \int d\mathbf{b}_1 dz_1 dz_1' \, \rho(\mathbf{r}_1, \mathbf{r}'_1) \, \exp\{i(p_z' - q_z)(z_1 - z_1')\} \tag{3}$$

$$\times \left[\theta(z_1 - z_1') \exp\{F(\mathbf{b}_1; z_1, z_1')\} + \theta(z_1' - z_1) \exp\{F^*(\mathbf{b}_1; z_1', z_1)\}\right],$$

where $F(\mathbf{b}_1; z_1, z_1')$ expresses the effect of the FSI (FSI-function),

$$F(\mathbf{b}_1; z_1, z_1') \equiv -(A-1) \int d\mathbf{r}_2 \, \rho(\mathbf{r}_2) \, g(|\mathbf{r'}_1 - \mathbf{r}_2|) \, g(|\mathbf{r}_1 - \mathbf{r}_2|)$$
$$\times \left(1 - \exp\{-\frac{i}{v} \int_{z_1'}^{z_1} dz_1'' \, V_{\mathrm{NN}}(\mathbf{r''}_1 - \mathbf{r}_2)\}\right), \qquad (4)$$

where $\rho(\mathbf{r}, \mathbf{r}')$ is the one-body density matrix with $\rho(\mathbf{r}) \equiv \rho(\mathbf{r}, \mathbf{r})$, and $\int d\mathbf{r} \, \rho(\mathbf{r}) = 1$. $V_{\mathrm{NN}}(\mathbf{r})$ is the N-N potential, which is constructed from the N-N scattering amplitude obtained experimentally. The path of integral in eq. (4) is taken to be along the z-axis which is the direction of the struck nucleon momentum, $z \parallel \mathbf{p}'$. $\mathbf{r}_1 = (\mathbf{b}_1, z_1)$, $\mathbf{r'}_1 = (\mathbf{b}_1, z_1')$, and $\mathbf{r''}_1 = (\mathbf{b}_1, z_1'')$. The velocity, $v = |\mathbf{p}'|/E_{p'}$, is that of the struck nucleon, and $|\mathbf{p}'|$ is determined by $\omega \simeq E_{p'} = \sqrt{|\mathbf{p}'|^2 + m_{\mathrm{N}}^2}$.

$g(|\mathbf{r}_1 - \mathbf{r}_2|)$ is the nuclear correlation function, which describes both dynamical and statistical two-body correlation between the nucleons, 1 and 2. That the correlation functions in eq. (4) appears in the off-diagonal form with respect to $\mathbf{r}_1$ and $\mathbf{r'}_1$ is a characteristic point of our formulation.

Please note that we have not used the expansion in terms of the N-N potential to derive the final expression, because the effect of the interaction with each nucleon cannot be assumed to be small. The expressions, eqs. (4), and (4), are our main results of this work.

3 Numerical Results and Discussion

Though we have obtained the closed form of the response function, we need to make some more approximations for numerical calculations. We examine two zero-range approximations for the N-N potential. One is the zero-range approximation in the z-direction only (ZR1), and another is the zero-range approximation in the x, y, z-direction (ZR2). We find out that the ZR1 is a good approximation for the finite-range interaction if we take into account the nuclear correlation [6].

The numerical results of the inclusive cross section are shown in Fig. 1. The incident electron has the energy, $\epsilon = 3.595$ [GeV], and its scattering angle is $\theta = 30°$. Since we are interested in the effects of the FSI and the nuclear correlation, other information, such as the one-body density matrix, is taken as the input. The PWIA result directly reflects the momentum distribution of a nucleon in the nuclear matter. Including only the FSI broadens the cross section, but

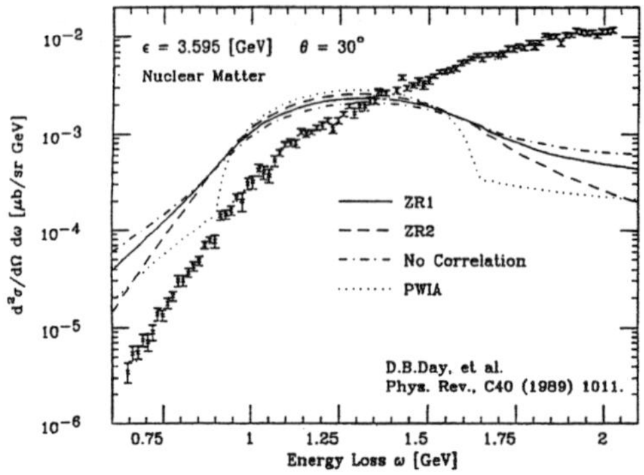

Figure 1: The cross sections of the inclusive scattering off nuclear matter with the experimental data [1]. The dotted curve is the case of PWIA. The dash-dotted curve is the case of only the FSI (no correlation effect). The solid curve (ZR1) and the dashed curve (ZR2) are the full calculation including both the FSI and the nuclear correlation with zero-range approximations.

adding the nuclear correlation (ZR1 and ZR2) reduces the broadening, that is, the effect of the FSI.

Although ZR1 includes the major part of the finite-range interaction, the cross section of ZR1 still overestimates the experimental cross section for low-ω region. Furthermore, ZR2 comes closer to the experimental data than ZR1 does, though ZR2 is more approximate expression than ZR1. One reason why this comes about is our use of the *on-shell* electron-nucleon cross section. These facts push us to study the *off-shell* cross section, and we leave the study in our future work.

4 Summary and Conclusion

We have shown a method to deal with the FSI in the high-energy (e, e') inclusive scattering. We have examined two zero-range approximations, and found that ZR1 is a good one for the finite-range interaction.

References

1. D.B.Day, *et al.*, *Phys. Rev.*, **C40** (1989) 1011.
2. A.Kohama, K.Yazaki, and R.Seki, *Nucl. Phys.*, **A551** (1993) 687.
3. O.Morimatsu and K.Yazaki, *Prog. Part. Nucl. Phys.*, **33** (1994) 679.
4. H.A.Gersch, *et al.*, *Phys. Rev.*, **A5** (1972) 1547.
5. O.Benhar, *et al.*, *Phys. Rev.*, **C44** (1991) 2328.
6. A.Kohama, K.Yazaki, and R.Seki, in preparation.

RELATIVISTIC FADDEEV DESCRIPTION OF BARYONS AND NUCLEON STRUCTURE FUNCTION IN THE NJL MODEL

W. BENTZ, H. MINEO, H. ASAMI, N. ISHII, K. YAZAKI
Department of Physics, Univ. of Tokyo
7-3-1 Hongo, Bunkyo-ku, Tokyo 113-0033, Japan
E-mail: bentz@tkynt2.phys. s.u-tokyo.ac.jp

In this work we use the Nambu-Jona-Lasinio (NJL) model as an effective quark theory based on QCD to describe the structure of baryons. Based on the solutions of the relativistic 3-quark Faddeev equation in the ladder approximation, we discuss the masses of the nucleon and the delta, the static properties of the nucleon, and the quark light cone momentum distributions in the nucleon.

The NJL model is an effective quark theory to describe the structure of hadrons, and is characterized by a chirally symmetric 4-fermion interaction lagrangian $\mathcal{L}_I$. Examples are the original NJL form [1] with scalar and pseudoscalar interaction terms in the $q\bar{q}$ channels, or the color current interaction lagrangian [2], which is given by eq. (1) below with $\kappa = 0$. By using Fierz transformations[3], we can extract the effective coupling constants in the pionic $\bar{q}q$ channel (g_π), the 'scalar' ($0^+, T = 0$) qq channel (g_s), and the 'axial vector' $1^+, T = 1$ qq channel (g_a). Introducing the ratios $r_s = \frac{g_s}{g_\pi}$, $r_a = \frac{g_a}{g_\pi}$, we have for for the color current lagrangian: $r_s = \frac{1}{2}$, $r_a = \frac{1}{4}$, while for the original NJL lagrangian [1] we have the much smaller values $r_s = \frac{2}{13}$, $r_a = \frac{1}{13}$. We see that if g_π is fixed by the pion mass, the strengths of the interactions in the 0^+ and 1^+ qq channels depend on the form of the interaction lagrangian.

To solve the relativistic Faddeev equation, we treat r_s and r_a as free parameters, reflecting different possible forms of the 4-fermi interaction lagrangian. The other parameters of the model are chosen to reproduce $m_\pi = 140MeV$, $f_\pi = 93MeV$, and a constituent quark mass of $M = 400MeV$. As an example [3] we show in fig.1 the baryon masses as functions of r_s for different values of r_a. (The delta mass depends only on r_a.)

While for the original NJL lagrangian [1] no 3-quark bound state exists, the color current interaction lagrangian gives a reasonable nucleon mass, but the delta is unbound. One can, however, construct a lagrangian which can describe simultaneously the masses of the pion, the nucleon and the delta [2]:

$$\mathcal{L}_I = -g \left[\sum_{c=1}^{8} \left(\bar{\psi}\gamma^\mu \frac{\lambda_c}{2} \psi \right)^2 + \kappa \left(\bar{\psi}\gamma^\mu \gamma_5 \psi \right)^2 \right]. \tag{1}$$

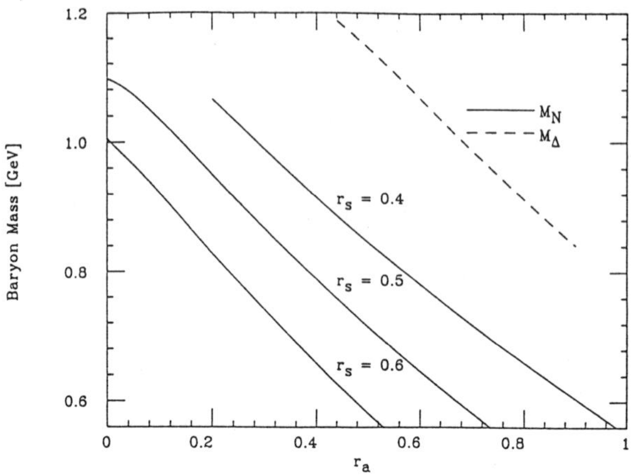

Figure 1: The solid lines show the nucleon mass as a function of r_a for three values of r_s, and the dashed line shows the delta mass as a function of r_a.

For $\kappa = 0.27$ ($r_s = 0.37$, $r_a = 0.44$), we obtain $M_\Delta = 1190MeV$, and $M_N = 920MeV$.

The main mechanism for the delta-nucleon mass difference is the interaction in the scalar diquark channel: Increasing r_s lowers the nucleon mass but leaves the delta mass unchanged. Our numerical results can be parametrized approximately by the relation $M_\Delta - M_N \simeq (1.1r_s - 0.06r_a - 0.18)\, GeV$.

Recently we have also estimated the effects of (composite) pion and sigma meson exchange between quarks by making a separable approximation to the meson exchange potentials and re-solving the relativistic Faddeev equation [4]. The result is that pion exchange gives a large attraction and leads to nucleon mass shifts of about 200 to 300 MeV, depending on the form of the interaction lagrangian. Before drawing conclusions, however, one should go beyond the separable approximation, and include also the effects of self energy corrections of the quarks due to the meson cloud.

The static properties of the nucleon are obtained by calculating bound state matrix elements of one-quark operators between the Faddeev wave functions. This amounts to calculating the Feynman diagrams shown in fig.2, and leads to the correct Ward-Takahashi identities [5] in the case of an external electromagnetic field. However, since we are using an Euclidean sharp cut-off, the charge conservation is violated by a small amount (typically 3%). Some of the results are shown in table 1. Here, for simplicity, we take into account only the scalar diquark channel ($g_a = 0$). For the case $r_s = 2/3$, which reproduces the correct nucleon mass, the axial vector coupling constant, the charge radii

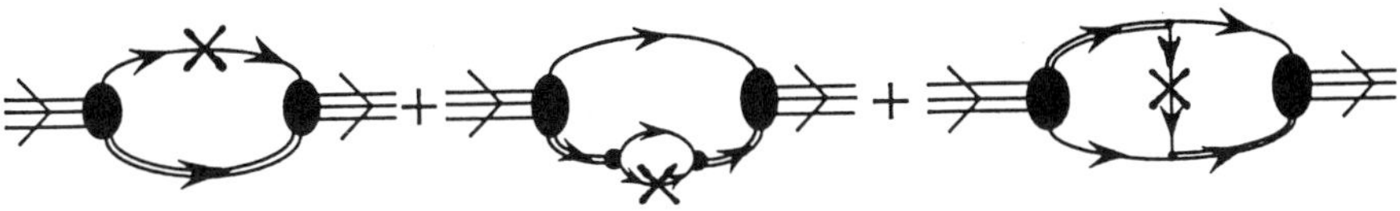

Figure 2: Feynman diagrams to calculate the expectation value of one-body operators (the cross) between 3-body wave functions.

Table 1: The mass, axial vector coupling constant, magnetic moments, and charge radii of the nucleon, as well as the pion nucleon sigma term, for three values of r_s and $r_a = 0$.

	$r_s = 1/2$	2/3	0.8	exp
M_N[MeV]	1096	934	764	939
g_A	1.23	1.33	1.40	1.25
μ_p [n.m.]	1.49	1.78	2.11	2.79
μ_n [n.m.]	-0.95	-1.34	-1.77	-1.91
$< r^2 >_p$ $[fm^2]$	3.39	0.806	0.534	0.74
$< r^2 >_n$ $[fm^2]$	-0.775	-0.075	-0.038	-0.12
$\Sigma_{\pi N}$ [MeV]	35.0	35.1	34.3	45 ± 7

and the pion nucleon sigma-term agree reasonably well with the experimental values. The magnetic moments are too small in magnitude, but one can expect that the interaction in the axial vector diquark channel gives further contributions.

We finally discuss the quark light cone momentum distributions in the nucleon, which are related to the structure functions F_1 and F_2. For this, one has to evaluate the same Feynman diagrams fig. 2, but with the momentum component $k_- = p_- x$ of the struck quark held fixed. (p is the nucleon momentum and x the Bjorken variable.) As a first step, we employ here the 'static approximation' to the Faddeev equation [2], in which the momentum dependence of the quark exchange kernel is neglected. In order to estimate also the sea quark distributions, we take into account the pionic cloud around the constituent quarks by using the familiar convolution formalism. We employ the 'invariant mass regularization', which can be formulated in terms of light cone variables [6]. The resulting distributions are shown in fig. 3 in comparison to the empirical distributions of ref. [7]. At the low energy scale ($Q_0^2 = 0.16 GeV^2$) the valence quarks carry 91% of the nucleon momentum. The pionic effects lead to the well known enhancement $\overline{d} > \overline{u}$, and to a reduction of the Gottfried sum from the naive value $\frac{1}{3}$ to 0.26, compared to the experimental value of 0.26.

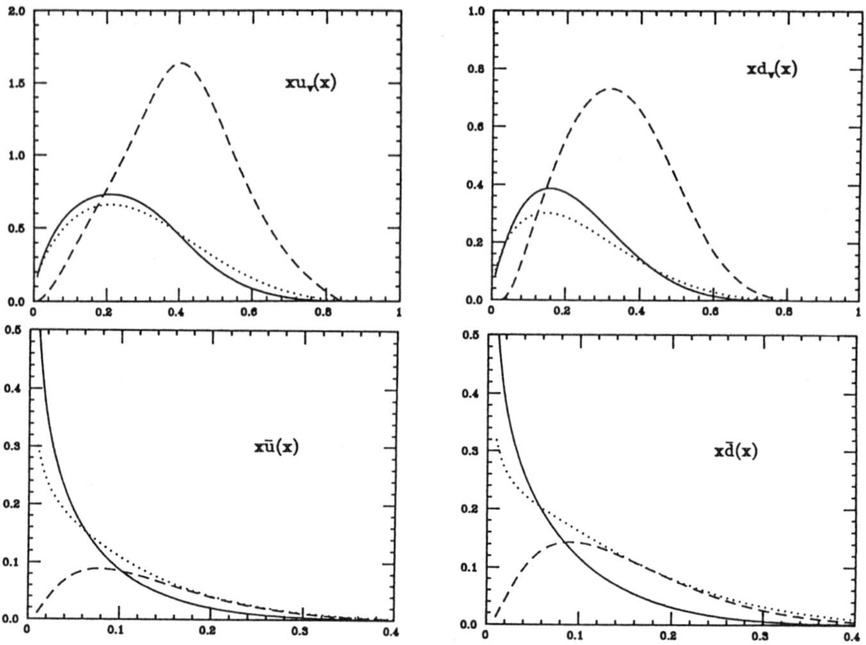

Figure 3: The valence quark (upper figures) and sea quark (lower figures) distributions in the nucleon. The dashed lines show the NJL input distributions ($Q_0^2 \equiv 0.16 GeV^2$), the solid lines the distributions evolved to $Q^2 = 4 GeV^2$, and the dotted lines the empirical distributions at $Q^2 = 4 GeV^2$.

The authors express their thanks to M. Miyama and S. Kumano for the Q^2 evolution program (ref. [8]), and A. W. Thomas for many helpful discussions.

1. Y. Nambu and G. Jona-Lasinio, *Phys. Rev.* **122**, 345 (1960); **124**, 246 (1961).
2. A. Buck, R. Alkofer and H. Reinhardt, *Phys. Lett.* B **286**, 29 (1992).
3. N.Ishii, W. Bentz and K. Yazaki, *Nucl. Phys.* A **587**, 617 (1995).
4. N. Ishii, *Phys. Lett.* B **431**, 1 (1998).
5. H. Asami, N. Ishii, W. Bentz and K. Yazaki, *Phys. Rev.* C **51**, 3388 (1995).
6. W. Bentz, T. Hama, T. Matsuki and K. Yazaki, NJL model on the light cone and pion structure function, subm. to *Nucl. Phys.* A.
7. A. D. Martin, W. J. Stirling and R. G. Roberts, *Phys. Rev.* C **50**, 6734 (1994).
8. M. Miyama and S. Kumano, *Comput. Phys. Commun.* **94**, 185 (1996).

Radiative muon capture in hydrogen in heavy baryon chiral perturbation theory

Shung-ichi Ando and Dong-Pil Min
Department of Physics and Center for Theoretical Physics, Seoul National University, Seoul 151-742, Korea

We calculate photon spectrum of radiative muon capture on hydrogen within heavy baryon chiral perturbation theory up to next-to-next-to leading order. We find that convergence of chiral expansion is very rapid, and the next-to-next-to leading order correction is less than five percent. Hence the recent experimental result of radiative muon capture on hydrogen, about 1.5 times larger pseudo-scalar coupling constant than that of PCAC prediction, cannot be explained by the higher order correction in the hadron sector.

1 Introduction

Recently, first measurement and analysis of data of radiative muon capture (RMC) on hydrogen ($p\mu \to n\nu\gamma$) was reported [1]. Surprisingly, the deduced pseudo-scalar coupling constant g_P was about 1.5 times larger than that of PCAC prediction, $g_P^{PCAC} = 8.42$, as well as other calculations by heavy baryon chiral perturbation theory (HBChPT)[2,3,4]. Moreover, the g_P deduced from the RMC experiment is also in disagreement with the value from ordinary muon capture (OMC) experiment on hydrogen ($p\mu \to n\nu$), $g_P^{OMC} = 8.7 \pm 1.9$ [5].

There are theoretical derivations of the RMC amplitude (See Ref.[6].), however, all of them give similar results, and small differences are found only in higher order terms [a]. Since one may question whether the loop order correction of hadronic amplitude could resolve the disagreement with the g_P, it is worthwhile to evaluate that order correction explicitly. In this talk, we report the result of the calculation of photon spectrum of the RMC within the HBChPT up to next-to-next-to leading order (NNLO), that is one-loop order[9].

2 Counting rule and effective lagrangian

The HBChPT is a low energy effective theory of QCD [10], where the effective lagrangian is constructed from $SU(2)_R \times SU(2)_L$, a symmetry of QCD lagrangian, and its spontaneous breakdown to $SU(2)_V$. The effective lagrangian is expanded in terms of derivatives and/or mass, m_π, of pion, the Goldstone boson. The most powerful feature of the HBChPT is a counting rule, in which

[a] In this workshop a new term to the RMC amplitude is suggested by I.-T. Cheon to explain the RMC experiment [7]. However, it breaks the gauge symmetry [8].

a physical quantity is expanded in powers of Q/Λ_χ, where Q denotes external small momentum and/or m_π, and Λ_χ is the chiral symmetry breaking scale, $\Lambda_\chi \simeq 4\pi f_\pi \simeq m_N \sim 1$ GeV with the pion decay constant f_π, and the nucleon mass m_N. Thus one can enjoy a systematic calculation up to a given order with the chiral counting rule. The order ν of a diagram with L loops and V_i vertices is given by [11]

$$\nu = 1 + 2L + \sum_i V_i \Delta_i, \quad \text{with} \quad \Delta_i = d_i + n_i/2 - 2, \tag{1}$$

where d_i is the number of derivative and m_π and n_i the number of nucleon legs in i-th vertex. Note that a one-loop diagram is two order higher, and infinities of the loop diagram is renormalized by counter terms of higher order lagrangian. The coefficients of the lagrangian are called as *low energy constants* (LECs), which contain information of high energy physics, and cannot be fixed by the symmetry but should be fixed by experiments and resonance saturation [12]. In our case the amplitude is expected to converge very rapidly, in powers of $Q/\Lambda_\chi \sim m_\mu/\Lambda_\chi \simeq 0.1$; m_μ is the muon mass.

Up to the NNLO the effective lagrangian being relevant to the RMC amplitude reads

$$\mathcal{L} = \mathcal{L}_0 + \mathcal{L}_1 + \mathcal{L}_2, \tag{2}$$

where $\mathcal{L}_0$, $\mathcal{L}_1$, $\mathcal{L}_2$ are leading order (LO), next-to leading order (NLO), NNLO lagrangian, respectively. Explicit form of them are given in Ref. [9]. Up to the NLO all coefficients of the effective lagrangian are determined by well-known constants; m_π, f_π, $g_A = 1.25$ for the LO lagrangian, and m_N, the anomalous magnetic moments of nucleon, $\kappa_p = 1.79$, $\kappa_n = -1.91$ for the NLO lagrangian. In the NNLO lagrangian we need seven independent LECs, $l_9^R + l_{10}^R$, c_3^R, $c_8^R + c_9$, c_9, c_{12}, c_{13}, where LECs with superscript R are renormalized ones and others are finite constants. Three of them are fixed by low energy structure of the weak nucleon current; c_3^V (c_{12}) is fixed by iso-vector vector (axial-vector) radius, and c_{14} by Goldberger-Treiman discrepancy. $l_9^R + l_{10}^R$ is fixed by a rare pion process, $\pi \to e\nu\gamma$ [13]. Remaining three are fixed by ρ and $\Delta(1232)$ saturation. Although the $\Delta(1232)$ saturation can be questioned because the mass difference $m_\Delta - m_N \simeq 300$ MeV would not be sufficient to obtain a natural value of the LECs and a convergence of the chiral power, we found that the spectrum is insensitive to the $\Delta(1232)$ and other resonances [14]. In fact, even though the c_9 term, fixed by the $\Delta(1232)$ saturation, are deliberately increased to a comparable size with a $1/m_N$ term, we find that the total NNLO correction is almost insensitive to the c_9. It cannot change more than 0.5 % of total rate with the doubled c_9 value.

3 Numerical result and conclusion

We plot photon spectra up to each chiral orders for statistical spin state of muonic atom in Fig. 1, and those for each spin states up to the NNLO in Fig. 2. We find that the convergence of the chiral power is very rapid, and the NNLO correction is less than five percent to the total rate. We also confirm the previous calculations[6] comparing the spectrums in Fig. 2. Thus the recent RMC experiment cannot be explained by the higher order correction. Hence our result suggests that no further doubts on hadron sector seem necessary.

References

1. G. Jonkmans *et al.*, *Phys. Rev. Lett.* **77**, 4512 (1996); D. H. Wright *et al.*, *Phys. Rev.* C **57**, 373 (1998).
2. V. Bernard *et al.*, *Phys. Rev.* D **50**, 6899 (1994).
3. H. W. Fearing *et al.*, *Phys. Rev.* D **56**, 1783 (1997); *Nucl. Phys.* A **631**, 735c (1998).
4. T. Meissner, F. Myhrer and K. Kubodera, *Phys. Lett.* B **416**, 36 (1998).
5. G. Bardin *et al.*, *Phys. Lett.* B **104**, 320 (1981).
6. M. Gmitro and A. A. Ovchinnikova, in *Advances in Nuclear Physics Vol. 18*, ed. J. W. Negele and E. Vogt (Plenum Press, New York, 1987), references therein.
7. I.-T. Cheon, Talk given in this workshop; I.-T. Cheon and M. K. Cheoun, nucl-th/9811009.
8. H. W. Fearing, nucl-th/9811027; J. Smejkal and E. Truhlík, nucl-th/9811080.
9. S. Ando and D.-P. Min, *Phys. Lett.* B **417**, 177 (1998).
10. S. Weinberg, *Physica* A **96**, 327 (1979); J. Gasser and H. Leutwyler, *Ann. Phys.* **158**, 142 (1984); *Nucl. Phys.* B **250**, 465 (1985).
11. S. Weinberg, *Phys. Lett.* B **251**, 288 (1990);*Nucl. Phys.* B **363**, 3 (1991).
12. G. Ecker *et al.*, *Nucl. Phys.* B **321**, 311 (1989).
13. J. F. Donoghue, E. Golowich and B. R. Holstein, Dynamics of standard model, (Cambridge University Press, 1992).
14. D. S. Beder and H. W. Feairng, *Phys. Rev.* D **35**, 2130 (1987); *Phys. Rev.* D **39**, 3493 (1989)

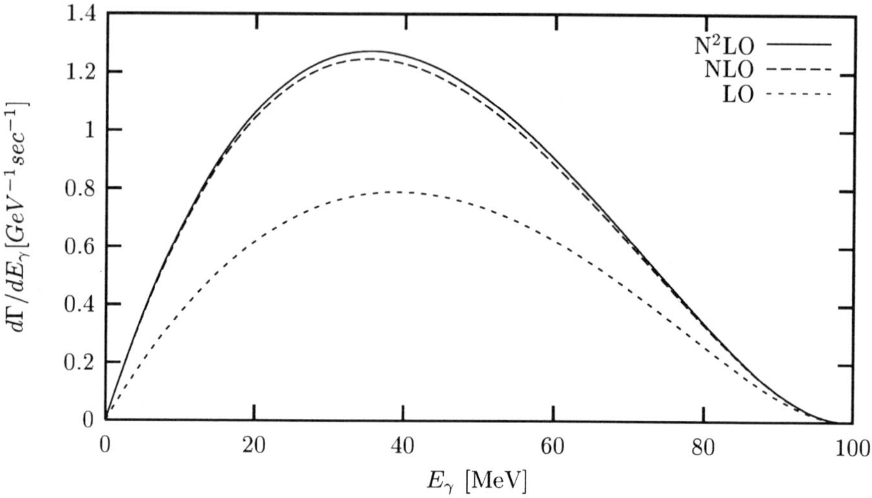

Figure 1: *Photon spectrums up to each chiral orders for statistical spin state.*

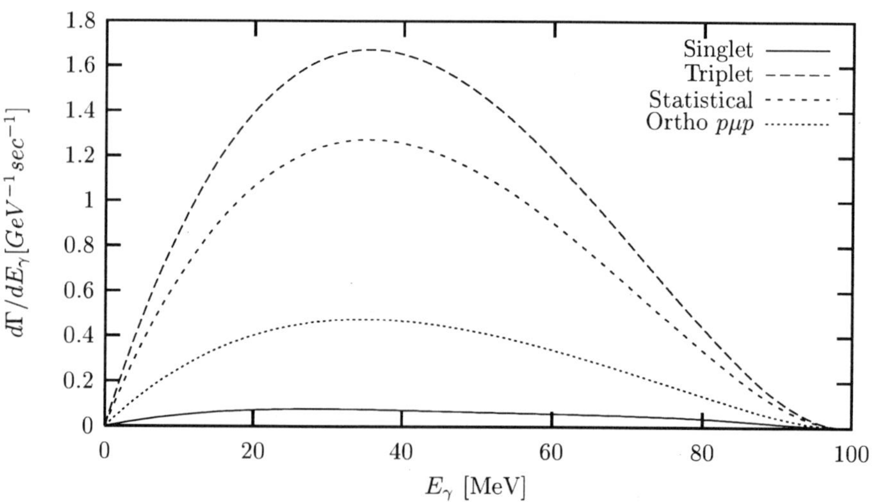

Figure 2: *Photon spectrums for each spin states within HBChPT up to NNLO.*

PHOTOPRODUCTION OF ψ' AS A TEST OF COLOR–OCTET MODEL

K. SUDOH
Graduate School of Science and Technology, Kobe University
1-1, Rokkodai, Nada, kobe, 657-8501, Japan
E-mail: sudou@radix.h.kobe-u.ac.jp

T. MORII and D. ROY
Graduate School of Science and Technology, Kobe University

We have calculated the cross section of ψ' photoproduction at small–p_T regions in polarized γp collisions for testing the color–octet model, which is based on NRQCD factorization formalism. We show that the measurement of the two–spin asymmetry of this process might be effective to constrain the value of NRQCD matrix elements.

1 Introduction

Recently, it has been reported that the cross sections of prompt J/ψ and ψ' production in unpolarized $p\bar{p}$ collisions measured by the CDF collaboration are largely inconsistent with the conventional color–singlet model calculation[1]. Far from that, these dramatic discrepancies could be explained well by taking a new color–octet model[2] into account. A rigorous formulation of the color–octet model has been presented in terms of a beautiful effective field theory called nonrelativistic QCD (NRQCD), in which the $\mathcal{O}(v)$ corrections of a relative velocity between the bound heavy quarks can be systematically calculated [3]. Physics of the color–octet model is now one of the most interesting topics for heavy quarkonium production at high energy.

In this work, as another test of the color–octet model we propose a different process,

$$\vec{\gamma} + \vec{p} \to \psi' + X, \tag{1}$$

at small–p_T regions which might be observed in the forthcoming HERA experiment, where both of an incident γ and target p are longitudinally polarized. The process is dominated by the s–channel photon–gluon fusion, so that the color–octet contribution is dominated at small–p_T regions. Hence we consentrate on only two lowest diagrams here. Note that in the case of J/ψ production, there are some additional processes like $\psi' \to J/\psi + X$, in addition to an original direct production of J/ψ, therefore the analysis must be complicated. Furthermore, the cross section and two–spin asymmetry are very sensitive to the gluon density in a proton, since the ψ' is dominantly produced by photon–gluon fusion. Thus one can get good information on the spin–dependent gluon distribution in a proton by analyzing the polarized process like this [4].

2 Two–Spin Asymmetry for $\gamma p \to \psi' X$

Let us introduce a two–spin asymmetry A_{LL} defined as

$$A_{LL} \equiv \frac{[d\sigma_{++} - d\sigma_{+-} + d\sigma_{--} - d\sigma_{-+}]}{[d\sigma_{++} + d\sigma_{+-} + d\sigma_{--} + d\sigma_{-+}]} = \frac{d\Delta\sigma}{d\sigma} , \qquad (2)$$

where $d\sigma_{+-}$, for instance, denotes that the helicity of one beam particle is positive and the other is negative.

The spin–independent and spin–dependent total cross sections via the color–octet state are given by [5]

$$\sigma(\gamma p \to \psi') = \frac{\pi^3 e_c^2 \alpha_{em} \alpha_S}{4m_c^5} xg(x, Q^2)\bigg|_{x=4m_c^2/s} \left\{ \langle \mathcal{O}_8^{\psi'}(^1S_0)\rangle + \frac{7}{m_c^2}\langle \mathcal{O}_8^{\psi'}(^3P_0)\rangle \right\} , \qquad (3)$$

$$\Delta\sigma(\gamma p \to \psi') = \frac{\pi^3 e_c^2 \alpha_{em} \alpha_S}{4m_c^5} x\Delta g(x, Q^2)\bigg|_{x=4m_c^2/s} \left\{ \langle \mathcal{O}_8^{\psi'}(^1S_0)\rangle - \frac{1}{m_c^2}\langle \mathcal{O}_8^{\psi'}(^3P_0)\rangle \right\} , \qquad (4)$$

where $g(x, Q^2)$ and $\Delta g(x, Q^2)$ are spin–independent and spin–dependent gluon distribution functions with the momentum fraction x at any Q^2, respectively. The factors $\langle \mathcal{O}_8^{\psi'}(^1S_0)\rangle$ and $\langle \mathcal{O}_8^{\psi'}(^3P_0)\rangle$ represent nonperturbative long distance parameters of NRQCD matrix elements and are associated with the production of a $c\bar{c}$ pair in a color–octet $^1S_0$ and $^3P_0$ states, respectively. The recent analysis on charmonium hadroproductions results in $\langle \mathcal{O}_8^{\psi'}(^1S_0)\rangle + (7/m_c^2)\langle \mathcal{O}_8^{\psi'}(^3P_0)\rangle \approx 5.2\times10^{-3}[\mathrm{GeV}^3]$ [6] and $\frac{1}{3}\langle \mathcal{O}_8^{\psi'}(^1S_0)\rangle + (1/m_c^2)\langle \mathcal{O}_8^{\psi'}(^3P_0)\rangle \approx (5.9 \pm 1.9) \times 10^{-3}[\mathrm{GeV}^3]$ [7], which lead to

$$\frac{\tilde{\Theta}}{\Theta} \equiv \frac{\langle \mathcal{O}_8^{\psi'}(^1S_0)\rangle - \frac{1}{m_c^2}\langle \mathcal{O}_8^{\psi'}(^3P_0)\rangle}{\langle \mathcal{O}_8^{\psi'}(^1S_0)\rangle + \frac{7}{m_c^2}\langle \mathcal{O}_8^{\psi'}(^3P_0)\rangle} \approx 3.6 \sim 8.0. \qquad (5)$$

Using this value and various parton distributions, we have numerically calculated the two–spin asymmetry A_{LL} originated from the color–octet state. Note that the A_{LL} in Eq. (2) has a simple form

$$A_{LL}^{\psi'}(\gamma p) = \frac{\Delta g(x, Q^2)}{g(x, Q^2)} \cdot \frac{\tilde{\Theta}}{\Theta}, \qquad (6)$$

and is directly propotional to the ratio of gluon densities and of NRQCD matrix elements in this process.

3 Numerical Results

We have calculated the A_{LL} at relevant HERA energies, which are presented in figures. Since the value of A_{LL} is rather large, we can sufficiently test the color–octet model in this reaction.

Let us discuss the sensitivity on our results. In order to examine the experimental accuracy of the forthcoming HERA experiments, we have estimated the experimental sensitivity of the A_{LL} for 100–day experiments at various $\sqrt{s}$ in the manner of Nowak [8], using the expected data of beam or target polarization ($P_B, P_T \sim 70\%$), the integrated luminosity ($\mathcal{L} \cdot \mathcal{T} \sim 66\text{pb}^{-1}$), and the combined trigger and reconstruction efficiency ($C \sim 50\%$) together with the values of unpolarized cross sections. As a result we found that the experimental sensitivity δA_{LL} is order of magnitude $\sim 10^{-3}$, which is very small. Hence we can neglect it safely and our predictions are expected to be actually tested in the HERA experiments.

However we can see in figures that the A_{LL} goes over 1 for small $\sqrt{s}$ or large x regions and become unphysical. To confirm that our results is reasonable, we attempt to explain such a odd behavior of the A_{LL} below. We have already seen that the A_{LL} is given by a product of $\Delta g(x)/g(x)$ and $\tilde{\Theta}/\Theta$. In the case of GS96 or GRSV96 parametrization which are widely used, the maximum value of $\Delta g(x)/g(x)$ is $0.2 \sim 0.35$, whereas the value of $\tilde{\Theta}/\Theta$ is $3.6 \sim 8.0$. If we rely on the value of the ratio of gluon distributions, the maximum value of the ratio of NRQCD matrix elements should be

$$\frac{\tilde{\Theta}}{\Theta}[\text{max}] \leq 5.0, \tag{7}$$

from the requierment that the A_{LL} should be less than 1. Actually, the uncertainty of matrix elements seems to be larger than the one of gluon distribution, because the value of matrix elements obtained from the Tevatron data does not include the contributions of higher order QCD corrections. Therefore we might be able to constrain the value of matrix elements by measuring the two–spin asymmetry in polarized γp collisions.

4 Summary and Discussion

We have proposed the photoproduction of ψ' at small–p_T regions in polarized γp collisions which might be available in the forthcoming HERA experiments. We have calculated two–spin asymmetry A_{LL} for various parameter regions $\tilde{\Theta}/\Theta = 3.6 \sim 8.0$, and found that the A_{LL} is rather large in the regions $\sqrt{s} = 10 \sim 50\text{GeV}$, $x = 0.005 \sim 0.1$. Therefore we conclude that we can sufficiently test the color–octet model in this process. In addition, we can

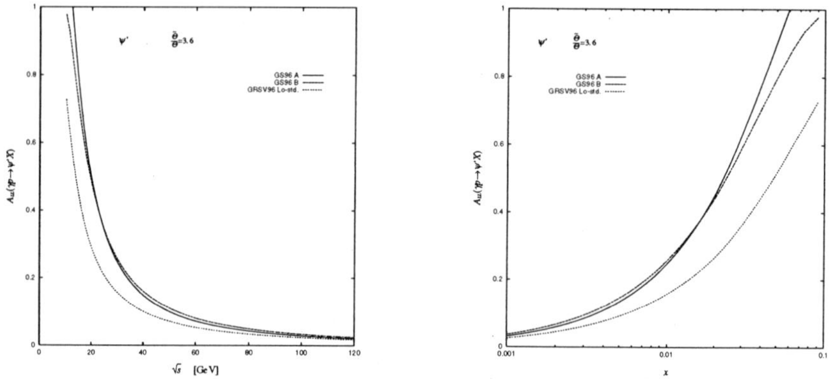

Figure 1: The two–spin asymmetry $A_{LL}(\gamma p \to \psi' X)$ with the parameter $\tilde{\Theta}/\Theta = 3.6$ as a function of $\sqrt{s}$ and momentum fraction x. The solid, dashed and dotted lines show the case of set A of ref.[9], set B of ref.[9] and the 'standard scenario' of ref.[10], respectively.

constrain the value of NRQCD matrix elements though there still remain big uncertainties for $\Delta g(x)$.

References

1. CDF Collaboration, F. Abe *et al.*, *Phys. Rev. Lett.* **69**, 3704 (1992); **71**, 2537 (1993).
2. E. Braaten and S. Fleming, *Phys. Rev. Lett.* **74**, 3327 (1995); P. Cho and A. K. Leibovich, *Phys. Rev.* D **53**, 150 (1996).
3. G. T. Bodwin, E. Braaten, and G. P. Lepage, *Phys. Rev.* D **51**, 1125 (1995).
4. T. Morii, D. Roy, K. Sudoh, Z. -X. Sun, S. Tanaka, and T. Yamanishi, *Phys. Rev.* D **57**, 4487 (1998); O. Teryaev and A. Tkabladze, *Phys. Rev.* D **56**, 7331 (1997).
5. S. Gupta and P. Mathews, *Phys. Rev.* D **55**, 7144 (1997).
6. M. Beneke and I. Z. Rothstein, *Phys. Rev.* D **54**, 2005 (1996).
7. A. K. Leibovich, *Phys. Rev.* D **56**, 4412 (1997).
8. W. D. Nowak, in *Trends in Collider Spin Physics*, Proceedings of the Adriatico Research Conference, Trieste Italy, 1995, ed. Y. Onel *et al.* (World Scientific, Singapore, 1997).
9. T. Gehrmann and W. J. Stirling, *Phys. Rev.* D **53**, 6100 (1996).
10. M. Glück, E. Reya, M. Stratmann, and W. Vogelzang, *Phys. Rev.* D **53**, 4775 (1996).

DILEPTON ENHANCEMENT FROM A NONPERTURBATIVE QUARK-GLUON PLASMA PHASE

Chang-Hwan LEE
Dept. of Physics and Astronomy, State University of New York at Stony Brook
Stony Brook, NY 11794, USA
E-mail : chlee@silver.physics.sunysb.edu

Assuming that gluon condensates are important even above the deconfining phase transition, we develop a model for the dilepton yield from a quark gluon plasma. Using a simple fire ball description of a heavy ion collision, and various estimates of the strengths of the gluon condensates, we compare our predicted dilepton yields with those observed in the CERES and HELIOS experiments at CERN. The simple model gives an adequate description of the data, and in particular it explains the observed considerable enhancement of the yield in the low mass region. This indicates that in the very early stages of a fire ball, quark gluon plasma phase and/or mixed phase, if exist, are consistent with hadronic phase as long as the dilepton enhancement is concerned.

Extensive lattice simulations have shown that QCD exhibits a phase transition at a temperature of the order of the infrared scale Λ. This transition is believed to be a confinement-deconfinement transition for heavy quarks, and a chiral transition for light quarks. High temperature lattice [1,2] and instanton simulations [3] seem to concur on the fact that in the vicinity of the critical temperature, T_c, the chromoelectric and chromomagnetic condensates are still substantial. A number of theoretical arguments, both perturbative [4] and nonperturbative,[5] suggests that in the Euclidean formulation, the high temperature phase of QCD exhibits strong $A_4 A_4$ correlations. It has even been argued that the global color symmetry might be spontaneously broken with the appearance of an A_4 condensate.[5,6] The value of the condensates can only be estimated using non-perturbative calculations such as lattice QCD, and at high temperature, the scale is expected to be set by the mass gap in the magnetic sector which is $\sim g^2 T$. Indeed, an analysis of the gluon-propagator in four-dimensions in Landau-gauge [5] gives $\beta \equiv \langle \frac{\alpha_s}{\pi} A_4^2 \rangle / T^2 \sim 0.4$.

In this work,[7] we present a calculation that suggests that the gluon condensates strongly affect the emission rates of low mass dileptons from the plasma. We will simply assume that the quark-antiquark annihilation processes that give rise to the dileptons occur in the presence of low momentum fluctuating background gluon fields. By averaging over the background fields, we get leading corrections to the emission rates proportional to the various condensates. Note that in doing so we take the concept of a fluctuating background field at face value and perform the calculations directly in the time-like region assuming that the temperature is high enough for perturbation theory to be useful. Our naive fire ball model combined with a simple parametrization of the production amplitudes based on the above ideas, and with the condensate parameters at the values suggested by the scale arguments, is quite successful in explaining the dilepton data from the CERES and HELIOS relativistic heavy ion experiments.[7]

In a translationally invariant, but non-perturbative, quark-gluon phase, the rate $\mathbf{R}$ of dileptons produced in a unit four volume is directly related to the electromagnetic current-current correlation function.[7,8,9] For a pair of leptons with mass m_l and momenta $p_{1,2}$, the rate per unit invariant momentum $q = p_1 + p_2$ is ,

$$\frac{d\mathbf{R}}{d^4q} = -\frac{\alpha^2}{6\pi^3 q^2}\left(1+\frac{2m_l^2}{q^2}\right)\left(1-\frac{4m_l^2}{q^2}\right)^{1/2}\frac{2}{e^{q^0/T}-1}\,\mathrm{Im}\,\mathbf{W}^R(q) \qquad (1)$$

where $\alpha = e^2/4\pi$ is the fine structure constant,

$$\mathrm{Im}\mathbf{W}^R = \frac{N_c}{4\pi}\left(\sum_{q=u,d,s} e_q^2\right) q^2\left[1+\frac{2T}{|\vec{q}|}\ln\left(\frac{n_+}{n_-}\right)\right] \quad , \qquad (2)$$

and N_c is the number of colors, e_q the charge of the quarks, and $n_\pm = 1/(e^{(q_0 \pm |\vec{q}|)/2T} + 1)$.

The contribution of the A_4A_4 condensate to the retarded function $\mathbf{W}^R$ can easily be calculated by picking up the term quadratic in the imaginary colored chemical potential A_4^a.[7] The result is

$$\mathrm{Im}\mathbf{W}_{11}^R = \frac{1}{2\pi}q^2\left[\left\langle\frac{\alpha_s}{\pi}A_4^2\right\rangle\left(\frac{4\pi^2}{T|\vec{q}|}\right)\left(n_+(1-n_+) - n_-(1-n_-)\right)\right] . \qquad (3)$$

Note that (3) has the same sign as (2) and will thus increase the dilepton yield. Below, we will see that this effect can be rather dramatic in the invariant mass region studied in the CERES experiments.

We will assume that the dileptons emanate from a simple fire ball composed of the nonperturbative plasma phase only.[7] The fire ball expansion is assumed to be homogeneous. In the left panel of Fig. 1, we show the evolved emission rates for CERES. The enhancement is sizable and goes in the direction of the data in the 500 MeV range in CERES. Again, the nonperturbative QCD results are indicated by the dotted line for the condensates we quoted ($\delta = 1$ indicates the inclusion of electro-magnetic condensates,[7]) the dashed-dotted

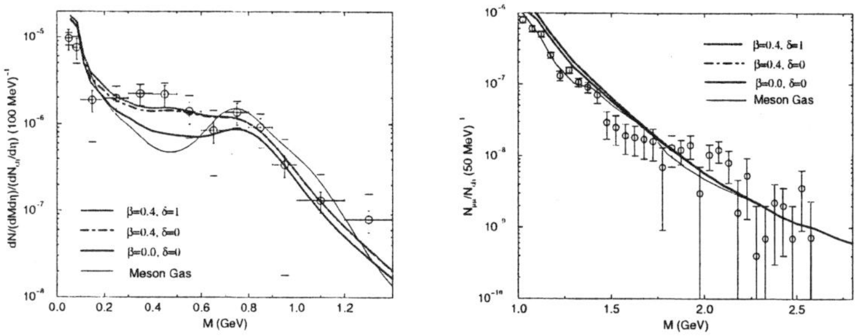

Figure 1: Dielectron rates(left panel) for CERES S-Au experiment and dimuon rates(right panel) for HELIOS-3 S-W experiment.

line (without the E- and B-condensates), while the perturbative QCD result is indicated by the solid line and the meson result by the thin solid line.

In the right panel of Fig. 1, we show the evolved emission rates for HELIOS. For this energy range the effect of chromoelectric and chromomagnetic condensates are small irrespective of the specific value for the thermal condensates.[7] The enhancement is compatible with the HELIOS data, and slightly overshoots both the data and the meson gas in the 1 GeV region.

Another important check on the nonperturbative emission rates can be obtained by using the recent p_t cut analysis by the CERES collaboration to probe the transverse momentum content of the dilepton emission rates in the fire-ball. Our results versus the data are shown in Fig. 2, with and without the chromoelectric and chromomagnetic condensates. The p_t dependent effects caused by the thermal condensates are most dramatic in the low p_t region, where there is a sizable condensate effect, mainly due to β. The condensate contributions are in the right direction around the low mass dielectron enhancement, and give important additions to the perturbative result. In the high p_t region, right panel of Fig. 2, the resulting condensate effects are rather small, and the curve is consistent with the present data. Therefore, the total transverse momentum content is quite accurately reproduced by the addition of the gluon condensates.

In summary, we find it encouraging that not only the dilepton emission rates, but also the observed low mass-region p_t enhancement is, at least partially, explained by our simple model, and we think that this further strengthens our proposal that nonperturbative condensate effects play a role in the plasma phase of heavy ion collisions.

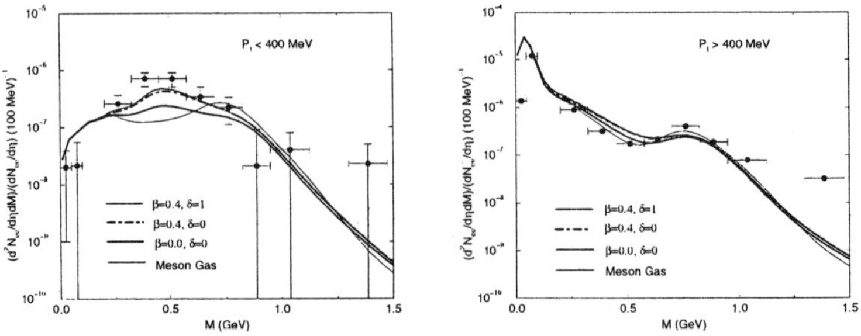

Figure 2: Dielectron rates for low and high p_t of CERES Pb-Au experiment.

Acknowledgments

This work was supported by the U.S. Department of Energy under grant no. DE-FG02-88ER40388. I am particularly thankful to I. Zahed with whom this work was done in collabaration. I would like to thank G.E. Brown, M. Rho, D.-P. Min for useful criticism and discussions.

References

1. M. Campostrini and A. Di Giacomo, Phys. Lett. **B197** (1987) 403; S.H. Lee, Phys. Rev. **D40** (1989) 2484.
2. C. Adami, T. Hatsuda and I. Zahed, Phys. Rev. **D43** (1991) 921; V. Koch and G.E. Brown, Nucl. Phys. **A560** (1993) 345;
3. T. Schaefer and E. Shuryak, Rev. Mod. Phys. **70** (1998) 323.
4. R. Anishetty, J. Phys. **G10** (1984) 423; K.J. Dahlem, Z. Phys. **C29** (1985) 553; V.M. Belayev and V.L. Eletsky, Z. Phys. **C45** (1990) 355.
5. J. Mandula and M. Ogilvie, Phys. Lett. **B201** (1988) 117.
6. S. Nadkarni, Phys. Rev. **D34** (1986) 3904.
7. C.-H. Lee, J. Wirstam, I. Zahed and T.H. Hansson, hep-ph/9809440.
8. L.D. McLerran and T. Toimela, Phys. Rev. **D 31**, 545 (1985).
9. H. A. Weldon, Phys. Rev. **D 42**, 2384 (1990).

ϕ-meson photoproduction and strangeness of the nucleon

Yongseok Oh
Center for Theoretical Physics, Seoul National University, Seoul 151-742, Korea

The polarization observables in ϕ photoproduction are suggested for probing the nucleon strangeness. Based on models for ϕ photoproduction, some double polarization observables are shown to be very sensitive to the strangeness content of the proton because of the different spin structures of the amplitudes associated with different mechanisms.

The possible existence of hidden strangeness in the nucleon has been one of the most controversial problems in hadron physics. Some experimental results have been interpreted as signatures of hidden strangeness of the nucleon[1,2], but it has been also claimed that these experiments could be understood with little or null strangeness in the nucleon[3].

It will be interesting, therefore, to study other processes that might be related directly to the hidden nucleon strangeness[4]. One of them is to use ϕ production from the proton. Since the ϕ is nearly pure $s\bar{s}$ state, its coupling to the proton is suppressed by the OZI rule. Then the idea is that studying the strange sea quark contribution through the OZI evasion processes may give informations about the hidden strangeness of the nucleon.

In this work, we discuss ϕ photoproduction process from the proton targets (Fig. 1). We define k, q, p, and p' as the four-momenta of the photon, ϕ-meson, target proton, and recoiled proton, respectively, and $t = (p - p')^2$ and $W^2 = (p + k)^2$. The scattering angle in c.m. frame is denoted by θ. In this case, the main contribution to the cross section is from the diffractive production process of vector-meson dominance (VDM), and the one-pion-exchange (OPE) process comes in as a correction to VDM[5]. This OPE process is possible because of the

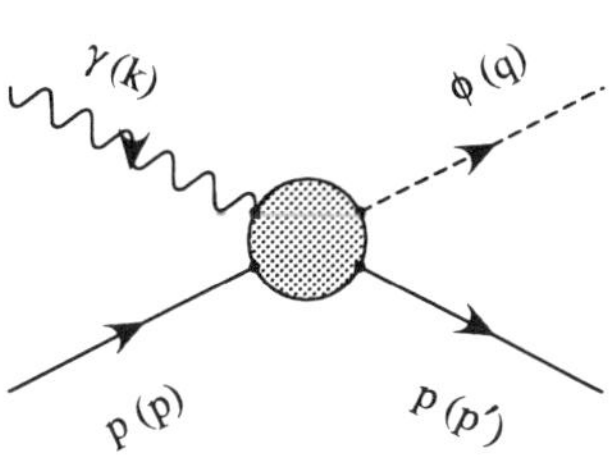

Figure 1: Kinematics for ϕ meson photoproduction from the proton, $\gamma p \to \phi p$.

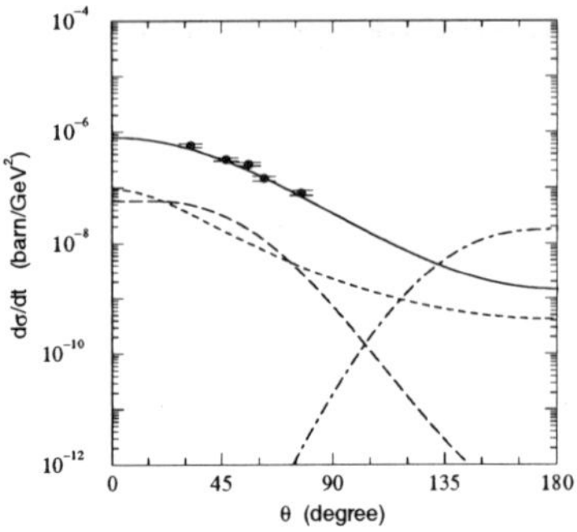

Figure 2: Unpolarized differential cross section $d\sigma/dt$ of ϕ photoproduction at $W = 2.155$ GeV. The solid, dotted, dashed, and dot-dashed lines give the cross section of VDM, OPE, $s\bar{s}$-knockout, and uud-knockout, respectively, with strangeness admixture $B^2 = 1\%$ assuming $|b_0| = |b_1|$. The experimental data are from Ref. 8.

ϕ-π-γ coupling, and represents the contributions from the (small) non-strange quark components of the ϕ-meson.

In addition to these processes, the contribution from the hidden strangeness of the proton arises through the direct knockout process. For this purpose, we write the proton wave function in Fock space as[6,7]

$$|p\rangle = A|[uud]^{1/2}\rangle + \sum_{j_{s\bar{s}}=0,1;j_c} b_{j_{s\bar{s}}} \, |[[[uud]^{1/2} \otimes [\mathbf{L}]]^{j_c} \otimes [s\bar{s}]^{j_{s\bar{s}}}]^{1/2}\rangle, \tag{1}$$

where the superscripts denote the spin of each cluster and (b_0, b_1) correspond to the amplitudes of the $s\bar{s}$ cluster with spin 0 and 1, respectively. Then the strangeness admixture of the proton is $B^2 = b_0^2 + b_1^2$. The symbol $\otimes$ represents vector addition of the cluster spin and the orbital angular momentum $\mathbf{L}$ ($\ell = 1$). The purpose of this study is to extract an information about B^2, if it is nonzero, through ϕ photoproduction from the proton targets.

In Ref. 6, Henley *et al.* calculated the contribution of the knockout process in ϕ electroproduction cross section using nonrelativistic harmonic oscillator quark model and found it comparable to that of VDM with an assumption of $B^2 = 10$-20%. This work was improved by Titov *et al.*[7] with the use of a relativistic harmonic oscillator quark model and leads to the conclusion that a theoretical upper bound of B^2 would be less than 5%. In Fig. 2, we give our results for the cross section for ϕ photoproduction. As can be seen from Fig. 2, it is not easy to distinguish the mechanisms from the cross section measurements because their respective contributions have similar dependence on the momentum transfer.

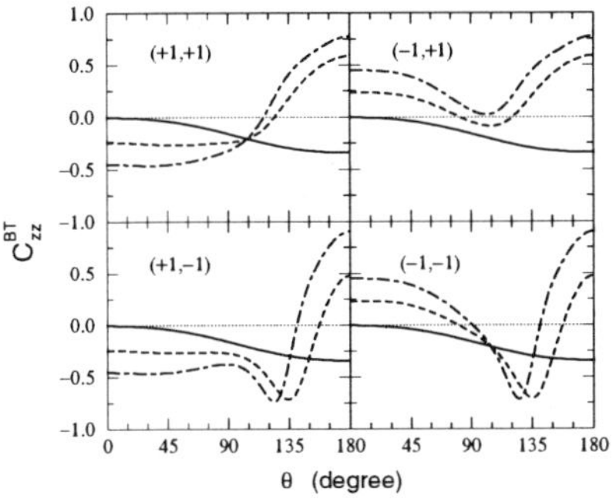

Figure 3: The double spin asymmetry $C_{zz}^{\rm BT}$ at $W = 2.155$ GeV with $B^2 = 0\%$ (solid lines), 0.25% (dashed lines), and 1% (dot-dashed lines) assuming $|b_0| = |b_1|$. The phases (η_0, η_1) of (b_0, b_1) are given explicitly in each graph.

It is, therefore, required to find other quantities which are very sensitive to the nucleon strangeness and we found that some double polarization observables in ϕ photoproduction could be very useful tools to study the nucleon strangeness[9]. Having the helicity amplitudes of each process, we can compute various polarization observables using the formalism developed, *e.g.*, in Ref. 10. In Fig. 3, as a typical example, we show our results for the double polarization observable $C_{zz}^{\rm BT}$ where both the photon and the target proton are polarized along the direction of the photon momentum;

$$C_{zz}^{\rm BT} \equiv \frac{d\sigma_{\uparrow\uparrow} - d\sigma_{\uparrow\downarrow}}{d\sigma_{\uparrow\uparrow} + d\sigma_{\uparrow\downarrow}}, \tag{2}$$

where $d\sigma$ represents $d\sigma/dt$ and the arrows denote the photon beam and target proton helicities. One can find that this quantity is very sensitive to the hidden nucleon strangeness B^2. At forward scattering angles, $C_{zz}^{\rm BT}$ approaches zero with $B^2 = 0$, but even with $B^2 = 1\%$ its magnitude is as large as 0.45.

This remarkable sensitivity of $C_{zz}^{\rm BT}$ on B^2 comes from the different spin structures of the associated amplitudes. At forward scattering angle limit, the helicity amplitudes read

$$H^{\rm VDM} \to -iM_0^{\rm VDM}\delta_{\lambda_f\,\lambda_i}\delta_{\lambda_\phi\,\lambda_\gamma}, \qquad H^{\rm OPE} \to -M_0^{\rm OPE}(2\lambda_i\lambda_\gamma)\delta_{\lambda_f\,\lambda_i}\delta_{\lambda_\phi\,\lambda_\gamma},$$
$$H^{s\bar{s}} \to -iM_0^{s\bar{s}}(2\lambda_i\lambda_\gamma)\delta_{\lambda_f\,\lambda_i}\delta_{\lambda_\phi\,\lambda_\gamma}, \tag{3}$$

where the helicities of the target and recoiled proton, photon beams, and ϕ-meson are denoted respectively by $\lambda_{i,f,\gamma,\phi}$. The relevant amplitudes of each process are given by $M_0^{\rm VDM}$, etc. The uud-knockout is suppressed at forward scattering angles and we can ignore it in this region. Then the above analysis gives

$$C_{zz}^{\rm BT} \simeq -2\eta_0\sqrt{\sigma^{s\bar{s}}/\sigma^{\rm VDM}}, \tag{4}$$

where η_0 is the phase of b_0. This shows that the above quantity is very sensitive to $\sigma^{s\bar{s}}$ as verified by numerical calculations given in Fig. 3.

As a summary, we found that some double polarization observables in ϕ photoproduction could be very useful for investigating the hidden strangeness of the nucleon. The optimal range of the initial photon energy to measure B^2 would be around 2-3 GeV, which can be reached by the current electron facilities, and several experiments have been proposed at JLab and RCNP to measure these quantities.

This work was done in close collaboration with A. I. Titov, S. N. Yang, and T. Morii. I am grateful to them for valuable discussions. I also thank D.-P. Min and C.-R. Ji for encouragements. This work was supported in part by the Korea Science and Engineering Foundation through the CTP of Seoul National University.

1. EM Collaboration, J. Ashman *et al.*, Nucl. Phys. **B328**, 1 (1989).
2. J. Ellis, E. Gabathuler, and M. Karliner, Phys. Lett. B **217**, 173 (1989).
3. M. Anselmino and M. D. Scadron, Phys. Lett. B **229**, 117 (1989); H. J. Lipkin, *ibid.* **256**, 284 (1991).
4. See, *e.g.*, M. J. Musolf *et al.*, Phys. Rep. **239**, 1 (1994).
5. D. G. Cassel *et al.*, Phys. Rev. D **24**, 2787 (1981); A. Donnachie and P. V. Landshoff, Nucl. Phys. **B244**, 322 (1984); M. A. Pichowsky and T.-S. H. Lee, Phys. Lett. B **379**, 1 (1996).
6. E. M. Henley, G. Krein, and A. G. Williams, Phys. Lett. B **281**, 178 (1992).
7. A. I. Titov, Y. Oh, and S. N. Yang, Chin. J. Phys. **32**, 1351 (1994); A. I. Titov, S. N. Yang, and Y. Oh, Nucl. Phys. **A618**, 259 (1997).
8. H. J. Besch *et al.*, Nucl. Phys. **B70**, 257 (1974).
9. A. I. Titov, Y. Oh, and S. N. Yang, Phys. Rev. Lett. **79**, 1634 (1997); A. I. Titov, Y. Oh, S. N. Yang, and T. Morii, Phys. Rev. C **58**, 2429 (1998).
10. C. Bourrely, E. Leader, and J. Soffer, Phys. Rep. **59**, 95 (1980); M. Pichowsky, C. Savkli, and F. Tabakin, Phys. Rev. C **53**, 593 (1996).

DISSIPATIVE FIELD THEORY WITH CALDEIRA-LEGGETT METHOD AND ITS APPLICATION TO NEUTRAL/CHARGED DCC STATES

Hiroyuki YABU
Department of Physics, Tokyo Metropolitan University, Hachioji Tokyo 192, Japan
E-mail: yabu@phys.metro-u.ac.jp

The effective field theory including the dissipative effect is developed based on the Caldeira-Leggett theory at the classical level. Application to the linear sigma model is given, and the decay process of the chiral condensate is calculated with it, both analytically in the linear approximation and numerically. With these results, we discuss the stability of chiral condensates within the quenched approximation. For details, see H. Yabu, K. Nozawa and T. Suzuki, Phys. Rev. **D57**, 1687 (1998).

1. Dissipative Field Equation with Caldeira-Leggett Theory To illustrate the application of the Caldeira-Leggett theory to the field equation, we take a simple system of the order-parameter $\Phi(x)$ and the background field $f(x)$ that describes the particle with the mass m:

$$L = \frac{1}{2}(\partial\Phi)^2 - V(\Phi) - F(\Phi)f + \frac{1}{2}[(\partial f)^2 - m^2 f^2]. \tag{1}$$

where $V(\Phi)$ is the effective potential for Φ and $F[\Phi]f$ gives the interaction between Φ and f. Taking variations with Φ and f, we obtain the field equations from (1):

$$\partial^2\Phi + \frac{\delta V}{\delta\Phi} + \frac{\delta F}{\delta\Phi} f = 0, \qquad (\partial^2 + m^2)f + F[\Phi] = 0. \tag{2}$$

The second equation of (2) is formally solved with the Green functions $G(x - y; m)$ that satisfies $(\partial^2 + m^2)G(x-y) = -\delta^4(x-y)$, and we obtain the integro-differential equation for Φ

$$\partial^2\Phi(x) + \frac{\delta V[\Phi(x)]}{\delta\Phi} + \frac{\delta F[\Phi(x)]}{\delta\Phi}\int d^4y G(x-y;m)F(\Phi[y]) = 0. \tag{3}$$

In the Caldeira-Leggett theory, the imaginary component of $G(k; m)$ (the energy-momentum representation of $G(x - y; m)$) can be considered as the dissipative term caused by the emission of particles for the background field f. and, we obtain the dissipative field equation corresponding to (3):

$$\partial^2\Phi(x) + \frac{\delta V[\Phi(x)]}{\delta\Phi} + \frac{\delta F[\Phi(x)]}{\delta\Phi}\tilde{f} = 0, \tag{4}$$

where

$$\tilde{f}(x) = \int d^4y \int \frac{d^4k}{(2\pi)^4} i\mathrm{Im}G(k;m)e^{-ik(x-y)}F(\Phi[y]). \tag{5}$$

2. Application to Linear Sigma Model We apply the formulation given in the last chapter to the linear sigma model:

$$\mathcal{L} = \frac{1}{2}\left[(\partial\sigma)^2 + (\partial\pi)^2\right] - V(\sigma,\pi), \tag{6}$$

where

$$V(\sigma,\pi) = \frac{\lambda}{4}(\sigma^2 + \pi^2 - \nu^2)^2 - H\sigma. \tag{7}$$

where the last term breaks the chiral symmetry explicitly and produces the finite pion mass.

The fields σ and π can be divided into two parts:

$$\sigma = \langle\sigma\rangle + \delta\sigma, \quad \pi = \langle\pi\rangle + \delta\pi, \tag{8}$$

where $\langle\sigma\rangle$ and $\langle\pi\rangle$ corresponds to the order parameter of the chiral symmetry for the (disoriented) condensed state Ψ: $\langle\sigma\rangle = \langle\Psi|\bar{q}q|\Psi\rangle$ and $\langle\pi\rangle = \langle\Psi|\bar{q}\tau\gamma_5 q|\Psi\rangle$ with the quark field $q = (u,d)$. The $\delta\sigma$ and $\delta\pi$ are the fluctuations around them, and represent the sigma and pi meson degrees of freedom.

Now we consider the dynamical behaviors of the disoriented-condensed vacuum state $|\Psi\rangle \neq |0\rangle$, taking $\langle\sigma\rangle \neq f_\pi$ and $\langle\pi\rangle \neq 0$ as the field Φ in the last chapter and the fluctuations $\delta\sigma$ and $\delta\pi$ as the background field f. In this paper, we discuss the case where the condensates are on the chiral circle, so that $\langle\sigma\rangle$ and $\langle\pi\rangle$ can be parameterized with the chiral angle ϕ: $\langle\sigma\rangle = \nu\cos\phi$, $\langle\pi\rangle = \nu\sin\phi$.

Now we can apply the CL theory for ϕ, $\delta\sigma$ and $\delta\pi$ and get the dissipative fields equation:

$$\partial^2\phi(x) + \frac{H}{\nu}\sin\phi - \nu\sin\phi\,\partial^2\delta\tilde{\sigma} + \nu\cos\phi\,\partial^2\delta\tilde{\pi} = 0, \tag{9}$$

where

$$\partial^2\delta\tilde{\pi}(x) = \frac{1}{2}\int_{x_0}^{\infty} dy^0 \int d^3y\Delta(x-y;m_\sigma)\partial_y^2\sin\phi(y), \tag{10}$$

and similar for $\tilde{\sigma}$. The $\Delta(x-y;m)$ is the invariant delta function.

3. Discussions and Summary Analyzing that dissipative equation analytically and numerically, we can get the results:

1) Asymptotic behavior of solutions : Instead of $\phi(t,\mathbf{x})$, we use the Fourier component $\phi_{\mathbf{k}}(t)$: The order parameter $\phi_{\mathbf{k}}$ is assumed to decrease when $t \to \infty$,

so that it can be regarded small. Then we can expand the equation and approximate to the first order (the linear approximation). That linear equation can be solved with the form $\phi_{\mathbf{k}}(t) = e^{-a(\mathbf{k})t}$ where $a(\mathbf{k}) = p(\kappa) + iq(\kappa)$:

$$p(\kappa) = \frac{m_\pi}{2}\sqrt{2\rho(\kappa) - (2\kappa^2 + \frac{1}{2} + \alpha)}, \quad q(\kappa) = \frac{m_\pi}{2}\sqrt{2\rho(\kappa) + (2\kappa^2 + \frac{1}{2} + \alpha)}, \tag{11}$$

where $\rho(\kappa) = \sqrt{\kappa^4 + (1/2 + \alpha)\kappa^2 + \alpha}$. It can be shown easily that $p(\kappa)$ takes real and positive value when $\kappa \geq 0$, so that the asymptotic behavior of $\phi_{\mathbf{k}}$ is found to be

$$\phi_{\mathbf{k}}(t) \sim e^{-p(\kappa)t} \sin[q(\kappa)t + \delta]. \quad (t \to \infty) \tag{12}$$

with the constant phase. Summarizing the whole results, we obtain the asymptotic differential equation to (9):

$$\ddot{\phi}_{\mathbf{k}} + \mathbf{k}^2\phi_{\mathbf{k}} + \gamma\dot{\phi}_{\mathbf{k}} + \sqrt{\alpha}\sin\phi_{\mathbf{k}} \sim 0. \tag{13}$$

where the dissipative coefficients are $\gamma = 2p(\kappa)$.

0.1 Numerical Results

2) Uniform Case $\phi = \phi(t)$: In this case, eqs. (9) are reduced to the ordinary differential equations, which can be easily solved.

The dissipating behaviors can be read off in the phase diagram (Fig. 1), too, where each lines are the phase trajectories to these solutions; the spiral pattern around the origin show that they behave like the damped oscillator asymptotically.

For quantitative check, we consider the damped rigid oscillator:

$$\ddot{\phi} + \gamma\dot{\phi} + \sin\phi = 0, \tag{14}$$

with the damping coefficient γ consistent with the dissipative coefficient $p(k)$ in (11): $\gamma = 2p(0) = 0.7m_\pi$. The phase diagrams for (14) are given in Fig. 2. The trajectories in Fig. 1 are found to behave similar to those in Fig. 2, especially in the asymptotic region. In the non-asymptotic region (far from the origin), the trajectories in Fig. 1 have the modulations around those of the damped rigid oscillator. They come from the non-linear dissipating behavior in (9), which is more effective in the non-asymptotic region.

3) Life-time of disoriented chiral condensates (DCC) : As a phenomenological application, we discuss about the disoriented chiral condensate that is expected to appear after the high-energy hadron collision. Because of the explicit chiral-symmetry breaking, the DCC state has higher energy, and it should be observed

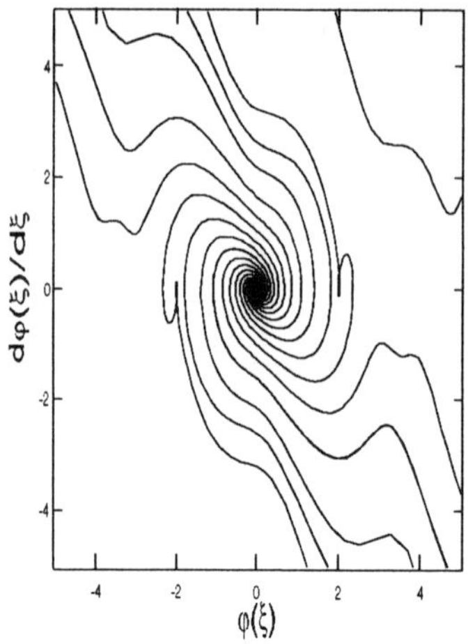

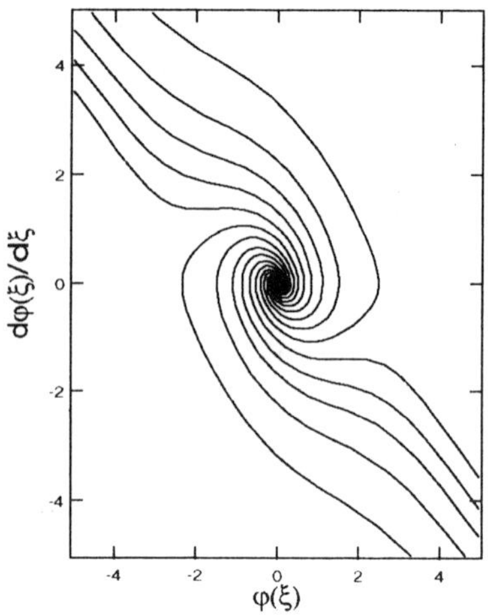

Figure 1: (left) The trajectories of the integro-differential equation (9) for the space-independent uniform case $\phi(x) \equiv \phi(t)$. The ϕ and $d\phi/d\xi$ are the chiral angle and the corresponding velocity with the scaled time $\xi = m \quad t$. The initial conditions of each trajectories are chosen as they behave asymptotically (12) with $\delta = \pi n/5$ $(n = 0, \pm 1, \cdots, \pm 5)$.

Figure 2: The trajectories of the damped rigid oscillator (14). The dissipative coefficient is chosen to be consistent with the asymptotic value in the integro-differential equation (9): $\gamma = 2p(0) = 0.7m \quad$. The ϕ and $d\phi/d\xi$ are the chiral angle and the corresponding velocity with the scaled time $\xi = m \quad t$. The initial conditions of each trajectories are chosen as in 1.

as a metastable state. We consider the neutral DCC state with $\langle \pi^0 \rangle \neq 0$ and $\langle \pi^\pm \rangle = 0$. The main decay process of this state should be the π^0-radiation, so that we can apply the above-developed formula with regarding π in (13) as π^0. The life-time τ_L can be estimated with the dissipative constant $p(0)$ in (11): $\tau_L = 1/p(0) \sim 3m_\pi^{-1}$. If we take the quenching scenario for the DCC formation, the formation time τ_R has been estimated to be $\tau_R \sim \sqrt{2}m_\sigma^{-1} \sim 0.3m_\pi^{-1}$. It tells us that the life-time τ_L is ten times longer than the formation time, so that the neutral DCC state will be enough metastable.

The dissipative decay of the charged DCC (with both *pi*- and γ-radiations) can be studied in a similar way, and will be published elsewhere.

About QCD sum rules for pion-nucleon coupling

Hungchong Kim
Department of Physics, Tokyo Institute of Technology, Tokyo 152-8551, Japan
E-mail: hckim@th.phys.titech.ac.jp

Abstract

We point out subtleties in the previous sum rule calculations of the πNN coupling associated with using either the PV or the PS coupling schemes in modeling the phenomenological side. We propose a sum rule which is independent of the coupling schemes used and has less uncertainty in the OPE. The obtained value is $g_{\pi N} = 9.76 \pm 2.04$ where the uncertainty mainly comes from the quark-gluon mixed condensate.

Since first introduced by Shifman, Vainshtein and Zakharov [1], QCD sum rule has been widely used to study the hadron properties such as masses or couplings of baryons [2]. QCD sum rule is a framework which connects the physical parameters with QCD parameters. In this framework, a correlation function is introduced in terms of interpolating fields constructed from quark and gluon fields. Then, the correlation function, on the one hand, is calculated by Wilson's operator product expansion (OPE) and, on the other hand, its phenomenological "ansatz" is constructed. A physical quantity of interest is extracted by matching the two descriptions in the deep Euclidean region ($q^2 = -\infty$) making use of the dispersion relation. The extracted value therefore should be independent of possible ansatz in order to be physically meaningful.

One important quantity to be determined in hadron physics is the pion-nucleon coupling constant, $g_{\pi N}$, whose empirical value is known to be around 13.4 but it is of interest to determine the coupling from QCD. QCD sum rule can be used for such purpose and indeed there are such calculations of $g_{\pi N}$ [2,3,4]. Reinders, Rubinstein and Yazaki [2] calculated $g_{\pi N}$ by retaining only the first nonperturbative term in the OPE. Later Shiomi and Hatsuda (SH) [3] improved the calculation by including higher order terms in the OPE. SH considered the two-point correlation function for the nucleon interpolating field J_N,

$$\Pi(q,p) = i \int d^4x e^{iq\cdot x} \langle 0|T[J_N(x)\bar{J}_N(0)]|\pi(p)\rangle \ , \qquad (1)$$

and evaluated the OPE in the soft-pion limit ($p_\mu \to 0$). More recently, Birse and Krippa (BK) [4] pointed out that the use of the soft-pion limit does not constitute an independent sum rule from the nucleon sum rule and they considered the sum rule beyond the soft-pion limit.

In both sum rules, the phenomenological side of the correlator is constructed by using the PS interaction Lagrangian,

$$\mathcal{L}_{ps} = g_{\pi N}\bar{\psi} i\gamma_5 \vec{\tau}\cdot\vec{\pi}\psi \ . \tag{2}$$

However, the PV interaction Lagrangian,

$$\mathcal{L}_{pv} = \frac{g_{\pi N}}{2m}\bar{\psi}\gamma_5\gamma_\mu \vec{\tau}\cdot\partial^\mu\vec{\pi}\psi \ , \tag{3}$$

can also be used which could change the results. This is because, in QCD sum rules, on-shell quantities are extracted from far Euclidean region where descriptions from the two Lagrangians are different.

To achieve the independence from the PS and PV coupling schemes, it is natural to go beyond the soft-pion limit. Then the method proposed by Ref. [4] seems to be appropriate. The $\gamma_5\, \not{p}$ structure considered in Ref. [4] uses the PS Lagrangion,

$$-\frac{i g_{\pi N}\lambda_N^2 m}{(q^2-m^2+i\epsilon)[(q-p)^2-m^2+i\epsilon]}+\cdots \ . \tag{4}$$

Then after some manipulation, the sum rule formula can be succinctly casted as

$$g_{\pi N} + AM^2 = f(M) \ . \tag{5}$$

$f(M)$ is a function of the Borel mass M. The unknown simple pole contribution A is associated with the transition $N \to N^*$ [5] and estimated to be small [4]. With the PV scheme, the phenomenological side takes the form,

$$-\frac{i g_{\pi N}\lambda_N^2/2m}{(q-p)^2-m^2+i\epsilon}-\frac{i g_{\pi N}\lambda_N^2 m}{(q^2-m^2+i\epsilon)[(q-p)^2-m^2+i\epsilon]}+\cdots \ . \tag{6}$$

Note that the double pole is the same as above but there is an additional simple pole of $N \to N$. It means that physical content of the single pole is coupling scheme-dependent. If this were used in BK's sum rule, then instead of Eq. (5), BK would have obtained

$$g_{\pi N}\left(1-\frac{M^2}{2m^2}\right)+AM^2 = f(M) \ . \tag{7}$$

Around $M \sim m \sim 1$ GeV, the new term is 0.5, impling that A is large within the prescriptions used in Ref. [4].

One more Dirac structure, $\gamma_5\sigma_{\mu\nu}$, contains the double pole only, independent of the coupling scheme. The common phenomenological side for this structure is given by

$$-\frac{g_{\pi N}\lambda_N^2 p^\mu q^\nu}{(q^2-m^2+i\epsilon)[(q-p)^2-m^2+i\epsilon]}+\cdots \ . \tag{8}$$

This structure is zero in the soft-pion limit, but beyond the soft-pion limit, this can provide a new sum rule for $g_{\pi N}$. Since there is no simple nucleon pole in this case, the simple pole structure comes only from $N \to N^*$.

To construct a sum rule for $\gamma_5\sigma^{\mu\nu}p_\mu q_\nu$ structure, we consider the correlation function

$$\Pi(q,p) = i\int d^4x e^{iq\cdot x}\langle 0|T[J_p(x)\bar{J}_n(0)]|\pi^+(p)\rangle \ . \tag{9}$$

Here J_p is the proton interpolating field of Ioffe [6],

$$J_p = \epsilon_{abc}[u_a^T C\gamma_\mu u_b]\gamma_5\gamma^\mu d_c \tag{10}$$

and the neutron interpolating field J_n is obtained by replacing $(u,d) \to (d,u)$. In the OPE, we will only keep the diquark component of the pion wave function and use the vacuum saturation hypothesis to factor out higher dimensional operators in terms of the pion wave function and the vacuum expectation value.

The calculation of the correlator, Eq. (9), in the coordinate space contains the following quark-antiquark component of the pion wave function,

$$\begin{aligned} D^{\alpha\beta}_{aa'} &\equiv \langle 0|u^\alpha_a(x)\bar{d}^\beta_{a'}(0)|\pi^+(p)\rangle \\ &= \frac{\delta_{aa'}}{12}\Big[(\gamma^\mu\gamma_5)^{\alpha\beta}\langle 0|\bar{d}(0)\gamma_\mu\gamma_5 u(x)|\pi^+(p)\rangle + (i\gamma_5)^{\alpha\beta}\langle 0|\bar{d}(0)i\gamma_5 u(x)|\pi^+(p)\rangle \\ &-\frac{(\gamma_5\sigma^{\mu\nu})^{\alpha\beta}}{2}\langle 0|\bar{d}(0)\gamma_5\sigma_{\mu\nu}u(x)|\pi^+(p)\rangle\Big] \ . \end{aligned} \tag{11}$$

These matrix elements can be written in terms of pion wave function. Since we are doing the calculation up to the first order of $p_\mu q_\nu$, we need only the overall normalization of the wave functions. In fact, to leading order in the pion momentum, the first and third matrix elements are given as [7],

$$\langle 0|\bar{d}(0)\gamma_\mu\gamma_5 u(x)|\pi^+(p)\rangle = i\sqrt{2}f_\pi p_\mu + O(x^2) \tag{12}$$

$$\langle 0|\bar{d}(0)\gamma_5\sigma_{\mu\nu}u(x)|\pi^+(p)\rangle = i\sqrt{2}(p_\mu x_\nu - p_\nu x_\mu)\frac{f_\pi m_\pi^2}{6(m_u+m_d)} \ . \tag{13}$$

In Eq. (12), $O(x^2)$ contains the twist 4 term which does not contribute to our sum rule up to the dimension we consider below. Note that in Eq. (13) the factor $f_\pi m_\pi^2/(m_u+m_d)$ can be written as $-\langle\bar{q}q\rangle/f_\pi$ by making use of the Gell-Mann$-$Oakes$-$Renner relation.

In our sum rule for the $\gamma_5\sigma^{\mu\nu}p_\mu q_\nu$ structure, the second matrix element in Eq. (11) does not contribute up to dimension 7. It is straightforward to calculate the OPE. We match the OPE with its phenomenological counterpart and take the Borel transformation. Then the unknown parameter, λ_N, which represents the coupling strength of the interpolating field to the physical nucleon, is eliminated by using the nucleon chiral odd sum rule. For more discussion, see Ref.[9]. The final expression for the sum rule is

$$g_{\pi N}(BM^2+1)\left(\frac{f_\pi}{m}\right) \\ = \frac{M^4E_0(x_\pi)/3 + 16\pi^2 f_\pi^2 M^2/3 + \pi^2\left\langle \frac{\alpha_s}{\pi}\mathcal{G}^2\right\rangle/54 - 2\pi^2 m_0^2 f_\pi^2/3}{M^4E_1(x_N) - \pi^2\left\langle\frac{\alpha_s}{\pi}\mathcal{G}^2\right\rangle/6}\,. \tag{14}$$

Here, B represents the unknown simple pole of $N \to N^*$, $x_\pi = S_{\pi N}/M^2$ with $S_{\pi N}$ being the continuum threshold, x_N is corresponding one for the nucleon sum rule, and $E_n(x) = 1-(1+x+\cdots+x^n/n!)e^{-x}$. In this sum rule, the main source for the error comes from m_0^2 which is related to quark-gluon mixed condensate via $\langle\bar{q}g_s\sigma\cdot\mathcal{G}q\rangle = m_0^2\langle\bar{q}q\rangle$. This contributes only to the highest dimensional term in the OPE so this will be suppressed in the Borel window chosen. According to Ref.[8], we take the range, $0.6 \le m_0^2 \le 1.4$ GeV^2 and see the sensitivity of our results.

We fit the RHS of Eq. (15) with a straight line within the Borel window chosen appropriately. Our result is $g_{\pi N} = 9.76 \pm 2.04$, where the error comes from the uncertainty in m_0^2. This result is independent of the coupling schemes and its error is relatively small.

Acknowledgments

This work is supported by Research Fellowship of the Japan Society for the promotion of Science.

References

1. M.A. Shifman, A.I. Vainshtein, and V.I. Zakharov, *Nucl. Phys.* B **147**, 385,448 (1979).
2. L.J. Reinders, H. Rubinstein and S. Yazaki, *Phys. Rep.* 1271 1985.
3. H. Shiomi and T. Hatsuda, *Nucl. Phys.* A **594**, 294 (1995).
4. M. C. Birse and B. Krippa, *Phys. Lett.* B **373**, 9 (1996); M. C. Birse and B. Krippa, *Phys. Rev.* C **54**, 3240 (1996).
5. B. L. Ioffe and A. V. Smilga, *Nucl. Phys.* B **234**, 109 (1984).
6. B. L. Ioffe, *Nucl. Phys.* B **188**, 317 (1981).
7. V. M. Belyaev, V. M. Braun, A. Khodjamirian and R. Rückl, *Phys. Rev.* D **51**, 6177 (1995).
8. A. A. Ovchinnikov and A. A. Pivovarov, *Yad. Fiz.* 481135 1988; M. Kremer and G. Schierholz, *Phys. Lett.* B **194**, 283 (1987); M. V. Polyakov and C. Weiss, *Phys. Lett.* B **387**, 841 (1996).
9. Hungchong Kim, Su Houng Lee and Makoto Oka, *Los Alamos Preprint nucl-th/9809004, Submitted to Physics Letters B.*

Effective $\Delta S = 1$ weak chiral Lagrangian from the instanton-induced chiral quark model

Mario Franz[(1)], Hyun-Chul Kim[(2)] , and Klaus Goeke[(1)]
[1] *Institute for Theoretical Physics II, P.O. Box 102148, Ruhr-University Bochum, D-44780 Bochum, Germany*

[2] *Department of Physics, Pusan National University, 609-735 Pusan, Republic of Korea*

We present the recent investigation of the $\Delta S = 1$ effective weak chiral Lagrangian within the framework of the instanton-induced chiral quark model. Starting from the effective four-quark operators, we derive the effective weak chiral action by integrating out the constituent quark fields. Employing the derivative expansion, we obtain the effective weak chiral Lagrangian to $\mathcal{O}(p^4)$ order with low energy constants.

Chiral perturbation theory (χPT), known as an effective field theory in very low-energy regime, explains low-energy phenomena in strong interactions very well. Based on its success in describing strong interactions, χPT was also applied to the nonleptonic processes [1,2]. However, the effective weak chiral Lagrangian of order $\mathcal{O}(p^4)$ brings in too many low energy constants (LECs) to be determined by experimental data, as the chiral corrections are considered. In order to proceed to make numerical calculations, it is inevitable to rely on models which provide the LECs.

Recently, Antonelli *et al.*[3] applied the χQM in order to obtain the effective weak chiral Lagrangian in the lowest order ($\mathcal{O}(p^2)$) in the chiral expansion. The large N_c expansion was also considered in orders $\mathcal{O}(N_c^2)$, $\mathcal{O}(N_c)$, and $\mathcal{O}(\alpha_s N_c)$. In their study, the correction from order $\mathcal{O}(\alpha_s N_c)$ plays an essential role in reaching the agreement with phenomenology. However, one should mention that treating order $\mathcal{O}(\alpha_s N_c)$ requires a special care. As noted by Ref.[4], $\mathcal{O}([\alpha_s N_c]^2)$ is in fact the same order as $\mathcal{O}(\alpha_s N_c)$ and might be nonnegligible.

More recently, Ref. [5] extended the work of Ref. [3] to the next-to-leading order in the chiral expansion. However, they calculated directly the transition amplitudes of the decay $K \to \pi\pi$ to order $\mathcal{O}(p^4)$, which is in a sense different from the original scheme of chiral perturbation theory.

In this work we rather want to furnish the LECs not only to order $\mathcal{O}(p^2)$ but also to order $\mathcal{O}(p^4)$ within the framework of the instanton-induced chiral quark model.

The effective chiral action with the $\Delta S = 1$ effective weak Hamiltonian can be written as follows:

$$\exp\left(-S_{\rm eff}^{\Delta S=1}\right) = \int \mathcal{D}\psi\mathcal{D}\psi^\dagger \exp\left[\int d^4x \left(\psi^\dagger D\psi - \mathcal{H}_{\rm eff}^{\Delta S=1}\right)\right], \qquad (1)$$

where D is the Dirac operator with the pseudo-Goldstone boson [6]

$$D = i\partial\!\!\!/ + iMU^{\gamma_5}. \qquad (2)$$

The constituent quark mass M is a momentum-dependent mass which characterizes the instanton-induced chiral quark model. The effect of the momentum dependence of the M will be considered in a later work [7]. U^{γ_5} denotes the chiral meson field. The effective weak quark Hamiltonian $\mathcal{H}_{\rm eff}^{\Delta S=1}$ consists of ten four-quark operators among which only seven operators are independent:

$$\mathcal{H}_{\rm eff}^{\Delta S=1} = -\frac{G_F}{\sqrt{2}} V_{ud}V_{us}^* \sum_i c_i(\mu)\mathcal{Q}_i(\mu) + \text{h.c.}. \qquad (3)$$

The Hamiltonian $\mathcal{H}_{\rm eff}^{\Delta S=1}$ can be found elsewhere (for example, see Ref. [8]). Since the Fermi constant G_F is very small, we can expand Eq.(1) with regard to it and obtain the following expression:

$$\mathcal{L}_{\rm eff}^{\Delta S=1} = -\frac{1}{\mathcal{N}}\int \mathcal{D}\psi\mathcal{D}\psi^\dagger \mathcal{H}_{\rm eff}^{\Delta S=1} \exp\left[\int d^4x \psi^\dagger D\psi\right]. \qquad (4)$$

By integrating out the quark fields and making the derivative expansion, we derive the effective chiral Lagrangian as follows:

$$\mathcal{L}_{\rm eff}^{\Delta S=1,\mathcal{O}(p^2)} = -\frac{G_F}{\sqrt{2}} V_{ud}V_{us}^* f_\pi^4 \Big[g_{\underline{8}} \langle \lambda_3^2 L_\mu L^\mu\rangle + g_{\underline{27}}\Big(\frac{2}{3}\langle\lambda_2^1 L_\mu\rangle\langle\lambda_1^3 L^\mu\rangle$$
$$+\langle\lambda_2^3 L_\mu\rangle\langle\lambda_1^1 L^\mu\rangle\Big)\Big] + \text{h.c.} = \mathcal{L}_{\underline{8}}^{(1/2)} + \frac{1}{9}\mathcal{L}_{\underline{27}}^{(1/2)} + \frac{5}{9}\mathcal{L}_{\underline{27}}^{(3/2)}, \qquad (5)$$

where $\mathcal{L}_{\underline{8}}^{(1/2)} = -\frac{G_F}{\sqrt{2}} V_{ud}V_{us}^* f_\pi^4 g_{\underline{8}} \langle\lambda_3^2 L_\mu L^\mu\rangle + \text{h.c.}$,

$$\mathcal{L}_{\underline{27}}^{(1/2)} = -\frac{G_F}{\sqrt{2}} V_{ud}V_{us}^* f_\pi^4 g_{\underline{27}}\Big(\langle\lambda_2^1 L_\mu\rangle\langle\lambda_1^3 L^\mu\rangle - \langle\lambda_2^3 L_\mu\rangle\langle\lambda_1^1 L^\mu\rangle$$
$$-5\langle\lambda_2^3 L_\mu\rangle\langle\lambda_3^3 L^\mu\rangle\Big) + \text{h.c.}, \quad \mathcal{L}_{\underline{27}}^{(3/2)} = -\frac{G_F}{\sqrt{2}} V_{ud}V_{us}^* f_\pi^4 g_{\underline{27}}\Big(\langle\lambda_2^1 L_\mu\rangle\langle\lambda_1^3 L^\mu\rangle$$
$$+2\langle\lambda_2^3 L_\mu\rangle\langle\lambda_1^1 L^\mu\rangle + \langle\lambda_2^3 L_\mu\rangle\langle\lambda_3^3 L^\mu\rangle\Big) + \text{h.c.}$$

Note that the isospin $T = 1/2$ part of the eikosiheptaplet is suppressed by the prefactor, so that one can treat the eikosiheptaplet part of the Lagrangian as

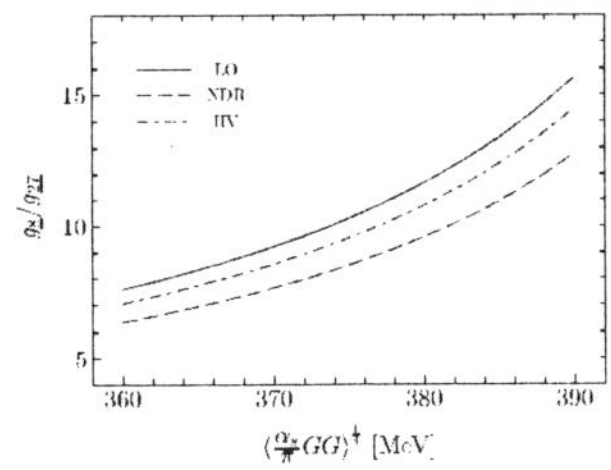

Figure 1: Dependence of the $g_{\underline{8}}/g_{\underline{27}}$ on the gluon condensate $\langle\frac{\alpha_s}{\pi}GG\rangle$. The solid curve denotes the LO renormalization scheme in Ref.[8], while the dashed curve and dot-dashed one stand for the NDR and the HV schemes, respectively. The value of the constituent quark mass $M = 300$ MeV is used and the quark condensate $\langle\bar{q}q\rangle/2 = -(250 \text{ MeV})^3$ is employed.

a pure $T = 3/2$ one. $g_{\underline{8}}$ and $g_{\underline{27}}$ can be extracted from the $K \to \pi\pi$ decay rate. At the tree level, the $\Delta T = 1/2$ enhancement is reflected in these constants.

We first revisit the recent results of Ref. [3] but we concentrate only on the low energy constants. To obtain the comparable results to the empirical data, Ref. [3] considered the effect of the gluon condensates coming from the external backgroud gluon fields [4] in order $\mathcal{O}(\alpha_s N_c)$. With orders $\mathcal{O}(N_c^2)$, $\mathcal{O}(N_c)$ and $\mathcal{O}(\alpha_s N_c)$ taken into account, one arrives at following results in the leading order $\mathcal{O}(p^2)$:

$$\begin{aligned} g_{\underline{8}}^{\delta_{\langle GG\rangle}} &= \left(-\frac{2}{5}+\frac{1}{N_c}(1-\delta_{\langle GG\rangle})\frac{3}{5}\right)c_1 + \left(\frac{3}{5}-\frac{1}{N_c}(1-\delta_{\langle GG\rangle})\frac{2}{5}\right)c_2 \\ &+\frac{1}{N_c}(1-\delta_{\langle GG\rangle})c_3 + \left(\frac{\langle\bar{q}q\rangle}{N_c f_\pi^2 M} - \langle\bar{q}q\rangle_2 + \frac{1}{N_c}(1-\delta_{\langle GG\rangle})\right)c_4 \\ &+\left(\frac{\langle\bar{q}q\rangle}{N_c f_\pi^2 M} - \frac{\langle\bar{q}q\rangle M}{8 f_\pi^4\pi^2}\right)c_5 + \left(\frac{\langle\bar{q}q\rangle}{f_\pi^2 M} - \frac{N_c\langle\bar{q}q\rangle M}{8 f_\pi^4 \pi^2}\right)c_6 \\ &+\left(-\frac{3}{5}+\frac{1}{N_c}(1-\delta_{\langle GG\rangle})\frac{2}{5}\right)c_9 + \left(\frac{2}{5}-\frac{1}{N_c}(1-\delta_{\langle GG\rangle})\frac{3}{5}\right)c_{10} \end{aligned} \qquad (6)$$

$$g_{\underline{27}}^{\delta_{\langle GG\rangle}} = \left(1+\frac{1}{N_c}(1-\delta_{\langle GG\rangle})\right)\left(\frac{3}{5}c_1+\frac{3}{5}c_2+\frac{9}{10}c_9+\frac{9}{10}c_{10}\right), \qquad (7)$$

where

$$\delta_{\langle GG\rangle} = \frac{N_c}{2}\frac{\langle\alpha_s GG/\pi\rangle}{16\pi^2 f_\pi^4}. \qquad (8)$$

Figure draws the dependence of the $g_{\underline{8}}/g_{\underline{27}}$ on the gluon condensate. As the gluon condensate increases, the ratio $g_{\underline{8}}/g_{\underline{27}}$ is improved remarkably, so that we are able to obtain the reasonable value of the ratio around $(380 \text{ MeV})^4$, which is acceptable with the counterterms being considered [9]. However, there is one important caveat. As was mentioned in Ref. [4], the $\mathcal{O}(\alpha N_c)^2$ corrections which are of order $\mathcal{O}(1)$ as well were neglected, based on the argument that they involve condensates of higher dimension requiring higher powers of the normalization scale. However, keeping in mind the fact that the $\mathcal{O}(\alpha_s N_c)$ corrections enhance the result of the ratio g_8/g_{27} so radically as shown above, the higher-order terms might give some contribution. We would rather regard the above results as a mere indication of the importance of gluon effects in the nonleptonic processes.

An effective weak chiral Lagrangian to order $\mathcal{O}(p^4)$ in χPT with a minimal set of independent terms was given by Ecker *et al.*[10] and Esposito-Farèse[2] (presented in Minkowski space). A part of the corresponding low energy constants are obtained from the present study:

$$\begin{aligned}
\mathcal{L}_{\rm eff}^{\Delta S=1,\mathcal{O}(p^4)} &= -\frac{G_F}{\sqrt{2}} V_{\rm ud} V_{\rm us}^* f_\pi^2 \Big[\Big(N_1^{(8)} \langle \lambda_6 L_\mu L^\mu L_\nu L^\nu \rangle + N_2^{(8)} \cdot \langle \lambda_6 L_\mu L^\nu L_\nu L^\mu \rangle \\
&+ N_3^{(8)} \langle \lambda_6 L_\mu L_\nu \rangle \langle L^\mu L^\nu \rangle + N_4^{(8)} \langle \lambda_6 L_\mu \rangle \langle L^\mu L_\nu L^\nu \rangle + N_{28}^{(8)} i\,\epsilon_{\mu\nu\rho\delta} \langle \lambda_6 L^\mu \rangle \langle L^\nu L^\rho L^\delta \rangle \Big) \\
&+ \frac{5}{9} t_{ijkl} \Big(N_1^{(27)} \langle \lambda_{ij} L_\mu L^\mu \rangle \langle \lambda_{kl} L_\nu L^\nu \rangle + N_2^{(27)} \langle \lambda_{ij} L_\mu L_\nu \rangle \langle \lambda_{kl} L^\mu L^\nu \rangle \\
&+ N_3^{(27)} \langle \lambda_{ij} L_\mu L_\nu \rangle \langle \lambda_{kl} L^\nu L^\mu \rangle + N_4^{(27)} \langle \lambda_{ij} L_\mu \rangle \langle \lambda_{kl} L_\nu L^\mu L^\nu \rangle \\
&+ N_5^{(27)} \langle \lambda_{ij} L_\mu \rangle \langle \lambda_{kl} \{ L^\mu, L_\nu L^\nu \} \rangle + N_6^{(27)} \langle L_\mu L^\mu \rangle \langle \lambda_{ij} L_\nu \rangle \langle \lambda_{kl} L^\nu \rangle \\
&+ N_{20}^{(27)}\, i\,\epsilon_{\mu\nu\rho\delta} \langle \lambda_{ij} L^\mu L^\nu \rangle \langle \lambda_{kl} L^\rho L^\delta \rangle + N_{21}^{(27)} i\,\epsilon_{\mu\nu\rho\delta} \langle \lambda_{ij} L^\mu \rangle \langle \lambda_{kl} L^\nu L^\rho L^\delta \rangle \Big) \Big] . \qquad (9)
\end{aligned}$$

As noted by G. Ecker *et al.*[10], the LECs $N_1^{(8)}$, $N_2^{(8)}$, $N_3^{(8)}$, and $N_4^{(8)}$ contribute to the process $K \to 3\pi$ while $N_{28}^{(8)}$ does to the radiative K-decays. In particular, the $N_{28}^{(8)}$ is related to the chiral anomaly [11]. Note that all LECs in the eikosiheptaplet are degenerate.

Taking into account the N_c corrections, the LECs are derived as follows:

$$\begin{aligned}
N_1^{(8)} &= \left(-\frac{N_c^2 M^2}{128\pi^4 f_\pi^2} + \frac{N_c}{8\pi^2} - \frac{f_\pi^2}{2M^2} \right) c_6 + \left(-\frac{N_c M^2}{128\pi^4 f_\pi^2} + \frac{1}{8\pi^2} - \frac{f_\pi^2}{2N_c M^2} \right) c_5, \\
N_2^{(8)} &= \frac{N_c}{60\pi^2} \left(\left(-2 + \frac{1}{N_c} 3 \right) c_1 + \left(3 - \frac{1}{N_c} 2 \right) c_2 + \frac{1}{N_c} 5 c_3 + 5 c_4 \right. \\
&+ \left. \left(-3 + \frac{1}{N_c} 2 \right) c_9 + \left(2 - \frac{1}{N_c} 3 \right) c_{10} \right), \quad N_3^{(8)} = 0, \\
N_4^{(8)} &= \frac{N_c}{60\pi^2} \left(\left(\frac{3}{2} - \frac{1}{N_c} \right) c_1 + \left(-1 + \frac{1}{N_c} \frac{3}{2} \right) c_2 + \frac{5}{2} c_3 + \frac{1}{N_c} \frac{5}{2} c_4 \right.
\end{aligned}$$

$$-\frac{5}{2}c_5 - \frac{1}{N_c}\frac{5}{2}c_6 + \left(1 - \frac{1}{N_c}\frac{3}{2}\right)c_9 + \left(-\frac{3}{2} + \frac{1}{N_c}\right)c_{10}\Big),$$

$$N_{28}^{(8)} = \frac{N_c}{60\pi^2}\left(\left(-\frac{3}{2} + \frac{1}{N_c}\right)c_1 + \left(1 - \frac{1}{N_c}\frac{3}{2}\right)c_2 - \frac{5}{2}c_3 - \frac{1}{N_c}\frac{5}{2}c_4\right.$$

$$\left.-\frac{5}{2}c_5 - \frac{1}{N_c}\frac{5}{2}c_6 + \left(-1 + \frac{1}{N_c}\frac{3}{2}\right)c_9 + \left(\frac{3}{2} - \frac{1}{N_c}\right)c_{10}\right),$$

$$N_1^{(27)} = N_5^{(27)} = N_6^{(27)} = N_{20}^{(27)} = 0, \quad N_2^{(27)} = -N_3^{(27)} = -N_4^{(27)} = N_{21}^{(27)}$$

$$= \frac{N_c}{60\pi^2}\left(1 + \frac{1}{N_c}\right)\left(-3c_1 - 3c_2 - \frac{9}{2}c_9 - \frac{9}{2}c_{10}\right). \tag{10}$$

The corresponding numerical results and further discussions such as the contribution of the quark axial-vector constant g_A and the momentum-dependence of the constituent quark mass will appear soon [7].

In summary, we have determined a part of the LECs for the $\mathcal{O}(p^4)$ and discussed the relevance of the gluon condensate to the LECs in the leading order.

The work is supported in part by COSY, DFG, and BMBF. HCK acknowledges the financial support of the Korean Research Foundation made in the program year of 1998.

References

1. J. Kambor, J. Missimer, and D. Wyler, *Nucl. Phys.* **B346** (1990) 17; *Phys. Lett.* **261B** (1991) 496.
2. G. Esposito-Farèse, *Zeit. f. Phys.* **C50** (1991) 255.
3. V. Antonelli, S. Bertolini, J.O. Eeg, M. Fabbrichesi and E.I. Lashin, *Nucl. Phys.* **B** (1996).
4. A. Pich and E. de Rafael, *Nucl. Phys.* **B358** (1991) 311.
5. S. Bertolini, J.O. Eeg, M. Fabbrichesi, and E.I. Lashin, *Nucl. Phys.* **B514** (1998) 63; *ibid.* (1998) 93.
6. D. Diakonov and V. Petrov, *Phys. Lett.* **B147** (1984) 351;, *ibid.* **B272** (1986) 457; D. Diakonov, V. Petrov and P. Pobylitsa, *Nucl. Phys.* **B272** (1988) 809.
7. M. Franz, H.-Ch. Kim, and K. Goeke, in preparation.
8. G. Buchalla, A.J. Buras, M.E. Lautenbacher, *Rev. Mod. Phys.* **68** (1996) 1125 (and references therein).
9. J. Kambor, J. Missimer, and D. Wyler, *Phys. Lett.* **261B** (1991) 496.
10. G. Ecker, J. Kambor, and D. Wyler, *Nucl. Phys.* **B394** (1993)101.
11. J. Bijnens, G. Ecker, and A. Pich, *Phys. Lett.* **B286** (1992) 341.

CHIRAL SYMMETRY FOR POSITIVE AND NEGATIVE PARITY NUCLEONS

D. JIDO

Department of Physics, Tokyo Institute of Technology, Tokyo, Japan

We investigate the role of chiral symmetry in the properties of positive and negative parity nucleons (N and N^*). When we consider two kinds of nucleons, two distinctive types of chiral assignment for N^* can be introduced. There are interesting differences between the two assignments in the properties of nucleons, which are studied using linear sigma models based on the two assignments. Behavior of N and N^* in the nuclear medium is also investigated using these linear models.

Chiral symmetry and its spontaneous breakdown are very important to understand properties of hadrons at low energy. However, the role of chiral symmetry for baryon is less understood as compared to the meson case. It should become very important in the study of, for instance, QCD phase transitions. In this report we consider the $SU(2)_R \times SU(2)_L$ chiral symmetry in the chiral limit in the following discussion. This report is based on the references[1,2].

We consider two kinds of nucleons N_1 and N_2. Then it is possible to introduce two different chiral assignments which we call the naive and mirror assignment. They are distingushed by the chiral transformations:

$$\text{Naive} \qquad N_{2R} \longrightarrow RN_{2R} \,, \qquad N_{2L} \longrightarrow LN_{2L} \,, \tag{1}$$

$$\text{Mirror} \qquad N_{2R} \longrightarrow LN_{2R} \,, \qquad N_{2L} \longrightarrow RN_{2L} \,, \tag{2}$$

where the chiral transformation rule for N_1 is defined as $N_{1R} \to RN_{1R}$ and $N_{1L} \to LN_{1L}$ in both cases.

From symmetry consideration, we can write down chiral invariant Status: O

Lagrangians for the two assignments. They are up to Yukawa terms,

$$\begin{aligned}
\mathcal{L}_{\text{naive}} &= \bar{N}_1 i\partial\!\!\!/ N_1 + \bar{N}_2 i\partial\!\!\!/ N_2 + a\bar{N}_1(\sigma + i\gamma_5 \vec{\tau}\cdot\vec{\pi})N_1 + b\bar{N}_2(\sigma + i\gamma_5\vec{\tau}\cdot\vec{\pi})N_2 \\
&\quad + c\{\bar{N}_2(\gamma_5\sigma + i\vec{\tau}\cdot\vec{\pi})N_1 - \bar{N}_1(\gamma_5\sigma + i\vec{\tau}\cdot\vec{\pi})N_2\} + \mathcal{L}_M \,, \qquad (3) \\
\mathcal{L}_{\text{mirror}} &= \bar{N}_1 i\partial\!\!\!/ N_1 + \bar{N}_2 i\partial\!\!\!/ N_2 + m_0(\bar{N}_2\gamma_5 N_1 - \bar{N}_1\gamma_5 N_2) \\
&\quad + a\bar{N}_1(\sigma + i\gamma_5\vec{\tau}\cdot\vec{\pi})N_1 + b\bar{N}_2(\sigma - i\gamma_5\vec{\tau}\cdot\vec{\pi})N_2 + \mathcal{L}_M \,, \qquad (4)
\end{aligned}$$

where a, b and c are coupling parameters. The second Lagrangian (4) was proposed first by DeTar and Kunihiro[3]. An important point here is that, in the mirror Lagrangian, chiral invariant (off-diagonal) mass terms are allowed with the mass parameter m_0. Therefore when chiral symmetry is restored, N

and N^* get degenerate with a finite mass m_0. The physical nucleons N and N^* are linear combinations of N_1 and N_2, when the Lagrangian has mixing terms.

We investigated properties for N and N^* using these Lagrangian (3) and (4) and obtained some interesting results.

- When chiral symmetry is restored, N and N^* get degenerate with being massless in the naive case, while they do with a finite mass m_0 in the mirror case.
- The πNN^* coupling in the soft pion limit vanishes in the naive case, while it remains finite in the second case. The result for the naive case is consistent with the QCD sum rule analysis[4,5] and seems to be qualitatively consistent with the small coupling of the $\pi NN(1535)$.
- The chiral structures of the naive case and mirror case are different. In the naive case N and N^* belong to the separate multiplets of the chiral group, while they do to the same multiplet in the mirror case, which is the reason why the N and N^* are the chiral partners.
- The relative sign between $g_A^{N^*N^*}$ and g_A^{NN} is positive in the first case, while it is negative in the second case. With the Goldberger-Treiman relation, same conclusion is obtained for πNN and πN^*N^* couplings in the strong interaction.

As an application of the two models, we investigate behaviors of N and N^* in the nuclear medium. From now we assume that N^* can be identified with the observed nucleon resonance $N(1535)$. We extend these models minimally so that these models reproduce the observed axial vector coupling of N. Namely, we add the minimally possible terms with one derivative to each model:

$$d_1\bar{N}_1\Pi_2N_1 + d_2\bar{N}_2\Pi_2N_2 + d_3(\bar{N}_1\Pi_2N_2 + h.c.) \quad \text{for naive case,} \tag{5}$$

$$d_1\bar{N}_1\Pi_2N_1 + d_2\bar{N}_2\Pi_2N_2 + d_3(\bar{N}_1\Pi_1N_2 + h.c.) \quad \text{for mirror case,} \tag{6}$$

where $\Pi_1 \equiv i\partial\!\!\!/\sigma\gamma_5 + \vec{\tau}\cdot\partial\!\!\!/\vec{\pi}$ and $\Pi_2 \equiv (\vec{\tau}\cdot\vec{\pi}\partial\!\!\!/\sigma - \sigma\vec{\tau}\cdot\partial\!\!\!/\vec{\pi})\gamma_5 - \vec{\tau}\cdot(\vec{\pi}\times\partial\!\!\!/\vec{\pi})$. The parameters d_1, d_2 and d_3 are determined phenomenologically. One important consequence of those terms is that physical parameters such as masses or couplings are functions of σ_0. As σ_0 is expected to change in the medium, our formalism could reveal interesting density dependence of those parameters.

We consider nuclear matter with $k_F = 1.36 fm^{-1}$ and calculate the energy density in the relativistic Hartree approximation,

$$\mathcal{E} = \lambda\left(\sigma_0^2 - f_\pi^2\right)^2 + \mathcal{E}_v^+ + \mathcal{E}_v^- + 4\int_0^{k_F}\frac{d^3k}{(2\pi)^3}\sqrt{\mathbf{k}^2 + M_+^{*2}} \tag{7}$$

Table 1: Comparison of the naive and mirror construction

	naive assignment	mirror assignment
mass at symmetric limit	0	m_0 (finite)
$\pi N N^*$ coupling	0	$\frac{2c(a+b)}{\sqrt{(a+b)^2+4c^2}}$
chiral partner	$N \leftrightarrow \gamma_5 N$, $N^* \leftrightarrow \gamma_5 N^*$	$N \leftrightarrow N^*$
$g_A^{NN} g_A^{N^*N^*}$	positive	negative
g_A^{NN} in nuclei	suppressed	suppressed
$g_A^{NN^*}$ in nuclei	suppressed	enhanced

where M_+^* denotes the mass of N in the medium and $\mathcal{E}_v^+$ and $\mathcal{E}_v^-$ denote the vacuum contributions from N and N^* to the energy density. We take $m_\sigma = 635$ MeV in our calculations. Minimizing the energy density (7) we obtain the value of σ_0 in the medium and determine the masses for N and N^* and the axial charges, g_A^{NN} and $g_A^{NN^*}$ as functions of the density.

The results are shown in the Figure 1. We see that the masses of N and N^* decrease as the density increases in both the cases and that g_A^{NN} is suppressed in nucleus. The notable difference between the naive and mirror models is the behavior of $g_A^{NN^*}$ in the medium. We get the suppression in the naive case while the enhancement for the mirror case. It is extremely interesting to observe the density dependence of $g_A^{NN^*}$, or equivalently $g_{\pi NN^*}$ as it indicates which model is realized in the real world.

In summary we have investigated the properties of the positive and negative parity nucleons from the viewpoint of chiral symmetry. There are two ways of assignment of chiral symmetry for the negative parity nucleon. Using linear sigma models we have seen how the differences between the two ways will manifest in the physical quantities not only in the free space but also in the finite density. The comparisons are summarized in the Table 1.

It is interesting if we can distinguish the naive and mirror models experimentally. One such candidate is to determine the sign of $\pi N^* N^*$ coupling relative to πNN. Consider the photo-π-η production process, $\gamma N \to N\pi\eta$ process. If the signs of $\pi N^* N^*$ and πNN coupling are same, the $\gamma N \to N^* \to \pi N^* \to \eta\pi N$ and $\gamma N \to N^* \to \eta N \to \pi\eta N$ diagrams interfere constructively, while otherwise, they would be added in the destructive way. Note that the η production probes the $N^*(1535)$ because it dominates the η production in this energy area. Another candidate is to see the behavior of the $\pi N N^*$ coupling in the nucleus as we have discussed above.

A more important subject is how the present analysis is related to QCD. How is the mass parameter m_0 in the mirror case represented in QCD ? Because

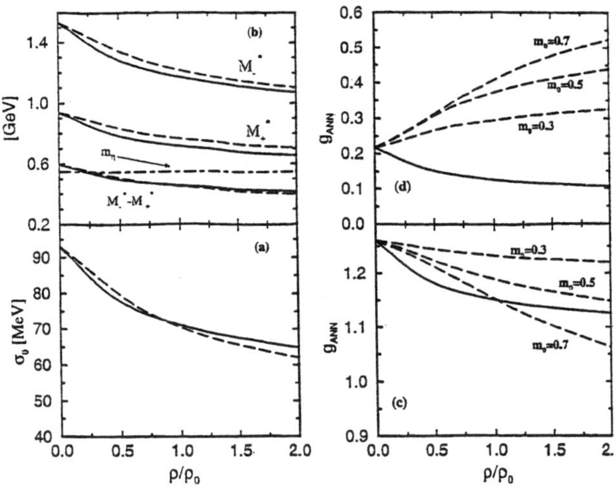

Figure 1: Figure (a) shows the density dependence of σ_0. The solid line is for the naive case and the dashed line is for the mirror case in all figures. Figure (b) shows our predictions for M_-^*, M_+^* as density increases. For the mirror case, $m_0 = 0.5$GeV is used. Figure (c) and Figure (d) show our prediction for axial charges. Figure (c) is for the nucleon axial charge and (d) is for the off-diagonal axial charge.

m_0 is the nucleon mass in the chirally symmetric phase, the quark condensate should not specify the scale of m_0. A candidate is the gluon condensate, because it is the dimensional parameter of QCD when chiral symmetry is restored. It is interesting and important to understand the mirror nucleons in the context of non pretubative QCD.

Acknowledgments

This work is supported by Japan Society for the Promotion of Science.

References

1. D. Jido, Y. Nemoto, M. Oka, and A. Hosaka, hep-ph/9805306
2. Hungchon Kim, D. Jido and M. Oka, *Nucl. Phys.* A **640**, 77 (1998)
3. C. DeTar and T. Kunihiro, *Phys. Rev.* D **39**, 2805 (1989).
4. D. Jido, N. Kodama and M. Oka, *Phys. Rev.* D **54**, 4532 (1996); D. Jido and M. Oka, hep-ph/9611322.
5. D. Jido, M. Oka and A. Hosaka, *Phys. Rev. Lett.* **80**, 448 (1998)

Optimized Perturbation Theory at Finite Temperature

S. Chiku
Institute of Physics, University of Tsukuba, Tsukuba, Ibaraki 305, Japan

T. Hatsuda
Physics Department, Kyoto University, Kyoto 606, Japan

The optimized perturbation theory (OPT) at finite temperature (T) recently developed by the present authors. The method resums automatically higher loops (including the hard thermal loops) at high T and simultaneously cures the problem of tachyonic poles at relatively low T. The advantages of this method is renormalizable at finite T and the existence the Nambu-Goldstone boson.

1 Introduction

Naive perturbation theory is known to break down at finite temperature (T). The two reasons are the existence of hard thermal loops (HTL) at high T [1] and the emergence of tachyonic poles at relatively low T [2]. If one adopts self-consistent resummation methods, HTL can be summed and tachyonic poles can be removed. However, most of the self-consistent methods proposed so far have difficulties of renormalization [3] and/or the violation of the Nambu-Goldstone theorem [4] at finite T.

In this talk [5], we show that a new loop-wise expansion at finite T, the optimized perturbation theory (OPT) [6,7], can solve these problems.

The organization of this talk is as follows. In section II, we introduce the OPT. The renormalization of UV divergences and the realization of the NG theorem in this method are also discussed. In section III, OPT is applied for the $O(4)$ σ model. Summary and concluding remarks are given in Section IV.

2 Optimized Perturbation at $T \neq 0$

2.1 New resummation method

We use modified perturbation theory at finite T for resummation. We start with the thermal effective action with an expansion parameter "δ":

$$\Gamma[\varphi^2] = \ln \int [d\phi] \exp \left[\frac{1}{\delta} \int_0^{1/T} d^n x \left[\mathcal{L}(\phi + \varphi; \mu^2) + J\phi \right] \right], \tag{1}$$

where $J \equiv -\partial \Gamma[\varphi]/\partial\varphi$ and $\int_0^{1/T} d^n x \equiv \int_0^{1/T} d\tau \int d^{n-1}x$.

We explain our method using ϕ^4 theory. We start with a renormalized Lagrangian with counter terms

$$\mathcal{L}(\phi;\mu^2) = \frac{1}{2}[(\partial\phi)^2 - \mu^2\phi^2] - \frac{\lambda}{4!}\phi^4 + \frac{1}{2}A(\partial\phi)^2 - \frac{1}{2}B\mu^2\phi^2 - \frac{\lambda}{4!}C\phi^4 + D\mu^4. \quad (2)$$

The $\overline{MS}$ scheme with the dimensional regularization is assumed in (2). Just for notational simplicity, the factor $\kappa^{(4-n)}$ to be multiplied to λ is omitted (κ is the renormalization point). The renormalization constants are completely fixed at $T = 0$ and $\mu^2 > 0$, in which A, B, C and D are expanded by δ

Then, we rewrite the Lagrangian (2) by introducing a new mass parameter m^2 following the idea of OPT [6]: $\mu^2 = m^2 - (m^2 - \mu^2) \equiv m^2 - \chi$. This identity should be used not only in the standard mass term but also in the counter terms [7], which is crucial to show the order by order renormalization in OPT:

$$\mathcal{L}(\phi;\mu^2) = \mathcal{L}_{\rm r} + \mathcal{L}_{\rm c} \quad (3)$$

$$\mathcal{L}_{\rm r} = \frac{1}{2}[(\partial\phi)^2 - m^2\phi^2] - \frac{\lambda}{4!}\phi^4 + \frac{1}{2}\chi\phi^2 \quad (4)$$

$$\mathcal{L}_{\rm c} = \frac{1}{2}A(\partial\phi)^2 - \frac{1}{2}B(m^2 - \chi)\phi^2 - \frac{\lambda}{4!}C\phi^4 + D(m^2 - \chi)^2. \quad (5)$$

Major difference between this expansion and the naive one is that $m^2 = O(\delta^0)$, $\chi = O(\delta)$. The physical reason behind this assignment is the fact that χ reflects the effect of interactions.

The final step is to find an optimal value of m by imposing physical conditions à la Stevenson [8] such as the following. (i) The principle of minimal sensitivity (PMS): this condition requires that a chosen quantity calculated up to L-th loops ($\mathcal{O}_L$) should be stationary by the variation of m: $\frac{\partial \mathcal{O}_L}{\partial m} = 0$. (ii) The criterion of the fastest apparent convergence (FAC): this condition requires that the perturbative corrections in O_L should be as small as possible for a suitable value of m. $\mathcal{O}_L - \mathcal{O}_{L-l} = 0$, where l is chosen in the range, $1 \leq l \leq L$.

2.2 Renormalization in OPT

We now prove the order by order renormalization in OPT. Let us first rewrite eq.(3) as

$$\mathcal{L}(\phi;\mu^2) = \mathcal{L}(\phi;m^2) + \frac{1}{2}\chi\phi^2 + \left[\frac{1}{2}B\chi\phi^2 + D\chi^2 - 2Dm^2\chi\right]. \quad (6)$$

The $\mathcal{L}(\phi; m^2)$ is renormalizable by definition. Although T-dependent divergences appear because of the T-dependent "resumed" mass $m^2(T)$, they are properly renormalized away since the counter terms (such as $Bm^2\phi^2$, Dm^2 and Dm^4) also acquires T-dependence through $m^2(T)$. In other words, the divergences arising from the resumed propagator is removed by the resumed counter terms.

The divergences arising from the insertion of $(1/2)\chi\phi^2$ can be shown to be removed by the last three counter terms in (6). This is obviously related to the renormalization of composite operators. For detailed proof of the renormalization, see Appendix A of ref.[5].

Thus, we can carry out renormalization order by order even within the self-consistent method.

2.3 Nambu-Goldstone theorem in OPT

The procedure and the renormalization in OPT discussed above do not receive modifications even if the Lagrangian has global symmetry. For $O(N)$ ϕ^4 theory, one needs to replace ϕ and ϕ^2 by $\vec{\phi} = (\phi_1, \cdots, \phi_N)$ and $\vec{\phi}^2$ respectively in all the previous formulas.

In OPT, V calculated up to L-th loops $V_L(\vec{\varphi}^2; m)$ has also manifest $O(N)$ invariance, because our decomposition used in (3) does not break $O(N)$ invariance. Therefore, the Nambu-Goldstone theorem is obviously satisfied.

3 Application of OPT

We apply OPT to the $O(4)\sigma$ model with FAC condition. The detailed analysis is reported by ref[5]. The result of the resumed masses and chiral condensate

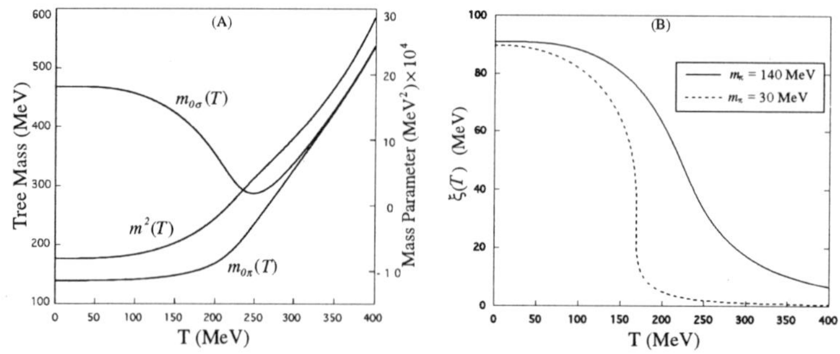

Figure 1: (A) Typical tree mass behaviors of the $m_{0\pi}(T)$ (NG-boson) and $m_{0\sigma}(T)$ (radial mode), and the mass parameter $m^2(T)$. (B) Chiral condensate $\xi(T)$ for some π masses.

are shown in fig.1. We can see that our method cures tachyon problem and HTL are correctly resummed at high T.

4 Summary

We have shown that the optimized perturbation theory (OPT) developed in [5] naturally cures the problems of the naive loop-expansion at finite T, namely, the breakdown of the naive perturbation at $T \gg T_c$ (due to the hard thermal loops) as well as at $T < T_c$ (due to the tachyonic poles). Furthermore, OPT has several advantages over other resummation methods proposed so far. First of all, the renormalization of the UV divergences can be carried out systematically. Secondly, the Nambu-Goldstone (NG) theorem is satisfied in any give order of the loop-expansion in OPT.

There are many directions where OPT at finite T may be applied. OPT may also have relevance to develop an improved perturbation theory for gauge theories in which the weak coupling expansion is known to break down in high orders [9].

Acknowledgments

This work was partially supported by the Grants-in-Aid of the Japanese Ministry of Education, Science and Culture (No. 06102004). S. C. would like to thank the Japan Society of Promotion of Science (JSPS) for financial support.

References

1. E. Braaten and R. D. Pisarski, *Nucl. Phys.* B **337**, 569 (1990); ibid. **B 339**, 310 (1990).
2. S. Weinberg, *Phys. Rev.* D **9**, 3357 (1974).
3. H. Matsumoto, I. Ojima and H. Umezawa, *Ann. Phys.* **152**, 348 (1984). G. Baym and G. Grinstein, *Phys. Rev.* D **15**, 2897 (1977).
4. See, A. Okopińska, *Phys. Lett.* B **375**, 213 (1996); G. Amelino-Camelia, *Phys. Lett.* B **407**, 268 (1997), and references therein.
5. This talk based on S. Chiku and T. Hatsuda, *Phys. Rev.* D **58**, 076001 (1998).
6. T. Hatsuda, T. Kunihiro and T. Tanaka, *Phys. Rev. Lett.* **78**, 3229 (1997). A. Okopińska, *Mod. Phys. Lett.* A **12**, 1003 (1997), and references therein.
7. N. Banerjee and S. Mallik, *Phys. Rev.* D **43**, 3368 (1991).
8. P. M. Stevenson, *Phys. Rev.* D **23**, 2916 (1981).
9. A. D. Linde, *Phys. Lett.* B **96**, 289 (1980).

CHIRAL PHASE TRANSITION AT FINITE TEMPERATURE AND DENSITY IN THE QCD-LIKE GAUGE THEORY

O. KIRIYAMA, M. MARUYAMA AND F. TAKAGI
Department of Physics, Tohoku University, Sendai 980–8578, Japan
E-mail : kiriyama@nucl.phys.tohoku.ac.jp

The chiral phase transition at finite temperature and/or chemical potential is studied. We calculate an effective potential in the QCD-like gauge theory and investigate the phase structure and have the tricritical point.

1 Introduction

At zero temperature and zero baryon density, the chiral symmetry is dynamically broken. It is generally believed, however, that at high temperature and/or density the QCD vacuum undergoes a phase transition chirally symmetric phase. This chiral phase transition plays an important role in the physics of neutron stars and the early universe and it may be realized in heavy-ion collisions. In this proceeding we study the phase structure of the QCD using the QCD-like gauge theory. We calculate the Cornwall–Jackiw–Tomboulis (CJT) effective potential[1] at finite temperature and/or chemical potential and investigate the chiral phase transition. Note that in this proceeding we fix the mass scale of our model by the condition of $\Lambda_{\rm QCD} = 1$, otherwise stated.

2 CJT Effective Potential at Finite T/μ

At $T = \mu = 0$, the CJT effective potential in the improved ladder approximation and in the Landau gauge is expressed as follows up to two loop level [2]

$$
\begin{aligned}
V_{\rm CJT}[\Sigma(p_{\rm E})] &= V_1[\Sigma(p_{\rm E})] + V_2[\Sigma(p_{\rm E})] \\
V_1[\Sigma(p_{\rm E})] &= -2\int^{\Lambda} \frac{d^4 p_{\rm E}}{(2\pi)^4} \ln \frac{\Sigma^2(p_{\rm E}) + p_{\rm E}^2}{m^2(\Lambda) + p_{\rm E}^2} \\
&\quad +4\int^{\Lambda} \frac{d^4 p_{\rm E}}{(2\pi)^4} \frac{\Sigma(p_{\rm E})(\Sigma(p_{\rm E}) - m(\Lambda))}{\Sigma^2(p_{\rm E}) + p_{\rm E}^2} \qquad (1) \\
V_2[\Sigma(p_{\rm E})] &= -6C_2 \int^{\Lambda}\int^{\Lambda} \frac{d^4 p_{\rm E}}{(2\pi)^4}\frac{d^4 q_{\rm E}}{(2\pi)^4} \frac{\bar{g}^2(p_{\rm E}, q_{\rm E})}{(p_{\rm E} - q_{\rm E})^2} \\
&\quad \times \frac{\Sigma(p_{\rm E})}{\Sigma^2(p_{\rm E}) + p_{\rm E}^2}\frac{\Sigma(q_{\rm E})}{\Sigma^2(q_{\rm E}) + q_{\rm E}^2} \qquad (2)
\end{aligned}
$$

where an overall factor $N_f N_c$ (the number of light quarks times colors) is omitted, $C_2 = (N_c^2 - 1)/(2N_c)$ is the quadratic Casimir operator for SU(N_c), and $\Sigma(p)$ is a dynamical mass function of quark. In the above equations we introduced the ultraviolet cutoff Λ in order to make bare mass well-defined. We abbreviate the subscript 'E' in the rest of this proceeding for simplicity. The extremum condition for $V_{\rm CJT}$ with respect $\Sigma(p)$ leads to the following Schwinger–Dyson (SD) equation

$$\Sigma(p) = m(\Lambda) + 3C_2 \int^{\Lambda} \frac{d^4q}{(2\pi)^4} \frac{\bar{g}^2(p,q)}{(p-q)^2} \frac{\Sigma(q)}{\Sigma^2(q) + q^2} \tag{3}$$

We use the effective running coupling [5]

$$\bar{g}^2(p) = 2\pi^2 a / \ln(p^2 + p_R^2) \ , \ a = 24/(11N_c - 2n_f), \tag{4}$$

where $n_f (= 3)$ controls the effective running coupling, and apply the Higashijima-Miransky approximation [2,3], $\bar{g}^2(p,q) = \bar{g}^2[\max(p^2, q^2)]$, for an argument of the running coupling, then we can derive the another expression of the CJT effective potential

$$V_{\rm CJT} = -2 \int \frac{d^4p}{(2\pi)^4} \ln \frac{\Sigma^2(p) + p^2}{p^2} - \frac{2}{3C_2} \int dp^2 \frac{1}{\frac{d}{dp^2}\mathcal{D}(p)} \left(\frac{d}{dp^2}\Sigma(p) \right)^2 \tag{5}$$

in the continuum limit ($\Lambda \to \infty$), where $\mathcal{D}(p) = \bar{g}^2(p)/p^2$. At finite temperature and chemical potential we use the imaginary time formalism,

$$\int \frac{dp_4}{2\pi} f(p_4) \to T \sum_{n=-\infty}^{\infty} f(p_4 = (2n+1)\pi T + i\mu) \ , \ n \in \boldsymbol{Z}. \tag{6}$$

We use the following "real" functions for $\mathcal{D}(p)$ and $\Sigma(p)$.

$$\mathcal{D}(p_4, \boldsymbol{p}) = \frac{2\pi^2 a}{\ln(\omega_n^2 + \boldsymbol{p}^2 + p_R^2)} \frac{1}{\omega_n^2 + \boldsymbol{p}^2}, \tag{7}$$

$$\begin{aligned}\Sigma(p_4, \boldsymbol{p}) &= m_R (\ln(\omega_n^2 + \boldsymbol{p}^2 + p_R^2))^{-a/2} \\ &+ \frac{\sigma}{\omega_n^2 + \boldsymbol{p}^2 + p_R^2} (\ln(\omega_n^2 + \boldsymbol{p}^2 + p_R^2))^{a/2-1}.\end{aligned} \tag{8}$$

where ω_n is the fermion Matsubara frequency, m_R is a renormalization group invariant quark mass and σ is an order parameter of chiral symmetry. We can determine the value of $\langle \bar{q}q \rangle$ through $\langle \bar{q}q \rangle = -(3/2\pi^2 a)\sigma_{min}$, where σ_{min} is the location of the minimum of $V_{\rm CJT}$. In this proceeding we consider the case of the chiral limit ($m_R = 0$). We do not take into account of μ dependence in $\mathcal{D}(p)$, because the gluon momentum squared is the most natural choice for the argument of $\mathcal{D}$ and therefore $\mathcal{D}$ should not depend on the quark chemical potential. We concentrate on the terms which represent the explicit chiral symmetry breaking and nonperturbative effect, therefore, do not include the terms arising through the perturbation and terms which proportional to γ_4. We expect that phase structure is not so affected by these terms. Using Eq. (5) with Eq. (6) and the ansatz Eqs. (7), (8), we can easily show that our effective potential is real different from the previous work [6] and calculate the effective potential as a function of σ,

$$V_{\mathrm{CJT}}(\sigma; T, \mu) = a_2(T,\mu)\sigma^2 + a_4(T,\mu)\sigma^4 + \cdots . \tag{9}$$

3 Numerical Results

At first, we put $\ln(p_R^2) = 0.1$ and study the two limiting cases. Figure 1 shows the temperature dependence of $\sigma_{min}/\Lambda^3_{\mathrm{QCD}}$. We found the second-order phase transition at $T_c = 129$ [MeV] [a]. On the other hand, μ dependence is shown in Figure 2, we have first-order phase transition at $\mu_c = 422$ [MeV].

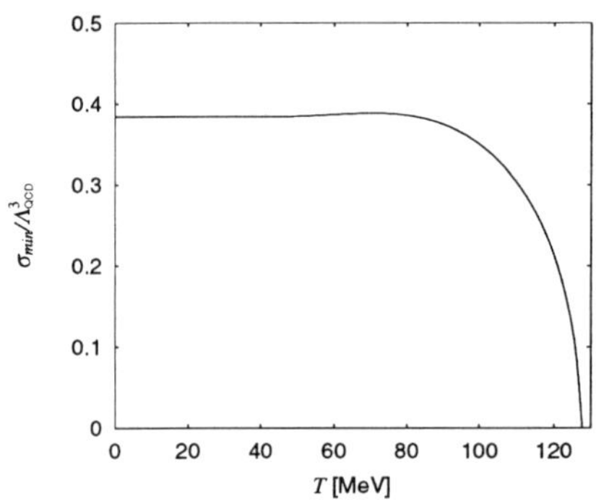

Figure 1: The temperature dependence of $\sigma_{min}/\Lambda^3_{\mathrm{QCD}}$. σ_{min} goes to zero continuously at $T_c = 129$ [MeV].

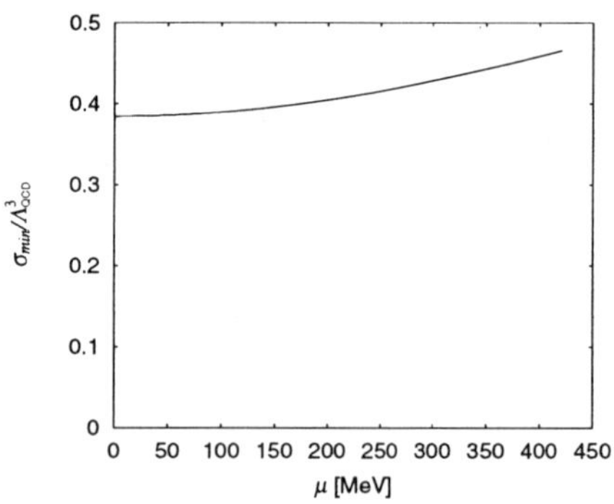

Figure 2: The chemical potential dependence of $\sigma_{min}/\Lambda^3_{\mathrm{QCD}}$. σ_{min} goes to zero discontinuously at $\mu_c = 422$ [MeV].

Next we study the simultaneous case and found the tricritical point 'P' at $(T_P, \mu_P) = (107, 210)$ [MeV]. This location of the tricritical point is determined

[a] We determine the value of Λ_{QCD} by the condition of $f_\pi = 93$ [MeV] at $T = \mu = 0$ and have $\Lambda_{\mathrm{QCD}} = 738$ [MeV]. Note that we use the Pagels–Stoker formula [4] for f_π.

by the condition

$$a_2(T_P, \mu_P) = a_4(T_P, \mu_P) = 0, \tag{10}$$

in Eq. (9).

4 Conclusions

In this proceeding we study the chiral phase transition at high temperature and/or density in the QCD-like gauge theory. We found the second-order phase transition along the $\mu = 0$ line and first-order phase transition along the $T = 0$ line. We also found the tricritical point on the phase diagram. The location of the tricritical point is approximately coincident with that found in the other models [7,8]. Recently it was pointed out [9] that we may be investigate the phase structure through the HBT effect. Consequently it is significant three different models shows the almost same results.

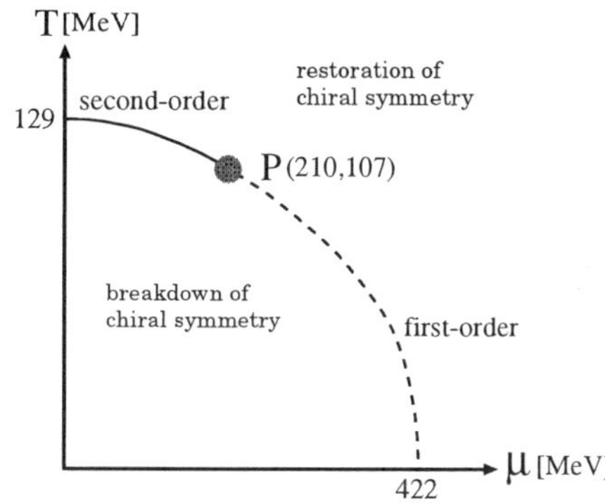

Figure 3: The cartoon of the phase diagram.

References

1. J.M. Cornwall, R. Jackiw and E. Tomboulis, Phys. Rev. **D10**, 2428 (1974).
2. K. Higashijima, Phys. Lett. **B124**, 257 (1983);
   see also O. Kiriyama, M. Maruyama and F. Takagi, Phys. Rev. **D58**, 116001 (1998).
3. V.A. Miransky, Sov. J. Nucl. Phys. **38**, 280 (1984).
4. H. Pagels and S. Stoker, Phys. Rev. **D20**, 2947 (1979).
5. K. Higashijima, Prog. Theor. Phys. Suppl. **104**, 1 (1991).
6. M. Harada and A. Shibata, hep-ph/9807408.
7. J. Berges and K. Rajagopal, hep-ph/9804233.
8. M.A. Halasz, A.D. Jackson, R.E. Shrock, M.A. Stephanov and J.J.M. Verbaarschot, Phys. Rev. **D58**, 96007 (1998).
9. M. Stephanov, K. Rajagopal and E. Shuryak, hep-ph/9806219

Mesons at Finite Temperature in the Linear Sigma Model by Cornwall-Jackiw-Tomboulis Effective Action

Y. Nemoto, K. Naito and M. Oka
Department of Physics, Tokyo Institute of Technology,
Meguro, Tokyo 152-8551 Japan
E-mail: nemoto@th.phys.titech.ac.jp

We study the $O(N)$ symmetric linear sigma model at finite temperature as the low-energy effective model of QCD using the Cornwall-Jackiw-Tomboulis(CJT) effective action for composite operators. We show that if one defines the propagator, which is determined from the Schwinger-Dyson(SD) equation, as the $O(N)$ symmetric form and the meson masses as the second derivatives of the effective potential, pions are always massless below the critical temperature of the chiral phase transition for finite N. We solve the SD-equations for the propagator and the sigma condensate numerically in the $N = 4$ case with the inclusion of the thermal effects. A problem about the derivation of the sigma meson mass at this approximation is pointed out.

1 Introduction

In QCD, existence of the nearly massless pions indicates that the chiral symmetry is spontaneously broken and the Nambu-Goldstone bosons appear. In increasing temperature, as shown in lattice QCD calculation, the chiral symmetry is believed to be restored at around $T_c = 100 \sim 300$ MeV. Behaviors of meson spectra in the course of chiral symmetry restoration may be modelled and studied in the $O(N = 4)$ linear sigma model.

Recently Amelino-Camelia *et.al.* investigated the $O(N)$ linear sigma model at finite temperature with the CJT action[1]. The CJT action, $\Gamma(\phi_c, G)$, has two arguments which correspond to a classical field and a two-point function and is a generating functional of the two particle irreducible vacuum graphs[2]. They conclude that in the CJT formalism the Goldstone theorem is violated, or the pions are massive at finite temperature in the Hartree-Fock approximation, i.e., the so-called super-daisy approximation and the case of finite values of N, while it is satisfied if and only if leading contributions in the $1/N$ expansion are taken. They also discuss the renormalization in the CJT formalism and insist that a consistent renormalization cannot be performed in the broken phase.

Here we re-analyze the $O(N)$ linear sigma model at finite temperature in the CJT formalism and show that the Goldstone theorem is always satisfied at any finite values of N in the Hartree-Fock approximation. The CJT action does not violate the $O(N)$ symmetry, so that the pions are always massless below the critical temperature if we define the meson masses by the curvature of the effective potential.

2 CJT Effective Action for the $O(N)$ Linear Sigma Model

We work with the following Euclidean linear sigma model:

$$\mathcal{L} = \frac{1}{2}\partial_\mu \phi^a \partial_\mu \phi^a + \frac{1}{2}m^2\phi^2 + \frac{\lambda}{6N}(\phi^2)^2 \tag{1}$$

where $\phi^2 = \phi^a \phi^a$, a runs over 1 to N and repeated indices are summed. We adopt the Hartree-Fock or super-daisy approximation, that is, we take into account only the "∞"-type as the two-loop part. The CJT effective potential is then

$$\begin{aligned} V(\phi_c, G) = V_0[\phi_c] &+ \frac{1}{2}\int_{\beta,k} \ln\det G^{-1}(k) + \frac{1}{2}\int_{\beta,k} \mathrm{tr}(\mathcal{D}^{-1}(\phi_c,k)G(k)) \\ &+ \frac{\lambda}{6N}\left[\left\{\int_{\beta,k} G_{aa}(k)\right\}^2 + 2\left\{\int_{\beta,k} G_{ab}(k)\right\}\left\{\int_{\beta,k} G_{ba}(k)\right\}\right] \end{aligned} \tag{2}$$

where

$$V_0[\phi_c] = \frac{1}{2}m^2\phi_c^2 + \frac{\lambda}{6N}(\phi_c^2)^2 \tag{3}$$

$$\mathcal{D}^{-1}_{ab}(\phi_c, k) = \left[k^2 + m^2 + \frac{2\lambda}{3N}\phi_c^2\right]\delta^{ab} + \frac{4\lambda}{3N}\phi_c^a\phi_c^b \tag{4}$$

and $\int_{\beta,k} \equiv \frac{1}{\beta}\sum_n \int \frac{d^3k}{(2\pi)^3}$ is a sum-integral used in the imaginary time formalism at finite temperature. Physical solutions demand minimization of the effective potential with respect to both ϕ_c and G:

$$\frac{dV(\phi_c, G)}{d\phi_c} = 0, \qquad \frac{dV(\phi_c, G)}{dG} = 0 \tag{5}$$

In order to solve Eq.(5), we need to parametrize G. Here we adopt the following form of G from the analogy of $\mathcal{D}$.

$$G^{-1}_{ab}(\phi_c, k) = (k^2 + m^2 + \frac{X(\phi_c^2)}{N}\phi_c^2)\delta^{ab} + \frac{Y(\phi_c^2)}{N}\phi_c^a\phi_c^b \tag{6}$$

where $X(\phi_c^2)$ and $Y(\phi_c^2)$ are unknown functions which are determined by solving the SD-equation. This form is crucial so as not to violate the $O(N)$ symmetry. We assume that X and Y are independent of the external momentum and also that no more parameters are needed.

The meson masses are defined as

$$m_\sigma^2 \equiv \left.\frac{d^2V}{d\sigma_c^2}\right|_{\sigma_c=\sigma_0,\pi_c=0}, \qquad m_\pi^2 \equiv \left.\frac{d^2V}{d\pi_c^2}\right|_{\sigma_c=\sigma_0,\pi_c=0} \tag{7}$$

where we have taken $\phi_c^a = (\sigma_c, \vec{\pi}_c)$. This definition of the masses is used in some literature[3,5]. In the CJT formalism, however, unless Eq.(6) is used in order to respect the $O(N)$ symmetry, the Goldstone theorem is violated even if one defines the masses in this way. With Eqs.(6) and (7), we conclude that below the critical temperature of the chiral phase transition the pions are massless and the sigma meson has the mass

$$m_\sigma^2 = 2\sigma_0^2 \left(A(\sigma_0^2) - \frac{4\lambda}{3N} \right), \tag{8}$$

$$\begin{aligned} A(\phi_c^2) = & \left(18\lambda N + 8\lambda^2(N+2)P_2[M_{X,N}]\right)\left(9N^2 + 6\lambda N(N+1)P_2[M_{X,N}]\right. \\ & \left. +18\lambda N P_2[M_{XY,N}] + 8(N+2)\lambda^2 P_2[M_{X,N}]P_2[M_{XY,N}]\right)^{-1}, \end{aligned} \tag{9}$$

and $P_2[M] \equiv \int_{\beta,k} \frac{1}{(k^2+M^2)^2}$, $M_{X,N} = m^2 + \frac{X}{N}\phi_c^2$ and $M_{XY,N} = m^2 + \frac{X+Y}{N}\phi_c^2$.

3 Numerical Results

We apply the present formalism to the study of finite temperature properties of the sigma and the pion, by taking $N = 4$. In this study, we ignore the quantum corrections which cause UV-divergence although a consistent renormalization procedure of the leading quantum correction is given elsewhere[4]. We show the effective potential for $m_\sigma(T = 0) = 600$ MeV in the chiral limit in Fig. 1. There occurs the first order phase transition at $T \sim 184$ MeV because in this level of approximation which neglects quantum correction, this result agrees with the mean field approach[5]. Temperature dependence of the sigma condensate, σ_0, for each pion mass is shown in Fig. 2. Finite pion masses can be obtained by adding the $c\sigma$ term in the lagrangin as usual. At $T = 0$, these values are 93 MeV, which is known as the pion decay constant. Because the first order phase transition takes place in the lower pion mass, σ_0 decreases and jumps to zero as temperature increases.

The meson masses for $m_\pi(T = 0) = 10$ and 138 MeV are shown in Fig. 3. For small $m_\pi(T = 0)$, there is discontinuity in the masses at the critical temperature. We also note that a peculiar behavior is seen near $T = 0$ for the small $m_\pi(T = 0)$ values. This results from the omission of the quantum corrections, so that the integrals in the sigma meson mass formula Eq.(8) have infrared singularity. In order to get rid of this problem, we need to include the quantum corrections[4].

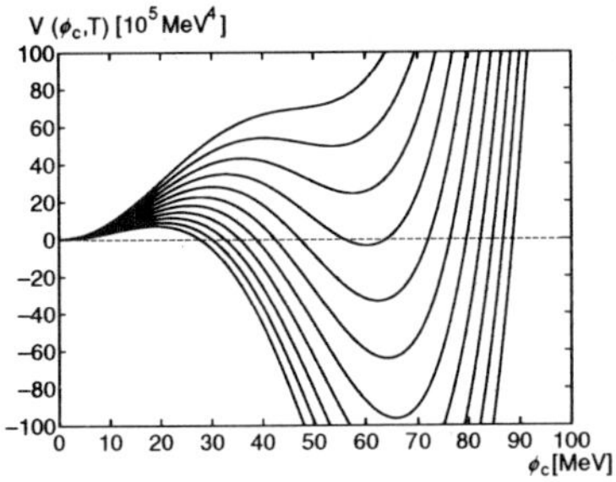

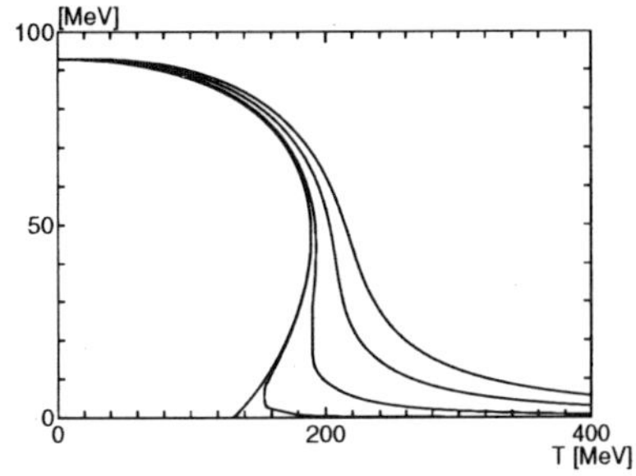

Figure 1: (LEFT) Effective potentials at $T = 190, 188, 186, \cdots, 170$ MeV from top to bottom in the chiral limit.

Figure 2: (RIGHT) Temperature dependence of the sigma condensate, σ_0. The pion mass at $T = 0, 10, 50, 100$ and 138 MeV from left to right.

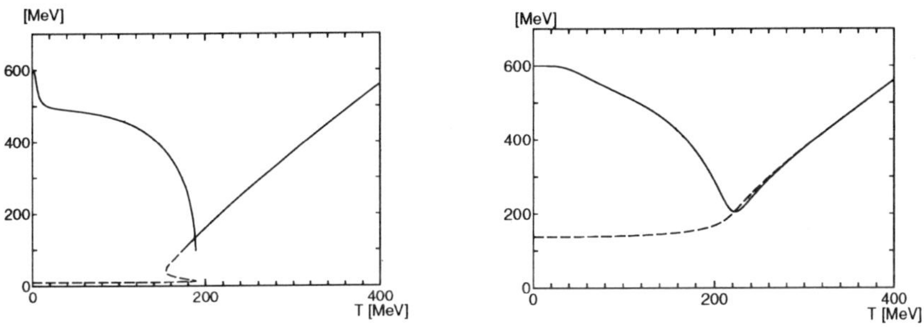

Figure 3: Temperature-dependence of the sigma meson mass(solid) and pion mass(dashed) for $m_\pi(T = 0) = 10$(LEFT) and 138(RIGHT) MeV.

References

1. G. Amelino-Camelia, *Phys. Lett.* B **407**, 268 (1976); N. Petropoulos, hep/ph-9807331.
2. J. M. Cornwall, R. Jackiw, and E. Tomboulis, *Phys. Rev.* D **10**, 2428 (1974).
3. A. Okopińska, *Phys. Lett.* B **375**, 213 (1996).
4. Y. Nemoto, Doctor thesis, Tokyo Inst. of Tech., 1999; Y. Nemoto, K. Naito and M. Oka, in preparation.
5. H.-S. Roh and T. Matsui, *Eur. Phys. J.* A **1**, 205 (1998).

PROVING QUARK CONFINEMENT IN QCD

Kei-Ichi Kondo
Department of Physics, Faculty of Science, Chiba University, Chiba 263, JAPAN
E-mail: kondo@cuphd.nd.chiba-u.ac.jp

I propose to reformulate the gauge field theory as the perturbative deformation of a novel topological quantum field theory. It is shown that this reformulation leads to quark confinement in QCD_4. Similarly, the fractional charge confinement is also derived in the strong coupling phase of QED_4. As a confinement criterion, we use the area decay of the expectation value of the Wilson loop.

1 Introduction

An idea of magnetic monopole and its condensation was used in the analytical proofs of "quark" confinement so far:

1975: Polyakov[1] for 3D compact U(1) gauge theory

1977: Polyakov[2] for 3D Georgi-Glashow model

1994: Seiberg-Witten[3] for 4D N=2 SUSY YM/QCD

How to prove quark confinement in QCD_4 analytically? Recent numerical simulations have established that also in QCD the magnetic monopole plays the dominant role in quark confinement (monopole dominance) under the Abelian projection [4]. To explain this fact analytically and then to prove quark confinement, we propose to use a novel type of topological quantum field theory (TQFT) [5] which describes 4D magnetic monopole, and to reformulate QCD as a perturbative deformation of the TQFT.

We can show:[6,7,8,9]

1. The TQFT is derived from QCD (without any approximation) in the maximal Abelian gauge (MAG).
2. The calculation of the Wilson loop in 4D YM theory is reduced to that in 2D nonlinear sigma model (NLSM) due to the dimensional reduction. This is an exact result.
3. Area decay of the Wilson loop is derived from 2D Instanton calculus in NLSM.

2 Basic Idea

We take into account the topological nontrivial background field Ω_μ which corresponds to 4D magnetic monopole current k_μ. The current k_μ obeys the topological conservation law $\partial_\mu k_\mu = 0$ and hence k_μ denotes a loop in 4D space-time. Usually, Ω_μ denotes an arbitrary, but fixed background (as a solution of classical field equation). In our case, we sum up all topological nontrivial background Ω_μ after integrating out quantum fluctuations Q_μ,

$$Z_{QCD} = \sum_{\Omega} \int [dQ_\mu] \exp\{-S_{QCD}[\Omega_\mu + Q_\mu]\}, \tag{1}$$

where

$$S_{QCD} = \int d^D x \left[-\frac{1}{2g^2} \mathrm{tr}_G(\mathcal{F}_{\mu\nu}\mathcal{F}_{\mu\nu}) + \bar{\psi}(i\gamma^\mu \mathcal{D}_\mu[\mathcal{A}] - m)\psi \right]. \tag{2}$$

We must regard this background to be different from the 4D instanton discussed in the topological YM theory (Witten[5]) and also from 3D magnetic monopole (Polyakov[2]).

3 Quantization of YM theory based on BRST formalism

Using the nilpotent BRST transformation δ_B, it is well known that the gauge fixing and the Faddeev-Popov (FP) ghost part is written in the form,

$$\mathcal{L}_{gf} = -i\delta_B G_{gf}, \quad G_{gf} := \mathrm{tr}_G \left\{ \bar{C} \left(F[\mathcal{A}] + \frac{\alpha}{2}\phi \right) \right\}, \tag{3}$$

where α is the gauge-fixing parameter and ϕ is the Lagrange-multiplier field. For the manifestly covariant Lorentz gauge, we choose $F[\mathcal{A}] = \partial^\mu \mathcal{A}_\mu$.

In the maximal Abelian gauge (MAG),[6,7] a suitable choice of gauge fixing parameter $\alpha = -2$ leads to

$$G_{gf} := \bar{\delta}_B \mathrm{tr}_{G/H} \left(\frac{1}{2}\mathcal{A}_\mu(x)\mathcal{A}_\mu(x) + iC(x)\bar{C}(x) \right). \tag{4}$$

where $\bar{\delta}_B$ is the (nilpotent) anti-BRST transformation. Using the decomposition,

$$\mathcal{A}_\mu(x) = \Omega_\mu(x) + Q_\mu(x) = \frac{i}{g} U(x)\partial_\mu U(x)^\dagger + U(x)\mathcal{V}_\mu(x)U^\dagger(x), \tag{5}$$

the action of the TQFT is given by

$$S_{TQFT} := \int d^D x \; i\delta_B \bar{\delta}_B \; \mathrm{tr}_{G/H} \left(\frac{1}{2}\Omega_\mu(x)\Omega_\mu(x) + iC(x)\bar{C}(x) \right), \tag{6}$$

and the partition function is given by

$$Z_{TQFT} = \int [dU][dC][d\bar{C}][d\phi] \exp\{-S_{TQFT}[\Omega_\mu]\}, \tag{7}$$

where U is an element of the gauge group G.

4 Strategy of a proof

We consider D-dim. QCD (with a gauge group G) for $D > 2$.

Step 1:[7] In MAG, D-dim. QCD (QCD_D) is reformulated as a perturbative deformation of TQFT_D. MAG is a partial gauge fixing which fixes the coset part G/H and leaves the maximal torus group H invariant.

Step 2:[7] It is shown that TQFT_D is equivalent to the coset G/H nonlinear sigma model (NLSM) in (D-2) dimensions. This is a consequence of Parisi-Sourlas dimensional reduction due to the hidden supersymmetry in TQFT (6). As extensively discussed more than 20 years ago, QCD_4 and NLSM_2 have common properties: renormalizability, asymptotic freedom ($\beta(g) < 0$), dynamical mass generation, existence of instanton solution, no phase transition for any value of coupling constant (one phase), etc. This similarity between two theories can be understood from this correspondence,

$$QCD_4 \supset TQFT_4 \Longleftrightarrow G/H\ NLSM_2 \tag{8}$$

For $G = SU(2)$, G/H NLSM is equivalent to O(3) NLSM. Existence of 2D instanton is guaranteed for any N, because $\Pi_2(SU(N)/U(1)^{N-1}) = \Pi_1(U(1)^{N-1}) = \mathbf{Z}^{N-1}$.

Step 3:[9] We define the full string tension σ by

$$\sigma := - \lim_{A(C)\to\infty} \frac{1}{A(C)} \ln\langle W^C[\mathcal{A}]\rangle_{YM_4}, \tag{9}$$

where $W^C[\mathcal{A}]$ is the non-Abelian Wilson loop with the area $A(C)$,

$$W^C[\mathcal{A}] := \mathrm{tr}\left[\mathcal{P}\exp\left(iq\oint_C \mathcal{A}_\mu^A(x)T^A dx^\mu\right)\right]. \tag{10}$$

On the other hand, the diagonal string tension σ_{Abel} is defined by

$$\sigma_{Abel} := - \lim_{A(C)\to\infty} \frac{1}{A(C)} \ln\left\langle W^C[a^\Omega]\right\rangle_{TFT_4}, \tag{11}$$

using the diagonal Wilson loop,

$$W^C[a^\Omega] = \exp\left[iq \oint_C dx^\mu a_\mu^\Omega(x)\right], \quad a_\mu^\Omega(x) := \Omega_\mu^3(x) := \mathrm{tr}(T^3\Omega_\mu(x)) \tag{12}$$

¿From the dimensional reduction, we find

$$\langle W^C[a^\Omega]\rangle_{TFT_4} = \left\langle W^C[a^\Omega]\right\rangle_{NLSM_2}. \tag{13}$$

Then it is shown that in the large Wilson loop limit

$$|\sigma - \sigma_{Abel}| \downarrow 0 \quad (A(C) \uparrow \infty). \tag{14}$$

This is derived from the non-Abelian Stokes theorem. For the large (non-intersecting planar) Wilson loop, the full string tension σ is saturated by the diagonal string tension σ_{Abel}. This explains the Abelian dominance and magnetic monopole dominance.

Step 4:[7] The whole problem is reduced to calculating

$$\left\langle W^C[a^\Omega]\right\rangle_{NLSM_2} = \left\langle e^{i\frac{q}{g}8\pi Q_S}\right\rangle_{NLSM_2}, \tag{15}$$

Note that the integrand of $Q_S = \frac{1}{8\pi}\int_S d^2z\ \epsilon_{\mu\nu}\mathbf{n}\cdot(\partial_\mu\mathbf{n}\times\partial_\nu\mathbf{n})$ is the instanton density in NLSM_2. Therefore, Q_S counts the number of instantons minus that of anti-instantons inside the area S bounded by the Wilson loop C. In order to perform the actual calculation, we adopt the the naive instanton calculus (dilute gas approximation). This leads to the area law of the diagonal Wilson loop and the non-zero diagonal string tension σ_{Abel} for the fractional charge q (i.e., when q/g is not an integer). More systematic instanton calculations enable us to identify the instanton solution with the Coulomb gas of vortices.

Note that all the above steps are exact except for the instanton calculus of the Wilson loop in NLSM. The above results imply that the non-zero string tension σ in QCD_4 follows from the non-zero diagonal string tension σ_A in NLSM_2. The problem of proving area law in QCD_4 is reduced to the equivalent problem in NLSM_2. The equivalence between the 4D TQFT and the 2D NLSM will be confirmed by direct numerical simulations.

Similar strategy is also applied to QED_4 to prove the existence of strong coupling confinement phase.[8] This follows from the existence of Berezinski-Kosterlitz-Thouless transition of the O(2) NLSM_2.

Another derivation of quark confinement in QCD_4 was given [10] from the viewpoint of Abelian-projected effective gauge theory of QCD.

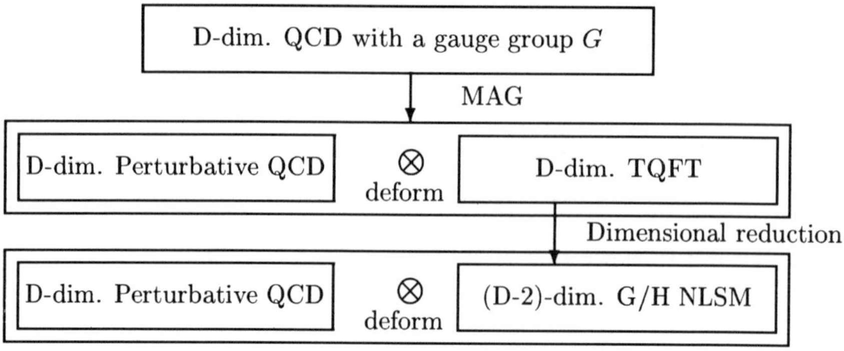

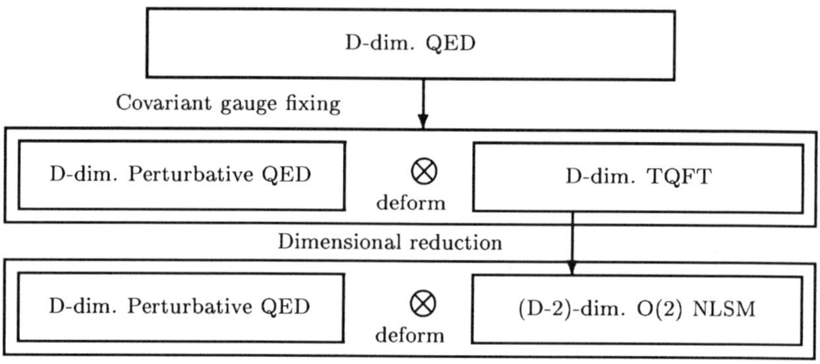

References

1. A.M. Polyakov, Phys. Lett. B **59**, 82 (1975).
2. A.M. Polyakov, Nucl. Phys. B **120**, 429 (1977).
3. N. Seiberg and E. Witten, Nucl. Phys. B**426**, 19 (1994); **431**, 484 (1994).
4. G. 't Hooft, Nucl. Phys. B **190** [FS3], 455 (1981).
5. E. Witten, Commun. Math. Phys. **117**, 353 (1988).
6. K.-I. Kondo, hep-th/9709109, Phys. Rev. D **57**, 7467 (1998).
   K.-I. Kondo, hep-th/9803063, Prog. Theor. Phys. Suppl., No. 131, 243 (1998).
7. K.-I. Kondo, hep-th/9801024, Phys. Rev. D **58**, 105019 (1998).
8. K.-I. Kondo, hep-th/9803133, Phys. Rev. D **58**, 085013 (1998).
9. K.-I. Kondo, hep-th/9805153, Phys. Rev. D **58**, 105016 (1998).
10. K.-I. Kondo, hep-th/9810167.

Finite Quark Mass Effects in the Improved Ladder Bethe–Salpeter Amplitudes

K. Naito, K. Yoshida, Y. Nemoto and M. Oka
Department of Physics, Tokyo Institute of Technology, Meguro, Tokyo 152-8551, Japan

M. Takizawa
Showa College of Pharmaceutical Sciences, Machida, Tokyo 194-8543, Japan

We study the finite quark mass effects of the low-energy QCD using the Schwinger–Dyson (SD) equation for the quark propagator and the Bethe–Salpeter (BS) equation for pion in the improved ladder approximation model of QCD. The observed values of the pion mass and the pion decay constant have been reproduced reasonably well and the empirical value of the quark condensate has been obtained with the rather large value of $\Lambda_{\rm QCD} \sim 600$ MeV. It may indicate that other non-perturbative interactions between quarks such as the instanton induced interaction also play an important role in the low-energy QCD. There are three parameters in our approach, namely, the current u,d-quark mass, $\Lambda_{\rm QCD}$ and the infrared cut-off below which the running coupling constant α_s stops running. The numerical studies show that the chiral perturbation seems to be reasonable up to the strange quark mass region.

1 Improved Ladder Approximation

The improved ladder approximation is an effective model of QCD proposed by Aoki et al.[1] Its interaction is given by

$$\mathcal{L}_{\rm int}[\psi,\overline{\psi}](x) := -\frac{1}{2}\int_{pp'qq'} \mathcal{K}^{mm',nn'}(p,p';q,q') \times \overline{\psi}_m(p)\psi_{m'}(p')\overline{\psi}_n(q)\psi_{n'}(q')e^{-i(p+p'+q+q')x}\,, \quad (1)$$

$$\mathcal{K}^{mm',nn'}(p,p';q,q') = \bar{g}^2\left((\frac{p_E - q'_E}{2})^2, (\frac{q_E - p'_E}{2})^2 \right) \times iD^{\mu\nu}\left(\frac{p+p'}{2} - \frac{q+q'}{2}\right)(\gamma_\mu T^a)^{mm'}(\gamma_\nu T^a)^{nn'} \quad (2)$$

where $\int_p$ denotes $\int \frac{d^4p}{(2\pi)^4}$ and p_E represents the Euclidean momentum. The Fourier transformations of fields are defined by $\overline{\psi}(p) = \int d^4x e^{ipx}\overline{\psi}(x)$ and $\psi(p) = \int d^4x e^{ipx}\psi(x)$. The indices $m, n, \cdots$ are combined indices $m := (a,i,f)$, $n := (b,j,g), \cdots$ with Dirac indices $a, b, \cdots$ and color indices $i, j, \cdots$ and flavor indices $f, g, \cdots$. T^a denotes the generator of the color $SU(N_C)$. According to

Higashijima and Miransky[2,3], we choose a particular set of the momenta that determines the running coupling constant, i.e.,

$$\bar{g}^2(p_E^2, q_E^2) = \theta(p_E^2 - q_E^2) g^2(p_E^2) + \theta(q_E^2 - p_E^2) g^2(q_E^2). \tag{3}$$

This is often called the Higashijima-Miransky approximation. The infrared cut-off t_0 and $t_{\rm IF}$ are introduced in the running coupling constant. Above $t_{\rm IF}$, $g^2(p_E^2)$ develops according to the one-loop result of the QCD renormalization group equation and below t_0, $g^2(p_E^2)$ is kept constant. These two regions are connected by the quadratic polynomial so that $g^2(p_E^2)$ becomes a smooth function. Here N_C is the number of colors and N_f is the number of active flavors. We use $N_C = N_f = 3$ in our numerical studies. The gluon propagator is given in the Landau gauge

$$iD^{\mu\nu}(k) = \left(g^{\mu\nu} - \frac{k^\mu k^\nu}{k^2} \right) \frac{-1}{k^2}. \tag{4}$$

The rainbow SD equation is derived as

$$iS_F^{-1}(q) - iS_0^{-1}(q) + C_F \int_p \bar{g}^2(q_E^2, p_E^2) iD^{\mu\nu}(p-q)\gamma_\mu S_F(p)\gamma_\nu = 0, \tag{5}$$

in which C_F denotes the second Casimir, $S_0(q)$ denotes the tree level quark propagator with the bare quark mass $m_0(\Lambda_{\rm UV}^2)$ and the integral is cut-off at $p_E^2 = -p^2 = \Lambda_{\rm UV}^2$.[4] The full quark propagator $S_F(q)$ is parametrized by two scalar functions as

$$S_F(q) = \frac{i}{\not{q} A(q^2) - B(q^2)}. \tag{6}$$

After the Wick rotaion, we obtain $A(q_E^2) \equiv 1$ and the integral equation for $B(q_E^2)$ is solved numerically. We employ the Kusaka's renormalization condition[5]

$$\left. \frac{\partial B(-\mu^2)}{\partial m_R(\mu^2)} \right|_{m_R(\mu^2)=0} = 1 \tag{7}$$

to obtain the renormalization constant $Z_m = Z_m(\Lambda_{\rm UV}^2, \mu^2)$ which relates the bare quark mass $m_0(\Lambda_{\rm UV}^2)$ and the current quark mass m_R renormalizaed at μ as

$$m_0(\Lambda_{\rm UV}^2) = Z_m^{-1} m_R(\mu^2). \tag{8}$$

The quark condensate in the finite quark mass system is not uniquely defined. Here we employ the following definition of the quark condensate.

$$\langle \bar{q}q \rangle_R(\mu^2) := Z_m^{-1} \langle \bar{q}q \rangle_0(\Lambda_{\rm UV}^2) \tag{9}$$

$$\langle\bar{q}q\rangle_0(\Lambda_{\rm UV}^2) := -\left(\int^{\Lambda_{\rm UV}} \frac{d^4q}{(2\pi)^4}{\rm tr}[S_F(q)] - \int^{\Lambda_{\rm UV}} \frac{d^4q}{(2\pi)^4}{\rm tr}[S_F(q)_{\rm pert}]\right) \quad (10)$$

$$S_F(q)_{\rm pert} := \left.\frac{\partial S_F(q)}{\partial m_R(\mu^2)}\right|_{m_R(\mu^2)=0} m_R(\mu^2). \quad (11)$$

In this definition the quark condensate does not include the term linear in $m_R(\mu^2)$ which causes a divergence.

The ladder BS equation is derived in the standard way and is written by

$$S_F^{-1}(q+P_B/2)\chi(q;P_B)S_F^{-1}(q-P_B/2)$$
$$= -iC_F\int_k \bar{g}^2(q_E^2,k_E^2)iD^{\mu\nu}(q-k)\gamma_\mu\chi(k;P_B)\gamma_\nu. \quad (12)$$

where $\chi(q;P_B)$ denotes the BS amplitude. After the Wick rotation, the integral region is $I = \{k_E | k_E^2 \pm k_E P_{BE} + P_{BE}^2/4 \le \Lambda_{\rm UV}^2\}$ which is consitent with the cut-off of the SD equation (5).[6] We treat Eq.(12) as the eigen equation by multiplying the eigen value λ on the left hand side for fixed total momentum P_B. We can solve this equation only in space like region $P_B^2 = -P_{BE}^2 < 0$. Then we must extrapolate the eigen value λ to the time like region to find the on-shell point $\lambda = 1$.

2 Result

We fix four of the model parameters as $\Lambda_{\rm QCD} = 600$[MeV], $t_{\rm IF} = -0.5, t_0 = -3.0$ and $\Lambda_{\rm UV} = 100$[GeV]. Note that the large part of t_0 dependence is absorbed in the $t_{\rm IF}$ degree of the freedom and the $\Lambda_{\rm UV}$ dependence of the model is very small for $\Lambda_{\rm UV} \gg \Lambda_{\rm QCD}$. The latter holds even in the finite quark mass case due to the Kusaka's renormalization condition (7). In order to study the current quark mass dependences of m_π and f_π, we vary the current quark mass $m_R(\mu^2 = 4.0[{\rm GeV}^2])$. In Fig.1 we show the estimation of pion masses. To obtain the decay constant, we normalize the BS amlitude in the standard way and employ the definition of f_π as

$$iP_{B\mu}f_\pi := \langle 0|J_{5\mu}(0)|\boldsymbol{P}\rangle \quad (13)$$

where $J_{5\mu}(x)$ is the Noether current in this model. $J_{5\mu}(x)$ deviates from the QCD axial-vector form $\overline{\psi}\gamma_\mu\gamma_5\psi$ because the interaction (1) is non-local. For this definition of f_π, the PCAC relation

$$\mathcal{R} := \frac{\langle 0|\partial^\mu J_{5\mu}(0)|\boldsymbol{P}\rangle}{2m_0\langle 0|J_5(0)|\boldsymbol{P}\rangle} = 1 \quad (14)$$

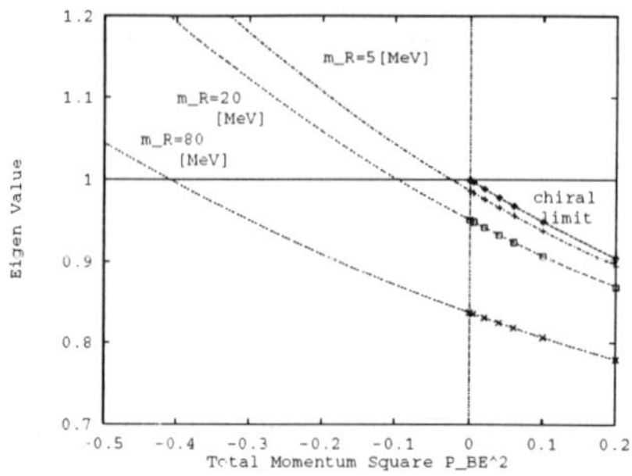

Fig.1 Extrapolation of the eigenvaule

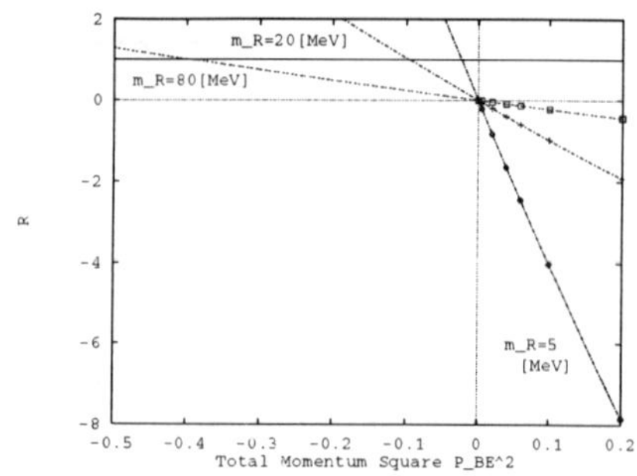

Fig.2 Ratios of the PCAC relation

holds.[6] In Fig.2 we plot and extrapolate $\mathcal{R}$ as a function of P_{BE}^2. One finds that $\mathcal{R}(P_{BE}^2) \simeq 1$ at the on-shell $P_{BE}^2 = -M_\pi^2$ which satisfies $\lambda(P_{BE}^2) = 1$. In Table 1 we summarize the numerical results. The meson mass square is almost propotional to the current quark mass and the decay constant depends on the current quark mass linearly. These results suggest that the chiral perturbation is reasonable up to the strange quark mass region.[7]

m_R[MeV]	0.0	5.0	10	20	40	80	120
$-\langle\bar{q}q\rangle^{1/3}$ [MeV]	259	259	259	258	257	250	239
M_π [MeV]	0	159	222	313	444	640	801
f_π [MeV]	86	88	90	93	101	116	132

Table 1: Numerical results

References

1. K.-I. Aoki, M. Bando, T. Kugo and M. Mitchard, Prog. Theor. Phys. **85** (1991) 355.
2. K. Higashijima, Phys. Rev. **D29** (1984) 1228.
3. V. A. Miransky, Sov. J. Nucl. Phys. **38** (1984) 280.
4. C. D. Roberts and A. G. Williams, Prog. Part. Nucl. Phys. **33** (1994) 477.
5. K. Kusaka, private communication.
6. K. Naito, K. Yoshida, Y. Nemoto, M. Oka and M. Takizawa to be published.(hep-ph/9805243)
7. K. Naito, K. Yoshida, Y. Nemoto, M. Oka and M. Takizawa in preparation.(hep-ph/9812502)

QCD SUM RULE ANALYSIS ON J/ψ-NUCLEON SCATTERING LENGTH AND THE MASS SHIFT IN NUCLEAR MATTER

ARATA HAYASHIGAKI
Yukawa Institute for Theoretical Physics, Kyoto University,
Kyoto 606-8502 Japan
E-mail: arata@yukawa.kyoto-u.ac.jp

We calculate the spin-averaged J/ψ-N scattering length $a_{J/\psi}$ by directly applying QCD sum rule to J/ψ-N forward scattering amplitude. Our result including twist-2 operator up to dimension-4, suggests the force is weakly attractive. If we apply $a_{J/\psi}$ to the effective mass of J/ψ in nuclear matter on the basis of the linear density approximation, it shows very slight decrease at normal matter density.

To investigate in-medium properties of charmonium in detail is a pressing need, because they are useful and simple probes in heavy-ion collisions. In this talk, we study the medium modification of J/ψ on the basis of the relation between J/ψ-N scattering length and the mass shift [3,4] through extending vacuum QCD sum rules (QSR's) [1,2] to finite density. That is, the first aim is to estimate the essential features of the interaction between J/ψ and N through the scattering length. In practice, by applying QSR to J/ψ-N forward scattering amplitude we calculate the scattering length. The scattering length is a physically very important quantity in free space, because it is the unique observable in J/ψ-N elastic scattering at low energy. If it is negative, then we could predict attractive interaction capable of binding J/ψ to nucleus, so that J/ψ could lead to a bound state with nucleus. The prediction of such exotic state will give exciting new directions in nuclear physics. The second aim is how the superposition of elementary J/ψ-N scattering at low energy affects the effective mass of J/ψ in nuclear matter. When we work in the dilute nucleon gas, we find that the mass shift is linearly dependent on the density and the scattering length in the framework of QSR.

In this approach with the Fermi gas model, in-medium correlation function (density ρ_N) is divided into vacuum part and static one-nucleon part near the normal matter density by applying OPE to the correlators at deep Euclidean region ($Q^2 = -q^2 > 0$). That is, in the framework of QSR, in-medium correlation function can be approximated reasonably well to the linear density of nuclear matter that all nucleons are at rest;

$$\Pi^{\rm NM}_{\mu\nu}(q) = i\int d^4x e^{iq\cdot x}\langle {\rm T}J_\mu(x)J^\dagger_\nu(0)\rangle_{\rho_N} \simeq \Pi^0_{\mu\nu}(q) + \frac{\rho_N}{2M_N}T_{\mu\nu}(q), \quad (1)$$

where $q^\mu = (\omega, \boldsymbol{q})$ is the 4-momentum of J/ψ ($J_\mu = \bar{c}\gamma_\mu c$). The one-nucleon part corresponds to the forward J/ψ-N scattering amplitude with the spin of nucleon averaged;

$$T_{\mu\nu}(\omega, \boldsymbol{q}) = i\int d^4x e^{iq\cdot x}\langle N(ps)|\mathrm{T}J_\mu(x)J_\nu^\dagger(0)|N(ps)\rangle, \tag{2}$$

where $|N(ps)\rangle$ denotes the nucleon state with $p = (M_N, \boldsymbol{p} = \boldsymbol{0})$ and spin s normalized covariantly as $\langle N(\boldsymbol{p})|N(\boldsymbol{p}')\rangle = (2\pi)^3 2p^0\delta^3(\boldsymbol{p}-\boldsymbol{p}')$. The QSR analysis on the forward scattering amplitude enables us to obtain the information for J/ψ-N interaction and in Eq. (1) the second term means the slight deviation from vacuum properties of J/ψ determined by $\Pi^0_{\mu\nu}$. By applying QSR to $T_{\mu\nu}$ directly, we relate the scattering length extracted from the QSR for $T_{\mu\nu}$ with the mass shift, $\delta m_{J/\psi} = 2\pi\frac{M_N + m_{J/\psi}}{M_N m_{J/\psi}}\rho_N a_{J/\psi}$. That is, using the dispersion relation for $T(\omega, \boldsymbol{q}) = T_\mu{}^\mu/(-3)$ we introduce three unknown parameters a, b and c defined in the spectral function $\rho(u, \boldsymbol{q} = \boldsymbol{0})$ $= a\,\delta'(u^2 - m^2_{J/\psi}) + b\,\delta(u^2 - m^2_{J/\psi}) + c\,\delta(u^2 - s_0)$, near the pole position of J/ψ. Here δ' is the first derivative of δ function with respect to u^2 and the parameter a is related to the spin-averaged scattering length $a_{J/\psi}$ as $a = 8\pi f^2_{J/\psi} m^4_{J/\psi}(M_N + m_{J/\psi})a_{J/\psi}$, where the coupling $f_{J/\psi}$ and the J/ψ mass $m_{J/\psi}$ are defined by $\langle 0|J_\mu\ |J/\psi^{(h)}(q)\rangle = f_{J/\psi} m^2_{J/\psi}\epsilon^{(h)}_\mu(q)$ with the polarization vector $\epsilon^{(h)}_\mu$. s_0 is regarded as the continuum threshold in vacuum. Moreover, among these parameters we impose the constraint from low energy theorem that in the low energy limit, $\omega \to 0$, $T(\omega, \boldsymbol{0})$ becomes equivalent to Born term $T^{\mathrm{Born}}(\omega, \boldsymbol{0})$ ($= 0$ for lack of charmed quarks inside a nucleon in this system). Thus, through the QSR we determine two unknown phenomenological parameters a and b after the parameter c is removed by the constraint $\frac{a}{m^4_{J/\psi}} + \frac{b}{m^2_{J/\psi}} + \frac{c}{s_0} = 0$.

On the other hand, we give the following OPE expression for n-th derivative for $T_{\mu\nu}$ with respect to q^2, $\widehat{T}^{(n)\ \mathrm{OPE}}(q^2)$ up to dimension-4 gluonic operators with new contributions from twist-2 operators included.

$$\frac{1}{n!}\left(\frac{d}{dq^2}\right)^n \frac{T^{\mathrm{OPE}}(q^2)}{q^2} = \frac{1}{3}\left[C_G^{(n)}(\xi)\left\{\left\langle\frac{\alpha_s}{\pi}G^2\right\rangle_N - 4\left\langle\frac{\alpha_s}{\pi}ST(G^a_{0\sigma}G^a_{0\sigma})\right\rangle_N\right\}\right.$$
$$\left. + \{D_1^{(n)}(\xi) - D_2^{(n)}(\xi) - D_3^{(n)}(\xi)\}\left\langle\frac{\alpha_s}{\pi}ST(G^a_{0\sigma}G^a_{0\sigma})\right\rangle_N\right] \tag{3}$$

These Wilson coefficients are calculated by fixed-point gauge method. Here we define the dimensionless parameter as $\xi = -q^2/4m_c^2$ (m_c; charmed quark mass) and $\rho = \xi/(1+\xi)$. $\mathcal{ST}$ means making the twist-2 operators symmetric and traceless in its Lorentz indices. In Eq. (3) each coefficient function is given using Gauss hypergeometric functions ${}_2F_1$ for arbitrary q^2 as follows;

$$D_1^{(n)}(\xi) = \frac{2^{n+3}(n+1)(n+1)!}{3(2n+3)!!}(4m_c^2)^{-(n+2)}(1+\xi)^{-(n+2)}$$
$$\times\left[2\,{}_2F_1\left(n+2,\frac{1}{2},n+\frac{5}{2};\rho\right) - \frac{2(n+2)}{1+\xi}\,{}_2F_1\left(n+3,\frac{1}{2},n+\frac{5}{2};\rho\right)\right.$$
$$\left. + \frac{3(n+2)^2}{(1+\xi)(2n+5)}\,{}_2F_1\left(n+3,\frac{1}{2},n+\frac{7}{2};\rho\right)\right] \quad (4)$$

$$D_2^{(n)}(\xi) = -\frac{2^{n+5}(n+1)(n+2)!}{3(2n+5)!!}(4m_c^2)^{-(n+2)}(1+\xi)^{-(n+2)}$$
$$\times\left[{}_2F_1\left(n+2,\frac{1}{2},n+\frac{7}{2};\rho\right) - \frac{n+2}{2(1+\xi)}\,{}_2F_1\left(n+3,\frac{1}{2},n+\frac{7}{2};\rho\right)\right] \quad (5)$$

$$D_3^{(n)}(\xi) = \frac{2^{n+3}(n+1)(n+1)!}{3(2n+5)!!}(4m_c^2)^{-(n+2)}(1+\xi)^{-(n+2)}\left[(n+2)\right.$$
$$\left.\times\,{}_2F_1\left(n+2,\frac{1}{2},n+\frac{7}{2};\rho\right) + 4(2n+5)\,{}_2F_1\left(n+2,\frac{1}{2},n+\frac{5}{2};\rho\right)\right] \quad (6)$$

The Wilson coefficient $C_G^{(n)}$ for scalar operator in Eq. (3) has already calculated in Ref.2 and Eq. (4), (5) and (6) are new contributions for twist-2 operator. Eventually we obtain the moment sum rule expressed as the form of the n-th derivative with respect to q^2, $\widehat{T}^{(n)\,\mathrm{ph}}(\xi\ ;\ a,b) = \widehat{T}^{(n)\,\mathrm{OPE}}(\xi)$. We must choose the reliable stability region of moment sum rule for the change of the both n and q^2. The moment sum rule for $T_{\mu\nu}$ is the method to investigate the deviation from vacuum properties obtained from $\Pi^0_{\mu\nu}$. So we should adopt the same regions of n and q^2 as $\Pi^0_{\mu\nu}$. We calculate the window of n for various values of ξ by applying the moment sum rule to vacuum correlation function $\Pi_0(\omega^2) = \Pi^{0\ \mu}_{\mu}(\omega, \boldsymbol{q}=\boldsymbol{0})\ /(-3\omega^2)$ in Eq.(1) and calculating $\widehat{\Pi}_0^{(n)}/\widehat{\Pi}_0^{(n-1)}$ in the same method as Ref.2. We select the appropriate range of n reproducing the experimental value of J/ψ bare mass for each ξ (ranging from 0.0 to 3.0 with 0.5 intervals) from this calculation, where we use $s_0 = 3.6^2$ [GeV^2] and $\langle \frac{\alpha_s}{\pi} G^2 \rangle_0 = 0.0126$ [GeV^4] [2].

By inserting the sets of ξ and n obtained thus into the moment sum rule, we can determine unknown parameters a and b simultaneously by fitting the left-hand side to the right-hand side. To calculate the scattering length we adopt the following values for other various parameters. In PH side, we use $m_{J/\psi} = 3.1$ [GeV] and $M_N = 0.94$ [GeV]. The coupling $f_{J/\psi}$ is determined from the experimental value of electronic width as $f^2_{J/\psi} = 1.7 \times 10^{-2}$. In OPE side, for

α_s and m_c we adopt the functions dependent on ξ given in Ref.2. The scalar nucleon matrix element $\langle \frac{\alpha_s}{\pi} G^2 \rangle_N = -(1.222 \pm 0.282)$ [GeV^2] is evaluated from the trace anomaly. The twist-2 part $\langle \frac{\alpha_s}{\pi} \mathcal{ST}(G^a_{0\sigma} G^a_{0\sigma}) \rangle_N = -(0.094 \pm 0.010)$ [GeV^2] is determined from the gluon distribution function of a nucleon, which is obtained by leading order parametrization to the experimental data of deep inelastic scattering [5].

To summarize, the direct application of moment sum rule to the forward J/ψ-N scattering amplitude supplies us the fascinating result for the J/ψ-N interaction. That is, the J/ψ-N scattering length $a_{J/\psi}$ indicates negative value (about -0.1 [fm]). This result suggests that the attractive J/ψ-N interaction is not sufficient to form a bound state with one nucleon, but it could make a bound state with nuclei. Our result is smaller than those obtained by Brodsky et al. [6] and de Téramond et al. [7], but is consistent with Ref.8. In this study we newly calculated the Wilson coefficients for twist-2 gluon operator (dimension-4) in the form with quark mass explicitly. The effect of twist-2 contribution makes the absolute value of scalar part larger about 30 %. In the linear density approximation, we can calculate J/ψ mass shift from $a_{J/\psi}$. The result gives very small decrease of mass (about -4 to -7 [MeV]), about 0.1 to 0.2 % at normal matter density. Since the slight mass shift is of the order of MeV, the change is sufficiently larger than the leptonic decay width of the order keV. So we consider that J/ψ is a good probe for the observation of medium effect. After this workshop, we submitted the full paper for this work [9].

References

1. M.A. Shifman, A.I. Vainshtein and V.I. Zakharov, Nucl. Phys. **B147** (1979), 385.
2. L.J. Reinders, H.R. Rubinstein and S. Yazaki, Phys. Rep. **127** (1985), 1; Nucl. Phys. **B186** (1981), 109.
3. Y. Kondo and O. Morimatsu, Phys. Rev. Lett. **71** (1993), 2855.
4. Y. Koike and A. Hayashigaki, Prog. Theor. Phys. **98** (1997), 631.
5. X. Jin, T.D. Cohen, R.J. Furnstahl and D.K. Griegel, Phys. Rev. **C47** (1993) 2882.
6. S.J. Brodsky and G.A. Miller, Phys. Lett. **B412** (1997), 125.
7. G.F. de Téramond, R. Espinoza and M. Ortega-Rodríguez, Phys. Rev. **D58** (1998), 034012.
8. After this talk, there appeared a preprint based on a similar QCD sum rule approach: F. Klingl, S. Kim, S.H. Lee, P. Morath and W. Weise, nucl-th/9811070 v2.
9. A. Hayashigaki, nucl-th/9811092, submitted to Prog. Theo. Phys..

Program of The APCTP-RCNP Joint International School on Physics of Hadrons and QCD

October 12-13, 1998
RCNP, Osaka University

October 12

Morning session I

9:30- 9:45 **H. Ejiri** (RCNP)
Opening Address

9:45-10:45 **J. Verbaarschot** (Stony Brook)
11:45-12:00 **J. Verbaarschot** (Stony Brook)
Random Matrix Theory and QCD at nonzero Temperature and Chemical Potential

(Lunch 12:00-14:00)

Afternoon session

14:00-15:00 **H. Toki** (RCNP)
Physics at RCNP

(Break 12:00-14:00)

Afternoon session II

15:20-16:20 **R. J. Perry** (Ohio State)
16:35-17:35 **R. J. Perry** (Ohio State)
Light-front QCD

(Banquet 18:00-)

October 13

Morning session

9:30-10:30 **M. Rho** (Saclay)
10:45-11:45 **M. Rho** (Saclay)
Chiral symmetry in nuclear physics

(Lunch 11:45-13:15)

Afternoon session

13:15-14:45 **T. Blum** (RIKEN-BNL)

Domain Wall Formalism in Lattice QCD

15:00-16:30 **Y. Koike** (Niigata)

Spin structure of the nucleon and perturbative QCD

16:30-16:45 **H. Toki** (RCNP)

Closing Address

Program of the 1998 YITP-Workshop on QCD and Hadron Physics

October 14-16, 1998
Auditorium of Yukawa Institute

October 14

Morning session I (Chairman : T. Matsui)

9:10- 9:15 **H. Yabu** (TMU)

Opening address

9:15-10:00 **J. Verbaarschot** (Stony Brook)

Thouless energy, partially quenched chiral perturbation theory and spectrum of the QCD Dirac operator (45)

10:00-10:30 **E.-M. Ilgenfritz** (Kanazawa)

Gauge invariant field strength correlators in RG smoothed lattices and in instanton/caloron models (30)

(Break 10:30-11:00)

Morning session II (Chairman : K. Yazaki)

11:00-11:45 **M. Rho** (Saclay)

Color anomaly and the Cheshire Cat phenomenon in the proton (45)

11:45-12:15 **T. Hatsuda** (Kyoto) and **T. Kunihiro** (Ryukoku)

Pre-Critical Chiral Fluctuations in Nuclear Medium (30)

(Lunch 12:15-13:45)

Afternoon session I (Chairman : M. Oka)

13:45-14:25 **A. Williams** (Adelaide)

Lattice calculation of the gluon propagator (40)

14:25-15:05 **D.-P. Min** (Seoul National)

Solar Proton Fusion Reaction by Chiral Perturbation Theory (40)

(Break 15:05-15:35)

Afternoon session II (Chairman : T. Hatsuda)

15:35-16:05 **M. Iwasaki** (Kochi)

Superconductivity in quark matter (30)

16:05-16:30 **T. Tanigawa** (Kyushu) and **M. Matsuzaki** (Fukuoka U. of Edu.)

Effects of Meson Mass Decrease on Superfluidity in Nuclear Matter (25)

16:30-16:55 **M. Yasuhira** and **T. Tatsumi** (Kyoto)

Equation of states for kaon condensed matter at finite temperature (25)

16:55-17:15 **N. Prasad** (Banaras Hindu)

Kaon Condensation in an Excluded Volume Mean Field Model with Meson Exchange Interaction (20)

17:15-17:45 **H. Kohno** (Saga)

Approximate discrete γ_5 symmetry in nuclear matter (20)

(19:00- Workshop Party in Shimogamo Prince Hotel)

October 15

Morning session I (Chairman : J. Kodaira)

9:00- 9:45 **R. Perry** (Ohio State)

A New Renormalization Group for Hamiltonian Field Theory (45)

9:45-10:15 **K. Harada** (Kyushu)

Theta in the LF massive Schwinger model (30)

10:15-10:35 **M. Taniguchi** (Nagoya)

Spontaneous Symmetry Breaking of the Light-Front ϕ^4 Theory (20)

(Break 10:35-10:55)

Morning session II (Chairman : K. Tanaka)

10:55-11:25 **C.-R. Ji** (North Carolina State)

Light-Cone Approach for Exclusive Reactions and Decays in QCD (30)

11:25-11:50 **D.-S. Hwang** (Sejong)

Light-cone wavefunction representation of electroweak matrix elements (25)

11:50-12:15 **B.-H. Lee** (Sogang)

Form Factor relations for B to light meson decay in Light Cone Approach (25)

(Lunch 12:15-13:45)

Afternoon session I (Chairman : K. Harada)

13:45-14:10 **K. Tanaka** (Juntendo)

Twist Three Light-Cone Distribution Amplitudes of Vector Mesons in QCD (25)

14:10-14:40 **N. Kochelev** (Joint Inst. for Nucl. Res.)

Proton spin problem in QCD (30)

14:40-15:00 **A. Kohama** (RIKEN)

Correlation Effects in High Energy Electron Inclusive Scattering (20)

15:00-15:20 **W. Bentz**, **H. Mineo** and **K. Yazaki** (Tokyo)

Nucleon structure function in the NJL model (20)

(Break 15:20-15:50)

Afternoon session II (Chairman : M. Matsuzaki)

15:50-16:10 **S.-I. Ando** (Seoul National)

Radiative muon capture in hydrogen in heavy baryon chiral perturbation theory (20)

16:10-16:30 **K. Sudoh** (Kobe)

Photoproduction of ψ' as a test of color-octet model (20)

16:30-17:10 **I.-T. Cheon** (Yonsei)

Induced pseudoscalar coupling constant (40)

17:10-17:30 **C.-H. Lee** (Stony Brook)

Dilepton Enhancement from Hadronic Phase and/or QGP (20)

October 16

Morning session I (Chairman : W. Bentz)

9:00- 9:20 **Y.-S. Oh** (Seoul National)

phi-meson photoproduction and strangeness of the nucleon (20)

9:20- 9:40 **H. Yabu** (TMU)

Dissipative Chiral Field Theory and its Application to neutral/charged DCC states (20)

9:40-10:00 **H. Kim** (TIT)

About QCD sum rules for pion-nucleon coupling (20)

10:00-10:25 **H.-C. Kim** (Pusan)

$\Delta S = 1$ effective weak chiral Lagrangian in the chiral quark model (25)

10:25-10:45 **D. Jido** (TIT)

Chiral Symmetry for Positive and Negative Parity Nucleons (20)

(Break 10:45-11:05)

Morning session II (Chairman : H. Kohno)

11:05-11:35 **H. Nakkagawa** (Nara)

Phase structure of the massive phi-4 model at finite temperature
—New resummation procedure a la RG improvement (30)

11:35-12:00 **S. Chiku** (Yukawa) and **T. Hatsuda** (Kyoto)

Optimized perturbation theory at finite temperature (25)

12:00-12:20 **O. Kiriyama** (Tohoku)

Chiral Phase Transition at Finite Temperature and Density in the QCD-like Gauge Theory (20)

(Lunch 12:20-13:50)

Afternoon session I (Chairman : T. Tatsumi)

13:50-14:30 **O. Morimatsu** (KEK)

Physics of JHP (40)

14:30-14:50 **Y. Nemoto** (TIT)

Chiral Mesons at Finite Temperature in Linear Sigma Model by Cornwall-Jackiw-Tomboulis Effective Action (20)

14:50-15:15 **K. Kondo** (Chiba)

Proving quark confinement in QCD (25)

(Break 15:15-15:40)

Afternoon session II (Chairman : H. Yabu)

15:40-16:00 **K. Naito** (TIT)

Finite Quark Mass Effects in the Improved Ladder Bethe–Salpeter Amplitudes (20)

16:00-16:20 **A. Hayashigaki** (Yukawa)

QCD Sum Rule Analysis on J/ψ-Nucleon Scattering Length and the Mass Shift in Nuclear Matter (20)

16:20-16:30 **D.-P. Min** (Seaul National) Closing remarks

List of Participants

Ando, Shung-ichi	Seoul Nat. Univ., Seoul, Korea
Bentz, Wolfgang	Univ. of Tokyo, Tokyo, Japan
Blum, Tom	RIKEN-BNL, NY, USA
Cheon, Il-Tong	Yonsei Univ., Seoul, Korea
Cho, Yong-Ming	APCTP, Seoul, Korea
Colanero, Klaus	Chinese Univ. of Hong Kong, Hong Kong, China
Chiku, Suenori	YITP, Kyoto Univ., Kyoto, Japan
Ejiri, Hiroyasu	RCNP, Osaka Univ., Osaka, Japan
Fukushima, Kenji	Univ. of Tokyo, Tokyo, Japan
Hatsuda, Tetsuo	Kyoto Univ., Kyoto, Japan
Harada, Koji	Kyushu Univ., Fukuoka, Japan
Hayashigaki, Arata	YITP, Kyoto, Japan
Hirano, Tetsufumi	Waseda Univ., Tokyo, Japan
Horie, Ryuuichi	Kyoto Univ., Kyoto, Japan
Hwang, Dae-Sung	Sejong Univ., Seoul, Japan
Ichie, Hiroko	Tokyo Inst. of Technology, Tokyo, Japan
Iida, Shinji	Ryukoku Univ., Siga, Japan
Ilgenfritz, Ernst-Michael	Univ. of Kanazawa, Kanazawa, Japan
Ishiguro, Katsuya	Kanazawa Univ., Kanazawa, Japan
Itakura, Kazunori	YITP, Kyoto, Japan
Iwasaki, Masaharu	Kochi Univ., Kochi, Japan
Ji, Chueng R.	North Carolina State Univ., NC, US
Jido, Daisuke	Tokyo Inst. of Technology, Tokyo, Japan
Kim, Hungchong	Tokyo Inst. of Technology, Tokyo, Japan
Kim, Hyun-Chul	Pusan National Univ., Pusan, Korea
Kiriyama, Osamu	Tohoku Univ., Sendai, Japan
Kochelev, Nikolai	Joint Inst. for Nucl. Res., Dubna, Russia
Kodaira, Jiro	Hiroshima Univ., Hiroshima, Japan
Kohama, Akihisa	RIKEN, Saitama, Japan
Kouno, Hiroaki	Saga Univ., Saga, Japan
Koike, Yuji	Niigata Univ., Niigata, Japan
Koma, Yoshiaki	RCNP, Osaka Univ., Osaka, Japan
Kondo, Kei-Ichi	Chiba Univ., Chiba, Japan
Kunihiro, Teiji	Ryukoku Univ., Seta, Japan
Lee, Bum-Hoon	Sogang Univ., Seoul,Korea
Lee, Chang-Hwan	SUNY at Stony Brook, US
Maedan, Shinji	Tokyo Nat. Coll. of Tech., Tokyo, Japan
Matsui, Tetsuo	YITP, Kyoto Univ., Kyoto, Japan

Matsuzaki, Masayuki	Fukuoka Univ. of Edu., Fukuoka ,Japan
Min, Dong-Pil	Seoul Nat. Univ., Seoul, Japan
Mineo, Hirobumi	Univ. of Tokyo, Tokyo, Japan
Morii, Kazushige	Hiroshima Univ., Hiroshima, Japan
Morii, Toshiyuki	Kobe Univ., Kobe, Japan
Morimatsu, Osamu	KEK-Tanashi, Tokyo, Japan
Morita, Kenji	Waseda Univ., Tokyo, Japan
Murata, Michika	Kanazawa Univ., Kanazawa, Japan
Naito, Ken-ichi	Tokyo Inst. of Technology, Tokyo, Japan
Nakamura, Hiroki	Waseda Univ., Tokyo, Japan
Nakkagawa, Hisao	Nara Univ., Nara, Japan
Nasuno, Takashi	Hiroshima Univ., Hiroshima, Japan
Nemoto, Yukio	Tokyo Inst. of Technology, Tokyo, Japan
Noaki, Junichi	Univ. of Tsukuba, Tsukuba, Japan
Nogi, Tetsurou	Osaka Univ., Osaka, Japan
Oh, Yongseok	Seoul Nat. Univ., Seoul, Korea
Ohkuma, Kazumasa	Kobe Univ., Kobe, Japan
Ohtani, Munehisa	Univ. of Tokyo, Tokyo, Japan
Ohyama, Satoshi	Kobe Univ., Kobe, Japan
Oka, Makoto	Tokyo Inst. of Technology, Tokyo, Japan
Perry, Robert J.	Ohio State Univ., OH, USA
Prasad, Nirvikar	Banaras Hindu Univ., Varanasi, India
Rho, Mannque	CE Saclay, France
Roy, Dayeeta	Kobe Univ., Kobe, Japan
Saito, Kenji	Tokyo Inst. of Technology, Tokyo, Japan
Sakai, Tsutomu	RCNP, Osaka Univ., Osaka, Japan
Sasaki, Kenji	Tokyo Inst. of Technology, Tokyo, Japan
Sudoh, Kazutaka	Kobe Univ., Kobe, Japan
Takahashi, Kazutaka	Ryukoku Univ., Seta, Japan
Takaishi, Tetsuya	Hiroshima Univ. of Economics, Hiroshima,Japan
Takayama, Miho	RCNP, Osaka Univ., Osaka, Japan
Tanaka, Kazuhiro	Juntendo Univ., Chiba,Japan
Tanigawa, Tomonori	Kyushu Univ., Fukuoka, Japan
Taniguchi, Masa-aki	Nagoya Univ., Nagoya, Japan
Tatsumi, Toshitaka	Kyoto Univ., Kyoto, Japan
Toki, Hiroshi	RCNP, Osaka Univ., Osaka, Japan
Tsukasaki, Yuichiro	Waseda Univ., Tokyo, Japan
Tsunemi, Tomohiro	Kanazawa Univ., Kanazawa, Japan
Verbaarschot, Jack J. M.	SUNY Stony Brook, NY, USA
Williams, Anthony G.	CSSM, Univ. of Adelaide, Adelaide, Australia

Yabu, Hiroyuki	Tokyo Metropolitan Univ., Tokyp, Japan
Yagisawa, Naohiko	Univ. of Tsukuba, Tsukuba, Japan
Yamamoto, Yuki	Kyushu Univ., Fukuoka, Japan
Yasuhira, Masatomi	Kyoto Univ., Kyoto, Japan
Yang, Jae-Choon	Chungnam Nat. Univ., Korea
Yazaki, Koichi	Univ. of Tokyo, Tokyo, Japan